Global Perspectives on Health Geography

Global Perspectives on Health Geography showcases cutting-edge health geography research that addresses pressing, contemporary aspects of the health-place interface. The bi-directional influence between health and place has been acknowledged for centuries, and understanding traditional and contemporary aspects of this connection is at the core of the discipline of health geography. Health geographers, for example, have: shown the complex ways in which places influence and directly impact our health; documented how and why we seek specific spaces to improve our wellbeing; and revealed how policies and practices across multiple scales affect health care delivery and receipt.

The series publishes a comprehensive portfolio of monographs and edited volumes that document the latest research in this important discipline. Proposals are accepted across a broad and ever-developing swath of topics as diverse as the discipline of health geography itself, including transnational health mobilities, experiential accounts of health and wellbeing, global-local health policies and practices, mHealth, environmental health (in)equity, theoretical approaches, and emerging spatial technologies as they relate to health and health services. Volumes in this series draw forth new methods, ways of thinking, and approaches to examining spatial and place-based aspects of health and health care across scales. They also weave together connections between health geography and other health and social science disciplines, and in doing so highlight the importance of spatial thinking.

Dr. Valorie Crooks (Simon Fraser University, crooks@sfu.ca) is the Series Editor of Global Perspectives on Health Geography. An author/editor questionnaire and book proposal form can be obtained from Publishing Editor Zachary Romano (zachary.romano@springer.com).

More information about this series at http://www.springer.com/series/15801

Ross Barnett • Tingzhong Yang
Xiaozhao Y. Yang
Editors

Smoking Environments in China

Challenges for Tobacco Control

Springer

Editors
Ross Barnett
School of Earth and Environment
University of Canterbury
Christchurch, New Zealand

Tingzhong Yang
School of Medicine
Zhejiang University
Hangzhou, China

Xiaozhao Y. Yang
School of Sociology and Anthropology
Sun Yat-sen University
Guangzhou, China

ISSN 2522-8005 ISSN 2522-8013 (electronic)
Global Perspectives on Health Geography
ISBN 978-3-030-76142-4 ISBN 978-3-030-76143-1 (eBook)
https://doi.org/10.1007/978-3-030-76143-1

This Springer imprint is published by the registered company Springer Nature Switzerland AG
The registered company address is: Gewerbestrasse 11, 6330 Cham, Switzerland

Acknowledgements

This book is a result of work by authors in both New Zealand and China. The New Zealand authors would like to acknowledge their colleagues at the Centre for Tobacco Control Research at Zhejiang University, in particular Professor Tingzhong Yang, for their long-term commitment to advancing the cause of tobacco control in China. The work of the Centre has been instrumental in raising the profile of public health and environmental perspectives on health, and for conducting tobacco control advocacy capacity building among medical students. Currently, this project covers all provinces, municipalities and autonomous regions in China, involving nearly 100 universities in more than 70 cities. The New Zealand authors thank the Centre for its international perspective and for enabling their visits to Zhejiang University, which provided the stimulus to write this book. Such visits were made possible by funding from the National Natural Science Foundation of China (Grant # 71490733).

The book's authors wish to thank Graham Moon, emeritus professor of human geography in the School of Geography and Environmental Science at the University of Southampton, for his critique of the original draft of the manuscript. Graham provided valuable comments and advice regarding the importance of health geography in improving understanding of smoking in China.

Dr Xiaozhao Y. Yang extends his gratitude on the part of the authors to Dr He Cai at Sun-Yat Sen University, who allowed, and arranged, access to the China Labour Force Dynamics Survey. Helpful inputs also came from Kecen Guo from the Royal Melbourne Institute of Technology.

The New Zealand authors thank the University of Canterbury Library for its support in obtaining research materials and the University for encouraging a research ethic and international perspective.

The authors thank Marney Brosnan, Mahi Pai Media Productions, Ōtautahi – Christchurch, for producing the graphics in the book and Pauline Barnett for assistance in editing the text.

Contents

About the Contributors

Pauline Barnett is an adjunct associate professor in the School of Health Sciences at the University of Canterbury. Her research focuses on public health policy and ways in which health services can be designed and managed to benefit the community. She has special interests in primary healthcare, mental health, disability and ageing.

Ross Barnett is an adjunct professor at the University of Canterbury. He is a senior international consultant for the Centre for Tobacco Control Research, Zhejiang University School of Medicine, Hangzhou, China. His research focuses on links between social and physical environments and health, health behaviour and policy making, with a particular focus on smoking and tobacco control.

Shuhan Jiang is a lecturer in the School of Humanities and Management at Zhejiang Chinese Medical University in Hangzhou. She is also a research member in the Centre for Tobacco Control Research, Zhejiang University. Her research interests focus on health behaviour, especially on tobacco use.

Sihui Peng is an instructor in the School of Medicine at Jinan University in Guangzhou and a researcher at the Centre for Tobacco Control, Zhejiang University. Her main research interests are tobacco control, smartphone addiction and uncertainty stress. Her research work underlines the importance of Chinese cultural influences on health behaviour.

Tingzhong Yang is the director of the Centre for Tobacco Control Research, Zhejiang University. His research has encouraged "tobacco control advocacy capacity" in medical schools in China. He served as an expert in updating the 2013–2020 version of the WHO Global Action Plan for the Prevention and Control of NCDs. Professor Yang's primary academic focus is health behaviour, but he also has research interests in tobacco control, mental stress and obesity.

Xiaozhao Y. Yang is a medical sociologist at Sun Yat-sen University. He studies at-risk health behaviours and health inequities across changing socioeconomic contexts. His research aims to reach out to policymakers and the public to inform about the fundamental social causes of health problems and their impacts on disadvantaged populations.

Weifang Zhang is a senior health administrator and academic. She is currently senior vice president of the Stomatology Hospital, Zhejiang School of Medicine, and previously served as the Vice President of the Children's Hospital, Zhejiang School of Medicine. She has been actively involved in establishing funds for child health and recently became the principal of the Zhejiang University Women's Health Foundation. Her research interests focus on social medicine, healthcare and health policy.

Chapter 1
Introduction

Ross Barnett

Abstract This chapter introduces the book. It examines smoking as a global issue and how the smoking burden has increasingly been felt in low- and middle-income countries. It highlights links between places and health, health behaviour and tobacco control, paying particular attention to western approaches to smoking research and providing a critique of these. The chapter discusses the importance of developing a greater understanding of the geography of smoking and tobacco control in China and provides a brief critique of past Chinese smoking research. This sets the platform for an outline of the key objectives of the book and the theoretical perspectives used in the investigation of Chinese smoking geographies.

Keywords China · Health geography · Places and health behaviour · Political determinants of health · Smoking geographies · Tobacco control

1.1 Introduction

1.1.1 Three Vignettes

1.1.1.1 Platform Smokers: Arrival in Xuzhou

The bullet train from Hangzhou was nearing Xuzhou where I was to visit the local university. I had been impressed with the speed of the journey, often looking up at the speedometer above the door of the carriage. We had glided along at 303 km per hour, sometimes a little faster. By the speedometer was a big sign which read 'No Smoking' (Fig. 1.1a). Smoking has been banned on China Railway high-speed trains for some years, and since 2018, harsher penalties, including 180-day travel

R. Barnett (✉)
School of Earth and Environment, University of Canterbury, Christchurch, New Zealand
e-mail: ross.barnett@canterbury.ac.nz

R. Barnett et al. (eds.), *Smoking Environments in China*,
Global Perspectives on Health Geography,
https://doi.org/10.1007/978-3-030-76143-1_1

bans, have been imposed on persons ignoring the regulations of smoking on board high-speed trains or in smoke-free areas of other trains (CATC 2018a). These, in theory, also apply to station waiting areas and platforms, but are poorly enforced (CATC 2018b, 2019). Relieved, I settled back in my seat and enjoyed the view, of the endless villages and fields flickering by as the train sped along. Soon the train began to slow and, with anticipation of my forthcoming visit to the university, I began to pack up the papers I had been reading. The typical Chinese urban landscape of multistorey high density housing blocks soon appeared yet the view was softened by many trees and I saw some lakes in the far distance. The train pulled into Xuzhou Station. Passengers crowded at the door ready to disembark. I wondered why so many people were getting off here, some with no luggage. Then when I stepped out on to the platform, I realised that many of the people disembarking were smokers leaving the train to have a quick puff before getting back on board (Fig. 1.1b). They were men of various ages, casually dressed. I stood there for a moment watching them until the announcement came asking people to get back on the train. They quickly returned to their carriages and the train, ever silent, pulled out of the station and, in a brief moment, too was gone.

1.1.1.2 Tobacco Advertising Still Matters: Friday Night on Nanjing East Road, Shanghai

As usual Shanghai was bustling. It was a warm Friday evening and everyone was out enjoying themselves. We took a short walk from the Grand Central Hotel, to Nanjing Road, one of the main pedestrian streets in Shanghai which stretches from the Bund to the city centre and beyond. Nanjing Road is one of Shanghai's famous

Fig. 1.1 (**a**) No Smoking sign on a bullet train (**b**) Platform smokers, Xuzhou. (Credit: Author)

shopping streets and once one of its busiest roads. In 2000, as a part of the local government development plan, Nanjing Road was renovated to be a pedestrian street and is popular with locals and tourists alike. You could spend years exploring this city, with its interesting colonial history and once the main producer of cigarettes in China. Of course in those days, foreign companies controlled the production and marketing of cigarettes, but today the state monopoly, the China National Tobacco Corporation, is dominant.

As we walked along, I noticed that few people were smoking, a welcome outcome of the smoking ban, affecting indoor spaces and many outdoor spaces, imposed in the city along with Beijing in 2016. We decided to window shop, admiring some of the brilliant silk scarves for which this part of the world is famous, along with its green tea. A lady beckoned us into a shop but we continued up the street, 'people watching' and just enjoying the ambiance. Soon a large neon lit tobacco store came into view (Fig. 1.2) and I stopped at the shop window to look at the cigarette packets, hundreds of them, all neatly arranged in rows with their prices clearly visible. There were no Rothmans here; all were Chinese, many with vibrant colours and designs including some with Chinese landscapes and architecture. I was mesmerised by the display. The shop was full of people but nevertheless I decided to go inside. I tried not to draw attention to myself, but no one took any notice of me. Most of the customers were young men, out for a good time and intent on buying their favourite packet or box of cigarettes. Most were cheap. One 20 pack only cost 10RMB, about a third of the cost of a cup of coffee back home. Some were over 100RMB, but most were in the 16–30RMB range. I watched the young men make their purchases and disappear out of the shop. They all looked affluent, well dressed, happy and ready for a good night out. Yes I thought this is definitely the new China....

Fig. 1.2 Specialist tobacco store, East Nanjing Road, Shanghai. (Credit: Author)

1.1.1.3 Retailing and Smoking Norms in a Hangzhou Neighbourhood

Hangzhou is famous for its Grand Canal, Westlake, tea plantations and production of silk. As in most large Chinese cities, many old hutongs have been destroyed and replaced by high density housing. However, some old neighbourhoods, such as Gudang, survive south of Zhejiang University where forested hills surround that part of the city. Here smaller apartment buildings of six storeys or thereabouts still remain, and their pleasant courtyards spill out onto leafy streets where the population goes about its business. The huge scale and geographic diversity of China can be overpowering, but I felt at home here as I walked along with my colleagues to buy food from a local market. The open air market was bustling with people selling their produce and was a popular meeting place for community residents. One of the traders, a middle-aged man selling fish, sat on a bucket enjoying the scene and, of course, his cigarette (Fig. 1.3a). His smoke drifted across to the next vendor who was selling naan bread, but he did not seem to mind being exposed to second-hand smoke. This, of course, was still quite legal as smoke-free legislation, but only banning smoking in indoor public places, workplaces and public transportation, did not come into effect until January 1, 2019.

We retraced our steps back down the street to the main commercial thoroughfare. While not as colourful as the market, there were many shops selling everything you could want, including tobacco. I saw many tobacco outlets, some large and selling alcohol as well, others very small just selling a few convenience goods (Fig. 1.3b). Coming from a country where tobacco products at the point of sale are highly regulated and generally hidden from view, I was struck by the sheer number of outlets present along the street and the visibility of the goods for sale. Across the street from one very large tobacco outlet was a school, a middle or high school, and close by was a primary school. What a pleasant neighbourhood I thought. How lovely for children to grow up here. But then I thought how normalised cigarette sales are in this community ….

This book is about smoking in China, the world's largest producer and consumer of cigarettes. Like many other low- and middle-income countries (LMICs) in the early stages of the smoking transition (Lopez et al. 1994), smoking rates have risen dramatically. In the early 1950s, on average, males smoked just one cigarette per day. By the early 1970s, this had risen to four cigarettes a day and by 1999 to fifteen cigarettes a day (Peto et al. 2009). During the next decade, despite a slight decline in male smoking prevalence, this rapid rise continued reaching almost 18 cigarettes per day in 2010 (Liu et al. 2017a). There are more than 300 million smokers in China, nearly one-third of the world's total. About one in every three cigarettes smoked in the world is smoked in China. In 2018, of a total retail volume of 5325.6 billion sticks, China accounted for 44.5%, dwarfing Indonesia, the next largest market, and the share of the top ten consumers (Table 1.1).

Given the rise of China within the global economy, these trends should come as no surprise. The rise in tobacco consumption mirrors the other rapidly expanding parts of the economy which have bought prosperity to China and helped it move from the ranks of poorer countries to become a middle-income nation, proud of both its heritage and recent accomplishments. Cigarette smoking, as in the case of food consumption and the rise of obesity, must be seen in the light of the lifestyle transitions that have accompanied socio-economic change. But they must also be seen in

Fig. 1.3 (**a**) Village market, Gudang, Hangzhou. (**b**) Small convenience store with cigarette counter, Gudang neighbourhood, Hangzhou. (Credit: Author)

Table 1.1 Major cigarette markets by retail volume (billion sticks), 2018

Country	Retail volume 2018 (billion sticks)	Percent of global total sales volume
China	2638.7	44.5
Indonesia*	307.1	5.8
USA	240.9	4.5
Russia	236.5	4.4
Japan	132.7	2.5
Turkey	118.5	2.2
Egypt	96.3	1.8
Bangladesh	91.6	1.7
India	82.5	1.5
Vietnam	80.9	1.5
Global Total	5325.6	100.0

Credit: Adapted from Fig. 4, Campaign for Tobacco Free Kids, 2019.
https://www.tobaccofreekids.org/assets/global/pdfs/en/Global_Cigarette_Industry_pdf.pdf
*Excluding hand-rolled kreteks

the cultural, economic and institutional context that has privileged certain parts of the economy in the interests of economic growth and political stability. As Matthew Kohrman (2018, p. 14) has emphasised:

> To primarily attribute this fifteen fold increase in (cigarette consumption) to lifestyle decisions is to fundamentally obfuscate the industry's own triumphalism regarding manufacturing. It is to obfuscate that a gigantic supply-chain investment made by government authorities, central and provincial, supercharged the surge in per capita smoking.

Thus to view tobacco smoking narrowly as a health behaviour is to misrepresent its significance as a cultural practice in China that, until recently, has been tolerated and encouraged by national and local political elites. The aim of this book, therefore, is to provide a deeper understanding of smoking in China by focusing on its significance within the cultural, economic, political and social environment.

In developing a geographic perspective on smoking in China we will provide a more in-depth appreciation of the importance of environmental factors and how they influence smoking prevalence and practices. Even though smoking has been practised in China since the sixteenth century with a large expansion of tobacco production, consumption and trade occurring up to the mid-eighteenth century (Benedict 2011), our focus is more contemporary. To achieve this focus, in the remainder of this chapter, we will do three things. First, we seek to briefly examine smoking as a global issue and how the smoking burden has increasingly been felt in low- and middle-income countries. Second, building on an earlier book, *Smoking Geographies* (Barnett et al. 2017), we highlight the importance of the geography of smoking and tobacco control, paying particular attention to western approaches in smoking research and providing a critique of these. In the third part of the chapter, we ask why it is important to develop a greater understanding of the geography of smoking and tobacco control in China and provide a brief critique of Chinese smoking research. This sets the platform where we outline the key objectives of the book. Finally we discuss the structure of the book. Here we focus on the theoretical perspectives which have guided our investigation of Chinese smoking geographies and how bringing together Chinese and western cultural and theoretical perspectives can add new insights to the Chinese smoking epidemic.

1.1.2 Smoking as a Global Issue

While the use of tobacco has a very long history and can be traced back to 5000 BC, the modern smoking epidemic began in the nineteenth century with the invention of automated cigarette-making machinery. Factories, primarily located in the southern USA, allowed for the mass production of cigarettes at low cost. In western countries, cigarettes became elegant and fashionable among society men as the Victorian era gave way to the Edwardian. Women soon followed as celebrities removed the stigma associated with smoking, and it became associated with glamour and independence. In the next few decades, the growth in prevalence was remarkable, as smoking became a normalised activity adopted by most social classes. Despite the emergence of scientific evidence from the 1930s that smoking was harmful to health, smoking prevalence continued to climb, peaking, in countries such as the UK, for example, in the 1940s for men and approximately two decades later for women (Cancer Research UK 2014). The effect of public health messages resulted in social differences in smoking prevalence, as more educated households were the first to quit quickly leading to the low levels of smoking we see among this group today. By contrast, the rapid diffusion of smoking from higher- to lower-income groups that occurred in the early twentieth century was not repeated by a similar rapid adoption of quitting among lower socio-economic status groups.

As smoking fell out of fashion, it changed from a normalised social activity to one that was increasingly stigmatised and concentrated among lower socio-economic status groups and certain ethnic minorities. These trends occurred at a time of recession in the global economy and the growth of concentrated poverty in old industrial cities especially in the USA and UK. These structural changes helped reinforce the idea that smoking was a socially unacceptable abnormal activity increasingly characteristic of more youthful members of an 'urban underclass' concentrated in social housing or in ghettoised conditions as in the USA (Barnett et al. 2017, pp. 42–43). The conservative political rhetoric of the time also emphasised the idea of 'cultures of poverty', which included smoking along with other 'deviant' abnormal behaviours such as domestic violence, alcoholism, female-headed households and general social anomie. Of course such social conditions reflected the negative effects of structural changes in the economy, heightened by stress and relative deprivation, which helped make such conditions worse. Under such situations smoking rates remained high.

Nevertheless, despite protestations by the tobacco industry, mounting public health evidence over the dangers of smoking and exposure to second-hand smoke resulted in the first major moves towards tobacco control in western countries. As the tobacco epidemic grew in significance, both in terms of increased prevalence and deaths attributable to smoking, so did public health demands for urgent action to control consumption. Beginning in the 1970s, countries such as Finland, Norway and Singapore instituted comprehensive national tobacco control legislation (Yach et al. 2007) that included a range of measures designed to reduce consumption. By contrast, in the USA, because of long-standing ties between the federal government and the tobacco industry, tobacco control was more likely to occur at state and local levels where concerted action and greater transparency of policy making provided many advantages for public health advocates when combatting the industry.

The increasing social concentration of demand in western countries also had an impact on cigarette manufacturers who were not immune to economic downturns and shrinking market share. In the face of such challenges, tobacco companies initially focused their attention on raising consumption among lower-income consumers (Brown-Johnson et al. 2014; Berg et al. 2018) and in poorer regions in richer countries (Dwyer-Lindgren et al. 2014). This, however, was a short-term strategy, with attention soon shifting to increasing demand, especially among women and younger people in Asia, Africa and Latin America where tobacco control laws were weak or non-existent.

However, the shift in tobacco marketing also reflected the effects of the rise of global neoliberalism during the 1970s and its impact particularly in many former Communist countries, now subject to a new world order. In much of Eastern Europe and the former Soviet Union, the collapse of former state enterprises led to opportunities for transnational tobacco companies (TTCs) to penetrate formerly closed economies (Holden and Lee 2009), leading to a subsequent rise in mortality rates particularly among women (Stefler et al. 2018). As pressures from global financial institutions encouraged these countries to liberalise their economies (Yach et al. 2007), US companies took advantage of lower tariffs with many of their subsidiaries taking over inefficient state-owned industries that produced high-price poor-quality products. In post-communist Poland, for example, former state-owned tobacco companies were quickly taken over by multinational companies which, by the end of the 1990s, owned 90% of the industry (Zatoński 2003). Forced privatisation imposed by the International Monetary Fund (IMF) and the weakening of state control resulted in the Polish government making large concessions to transnational tobacco companies. For example, limits placed on tobacco taxation meant that by the end of the 1990s, the average price of a packet of cigarettes was lower than the price of a loaf of bread (Wipfli and Samet 2016).

The increased globalisation of cigarette manufacturing resulted in a loss of local control over production and marketing and increased profits for western tobacco companies many of which had undergone a series of mergers to achieve greater economies of scale and access increased capital (Lee et al. 2016). For example, beginning in the 1980s, the US government, under the guise of encouraging free trade as part of WTO agreements, used the office of the US Trade Representative to open up Asian markets to US tobacco companies (Yach et al. 2007). Countries such as Japan, South Korea and Taiwan were pressured to allow American companies access to local markets, with cigarettes produced by local state run monopolies soon overtaken by American brands (Taylor et al. 2000). In South Korea, for example, TTCs identified different population groups more favourably inclined towards foreign brands and used different marketing tactics to encourage consumption. Demand rose rapidly, notably among youth and young women (Lee and Kamradt-Scott

2014). Other countries however, such as China and Thailand, resisted the incursion of American-owned TTCs. Globalisation thus was not a uniform process.

Thus the end of the twentieth century, with increasingly open international borders and a return to a more monopolistic global tobacco industry, saw a rapid shift in demand from richer to low- and middle-income countries. According to the World Bank classification of countries by income level, between 2000 and 2015, the greatest reductions in tobacco smoking were seen in high-income countries (WHO 2018). Although upper middle-income countries experienced a net increase in the number of smokers between 2001 and 2004, this was soon reversed with a net reduction occurring from 2005 to 2015. For high- and middle-income countries, it was estimated that there was a net reduction of 28.6 million smokers from 2000 to 2015 with the bulk of this (84.3%) occurring in high-income countries (−52.2 million). By contrast, in lower middle-income and low-income countries, there was a total net gain of 28 million smokers mostly in lower middle-income countries. Given the highly gendered nature of smoking in Asia, Africa and Latin America, this increase was mainly made up of increased numbers of male smokers (Table 1.2). The experience of China reflects these broader country trends. Although typical of other upper middle-income countries, which registered a slight absolute decline in the number of smokers, actual cigarette production has not, and, with few

Table 1.2 Fitted estimates of number of current smokers aged 15 years and over by World Bank income group of country and by sex, 2000–2015

	Estimated number of smokers (millions)					
Country income group	2000	2005	2010	2015	Change	% Change
	Total smokers					
All	1143	1134	1125	1114	−29	−2.5
High income	272	254	238	220	−52	−19.1
Upper middle income	502	507	505	497	−5	−1.0
Lower middle income	328	331	339	349	+21	+6.4
Low income	41	42	45	48	+7	+17.1
	Male smokers					
All	912	924	934	939	+27	+3.0
High income	165	155	145	135	−30	−18.2
Upper middle income	429	438	441	437	+8	−1.9
Lower middle income	285	296	310	325	+40	+14.0
Low income	33	35	38	42	+9	+27.3
	Female smokers					
All	230	210	192	175	−55	−23.9
High income	107	99	92	85	−22	−20.6
Upper middle income	73	69	64	60	−13	−17.8
Lower middle income	43	35	29	24	−19	−44.2
Low income	8	7	6	6	−2	−25.0

Credit: Adapted from: World Health Organization, 2018, Table 8 Public domain. https://www.who.int/tobacco/publications/surveillance/trends-tobacco-smoking-second-edition/en/

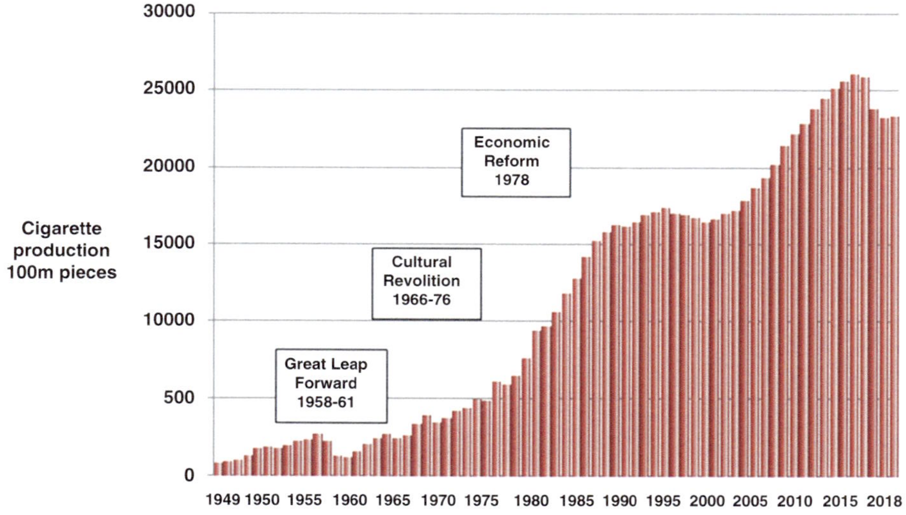

Fig. 1.4 Cigarette production in China, 1949–2018. (Credit: State Tobacco Monopoly Bureau, China Statistical Yearbook (various years). Public domain)

exceptions, during the Great Leap Forward (1958–1961), the Cultural Revolution (1966–1976) and since 2014, has continued to increase (Fig. 1.4). This is of concern since the greatest health risks are likely to occur in countries, like China, with high prevalence rates and high daily consumption (Ng et al. 2014).

1.1.3 Western Geographical Approaches to Smoking Research

The growth of the global smoking epidemic and attempts to curtail its spread have meant that cigarette smoking and tobacco control have received considerable academic attention. Beginning with the epidemiological studies of Sir Richard Doll in the 1950s, which demonstrated a causal link between smoking and poor health (Doll and Hill 1954), research into tobacco consumption and tobacco control has grown significantly across a large number of academic disciplines, including the medical and social sciences and humanities. This work has generated important insights into a wide range of topics including tobacco addiction, the effects of passive smoking on health, smoking and health inequalities, behavioural aspects of individual consumption, smoking and social class, and smoking in popular culture, including the use of social media. In addition, social sciences such as economics, political science and sociology have explored the costs of smoking as drivers of consumption and a range of topics related to cultures of smoking, tobacco control and its implementation. Collectively this work has provided important insights into smoking as a health behaviour and aided public policies designed to reduce tobacco consumption.

Geographers, until quite recently, paid little attention to smoking and other forms of health behaviour and were seldom concerned with wider issues of health inequality. This may have reflected the previous narrow focus of 'medical geography' which was largely concerned with disease distributions and healthcare provision. It was not until the 1990s that there was a gradual shift away from a predominantly biomedical model of health towards a more holistic framework that emphasised the political, social and cultural antecedents of health (Crooks et al. 2018). This shift led to a wider interest in social models and helped raise awareness of the role of place or environmental factors in shaping health.

The development of a wider social lens among health geographers in western countries was partly influenced by political and social trends of the time, including the effects of globalisation, greater awareness of environmental justice, as well as the negative social effects of neoliberal policies upon the social fabric of western societies. These trends were also evident to some degree in China where a series of economic and social reforms (*Gǎigékāifàng*), ushered in by Deng Xiaoping in 1978, led to the development of 'socialism with Chinese characteristics' and a 'socialist market economy'. One of the wider impacts of such reforms, both in China and the West, was to highlight economic differences within countries, in regions, cities and neighbourhoods adversely affected by the economic and social restructuring which followed. In a sense these wider processes sharpened the spatial lens of researchers as their adverse health effects became clear.

Within the geographical literature, therefore, the 1980s and 1990s were decades when the term 'place effects' entered the research vocabulary and characterised much of the subsequent research.

1.1.3.1 Places and Health

There is a great deal of evidence to indicate that the environment within which people live affects their health behaviour, health status and a variety of other social outcomes. Although context effects, especially in voting behaviour, educational aspirations and attainment and patterns of crime were first identified in the 1960s and 1970s, it was not until the 1990s that such work was extended to health. It was particularly through the work of health researchers such as Sally Macintyre, Ichiro Kawachi and Richard Wilkinson that the wider impacts of place effects on health began to be more fully appreciated. As a result, it was soon realised, in part as a result of the application of multilevel modelling (Duncan et al. 1996), that different processes operated at different scales and that structural processes leading to area disadvantage and high rates of social inequality were important. It also became clear that certain population groups, such as women, low-income households and some ethnic minorities, bore the brunt of these societal processes.

In the twenty-first century, many of these issues remain. Context still matters but in a variety of different ways, some very subtle, others less so. The emergence of interest in contextual or place effects since the 1970s is hardly surprising, reflecting the impact of global economic change as well as a reaction to the extremes of individualism and self-interest encouraged by neoliberal philosophies. This shift was present not only in the discipline of geography but also among researchers in social epidemiology, medical sociology and, more recently, medical anthropology who also expressed increased interest in contextual effects and the increased interdisciplinary nature of place-based research (Neely and Nading 2017).

Partly reflecting these trends, but also changes in the global tobacco industry and concern over the global health burden (GBD 2015 Tobacco Collaborators 2017), there has also been an increased interest in research in non-western contexts. As we shall see, this has also occurred in China and reflects a greater awareness of the need to explore the role of environmental factors, such as increased air pollution or high rates of urban stress, upon health. In western countries, however, the impetus came from the widespread effects of globalisation and deindustrialisation as well as the negative social impacts of neoliberalism on the urban fabric of many countries as their economies entered the more laissez faire post-industrial age. The emergence of increased inequality and entrenched disadvantage saw many neighbourhoods of old industrial cities, such as the South Bronx in New York City or Easterhouse in Glasgow, experience declining rather than increased life expectancy. Such trends led to a rising interest in urban regeneration and other area-based solutions to the many seemingly intractable social problems which characterised many post-industrial cities. However, the election of neoliberal governments, especially in Britain, the USA and New Zealand, saw such developments called into question as a new political climate encouraging individualism rather than state intervention became the norm. One consequence of this shift was that structural theories of health inequality became highly politicised, most notably in Britain following the publication of the Black Report (Whitehead et al. 1992). Health-related behaviours, such as smoking and obesity, were seen as outcomes of individual choice rather than being influenced by the characteristics of places where people lived. Poor individual choices were seen as the key causes of health inequalities rather than structural factors which emphasised differences in wealth, power and access to society's resources (Moon and Smith 2018). At the political level, this behavioural philosophy, stressing individual rather than state responsibility, provided a justification for the health service privatisation and the withdrawal of the state which occurred in many countries up to the end of the 1990s. Such moves, however, were highly contested, both politically and in academia, but it was not until the early 2000s that an environmental perspective began to reassert itself in the social sciences especially in academic publications on health inequalities and health behaviour.

Reaction to these broader societal processes set the scene for renewed academic interest in place effects. This was also aided by methodological developments such as multi-level modelling, which enabled a more critical assessment of the significance of the multilayered nature of environmental effects on health and health behaviour. By contrast earlier attempts to unpack the significance of geographical context on health were relatively rare or simple in approach (Twigg and Cooper 2009). Multi-level modelling enabled the specification of both individual and place effects as well as interactions between them and, for the first time, provided a more powerful method for estimating the independent effects of place characteristics on smoking and other health behaviours.

Since the 1990s, geographical research has expended a vast amount of effort examining the impact of place effects on health and health behaviour. A distinction is usually made between 'compositional' and 'contextual' effects (Macintyre et al. 2002); the former emphasising the individual characteristics of people living in different localities and how these affect health, while the latter focuses on the independent influence of place on health or behavioural outcomes. As many have pointed out, however, this context/compositional distinction, while analytically useful, is artificial. Individual characteristics are themselves likely to be influenced by context and thus individual and area characteristics may mutually reinforce each other (Cummins et al. 2007). For example, as many residents of social housing estates may testify, the presence of neighbourhood stigma may influence the likelihood of employment or the social mix of the local school may influence educational aspirations and academic performance. Similarly the length of time individuals are exposed to different types of neighbourhood environments will be important in affecting their life chances (Pearce 2018). Despite these caveats, most social scientists would conclude that 'place matters'. The challenge for researchers has been to attempt to understand important interactions; in other words, to clearly define for whom place matters most, under what circumstances it matters and processes at what spatial scales are most important in influencing health.

Places are complex. Place effects may operate at a variety of spatial scales, macro, meso or micro, all of which may be important in influencing both health behaviours, such as smoking, and overall population health. Places can be defined in a variety of ways. At the macro scale, they may reflect different national political cultures and different policy initiatives. At the meso scale, places may be different cultural regions, imbued with meaning and identity, or degree of rurality which, in many LMICs, will also be associated with higher levels of poverty. Or they may be politically defined spaces, such as states or provinces, whose public policies will differentially affect levels of service provision. At the more micro scale, city level policies may influence housing or green space policies or tobacco control ordinances which will impact well-being. Similarly at the local neighbourhood level, the provision of health damaging consumer goods will have a direct effect on consumption patterns as well as influence neighbourhood norms. The multi-scalar nature of these factors and their social and economic diversity makes it difficult to determine their health impacts.

Place effects are also made up of different kinds of environmental factors, whether these be cultural, economic, political or social, all of which influence health and health behaviour in numerous ways. This may occur as a result of easy access easy to unhealthy products, such as fast food, by restrictions on access to green and safe spaces for exercise and contemplation, or through social norms which may reinforce unhealthy lifestyles. Smoking is just one health behaviour which is dramatically affected by such mechanisms.

1.1.3.2 Places and Health Behaviour

Place effects are also important in understanding health behaviour. Smoking, along with obesity and alcohol consumption, has been of major concern, and there have been a number of reviews of the significance of place effects on these behaviours. Witten and Pearce (2010), for example, provide a comprehensive overview of environmental understandings of obesity while Ayuka and Barnett (2015) provide a more recent overview of place effects on alcohol consumption. Both contributions mainly focus on more affluent countries where lifestyle diseases have, until recently, been more prevalent.

Smoking is one health behaviour among many that is influenced by the context in which people live (Duncan et al. 1996, 1999). While individuals may choose to smoke, that choice will be strongly influenced by a range of external factors. As Cummins et al. (2007) have indicated, these factors come together to provide a geography of health behaviour. This geography reflects both individual characteristics and broader area-level determinants such as programmes to reduce (un)healthy behaviour and the interaction of compositional and contextual factors. Later work by Pearce et al. (2012) argued that at the broadest level two key pathways or domains were important in understanding links between places and smoking: place-based practices and place-based 'regulation'. Within these domains a variety of specific processes were likely to be important at different scales.

At a global level, differences in smoking prevalence between countries will reflect a variety of national characteristics. These will include a range of factors such as stage in the smoking epidemic, levels of economic development, the nature of patriarchal social relations, levels of female empowerment, the role of transnational companies in stimulating demand or their effects on limiting the development of effective tobacco control policies.

At the regional level, smoking prevalence will also reflect many of the above factors, but in places like China, local dependence upon the tobacco industry is also important. The presence of a 'beneficent' industry affects community norms and the social acceptance of smoking. In addition, community affluence will also be important in influencing awareness of the dangers of smoking and the presence of anti-tobacco lobbies. Many studies have found that both regional GDP per capita and individual per capita income have independent effects upon smoking prevalence. As community affluence increases, so too will changes occur in the social distribution of smoking, reflecting the importance of urban educational and other tobacco control programmes (Yang 2017b).

At a local level, different processes will come into play such as the availability of tobacco retailing surrounding homes and schools (Finian et al. 2019). In addition, characteristics of the local social environment, such as levels of social capital between neighbours and community, will affect social norms and the acceptability of smoking. Levels of local deprivation will also be important, and in western countries there is a large body of research which has shown that it is not just individual socio-economic disadvantage that matters. Residents of socially disadvantaged neighbourhoods, regardless of their individual SES, are much more likely to smoke. This effect appears to be strongest for current smokers and especially for females and may reflect the greater local embeddedness of women living in deprived areas (Barnett et al. 2017).

Understanding why unhealthy behaviour is present in some contexts more than others thus requires an understanding of the different types of context effects and how they relate to smoking and the effectiveness of tobacco control. It also raises questions about the nature of tobacco control and the extent to which processes of denormalisation and stigmatisation interact with other forms of stigma, ascribed by race, ethnicity, gender or sexuality (Anlin et al. 2017). As Moon and Smith (2018) have suggested, 'unhealthy behaviour can be entirely normal and understandable in some contexts, with the challenges being to effect change without demonizing people or communities where the behaviour is manifest and to understand why unhealthy behaviour can be a rational response to disadvantage'.

From a theoretical perspective, health geographers have drawn upon a large variety of approaches to better understand links between places and health and health behaviour. Many of these, such as theories of stress, relate to individual behavioural responses to external stimuli, such as increased deprivation and income inequality (Wilkinson and Pickett 2009) and its effect on smoking, while other approaches focus on social demands and personal subjective motivations arising from legislative changes affecting where people can smoke (Kasemets 2018). Many of these themes are bought together in Barnett et al. (2017) and point to a theoretical diversity in western smoking research. While the focus on contextual effects on smoking and other health behaviours has resulted in a quantitative focus of much recent smoking research, qualitative studies are also apparent and provide a different lens on how individuals interpret the act of smoking and what it means to them. While quantitative approaches, usually involving multi-level modelling, are seen as necessary in large scale national studies and to unpack the effects of forces operating at different spatial scales, it is important not to lose sight of more experiential and cultural perspectives (Thompson et al. 2007) that provide a different perspective and which are necessary to ensure a more nuanced interpretation of patterns revealed by quantitative research.

1.1.3.3 Places and Tobacco Control

The focus on place-based attributes also has relevance to understanding the geography of tobacco control. Most studies of tobacco control have focused on the individual decision to quit and the many ways in which this may take place, including unassisted attempts at quitting (Yang et al. 2019). Where context has been considered, it has usually been defined in terms of health service delivery in the case of support from a health professional during a doctor's consultation or receiving advice and support from a smoking quitline. However, as indicated in earlier work (Barnett et al. 2017), there is a distinct geography of tobacco control, with four themes identified as important here.

First, there has been an unequal development of tobacco control policy, both internationally and within particular countries where regional and local variations may be important. As indicated earlier in the chapter, tobacco control initiatives have proceeded furthest in richer countries. In LMICs the opposite has been the case often reflecting lack of financial resources, competing economic development and health imperatives and the interference of Big Tobacco in the policy process. Second, where comprehensive tobacco control policies have been introduced, while these have produced impressive reductions in smoking prevalence, they have been accompanied by unintended consequences – increased social inequalities in smoking and greater stigmatisation of smokers. Third, there is a need to stress the idea of intersectorality, that is, to link smoking to wider community initiatives, such as urban regeneration, aimed at increasing well-being. Finally, there is the idea of 'shrinking geographies of smoking' and the health issues, particularly those facing children and adolescents, as tobacco control policies increasingly move from public to private spaces such as the home. These are important for limiting exposure to second-hand smoke and restricting smoking initiation, but also have human rights implications.

These issues remain important in any geographic analysis of tobacco control policy. However, a particular challenge lies in identifying useful theoretical frameworks that can lead to greater understanding of the nature of tobacco control initiatives, where such moves take place, the extent to which policies are enforced, and the influence of local community characteristics in understanding the policy process and its outcomes. Traditional geographic theories which have considered the contextual and compositional factors are important but, in themselves, because they seldom consider national factors, are of less use for understanding policy development and its impact in different communities. As aspects of the wider political, social and cultural environment will be important, then a political economy approach, which emphasises cultural differences and power relationships, will be of more relevance to understanding the geography of tobacco control than local community or individual behavioural and personal differences. Here, the focus is less on the individual smoker, and more on the variety of local political elites, economic and other interests who may resist the implementation of tobacco control, compared to public health advocates and other important role models who will strongly support it. Thus, some research which has focused on international differences has used different theoretical approaches including agenda setting theory to understand how different political contexts facilitate policy changes (Albaek et al. 2007). Others have focused on the idea of corporatism where social partnerships with the tobacco industry have led governments to be reluctant to impose penalties on the tobacco industry (Studlar et al. 2011).

Because tobacco control is a politicised institutional process, it is more difficult to research than individual smoking behaviour. Understanding how and why political decisions are made with respect to policy changes is necessary as is developing a knowledge of the barriers or resistance to tobacco control efforts, whether this resistance comes from the tobacco industry, local governments or consumers themselves. It is also necessary to ascertain where policies have been implemented and the political cultures of the regions or cities that have been most active in tobacco control efforts. It is also essential to assess the effects of any such policies (such as raising cigarette prices) on reducing smoking prevalence in particular places.

1.1.3.4 Critique of Past Geographical Smoking Research

Since the seminal work of Macintyre et al. (Macintrye et al. 1993; Macintyre et al. 2002) geographers have invested a great deal of effort in understanding place effects on health and health behaviour. This has been a valuable addition to the smoking literature and has increased our understanding of the different place-related processes affecting smoking and its control. However, by relying upon an approach emphasising context and compositional effects, and interactions between them, it could be argued that the results of such research have provided only a partial view of the broader societal processes affecting smoking prevalence in different societies. To emphasise this point, a review of all papers published in the journal, *Health and Place*, foremost among academic journals that emphasise research on environments and health, was undertaken from 2000 to 2019. Only papers that considered smoking or smoking along with other health behaviours were considered. This yielded a total of 73 papers, with the following conclusions drawn. First, in terms of the scale of analysis, most research (55% of papers) focused on the local neighbourhood scale, either the residential neighbourhood or school catchment area with a further 14% focusing on home, work, recreation or other public spaces usually within cities. While some papers considered regional and rural-urban variations in smoking, only two, both in Europe, examined national factors, especially the role of political culture (Mackenbach 2014) in affecting variations in prevalence.

Second, the local emphasis of research was also reflected in the types of topics examined. The most popular topics were smoking prevalence studies (38%), often using a context-composition lens. Related to the above, but having a more specific focus, were studies of tobacco retailing (19%). These mainly dealt with the effects of outlet density on tobacco pricing, proximity to schools, youth smoking initiation and the socio-economic and ethnic residential location of retailers. A larger proportion (26%) of very diverse papers dealt with aspects of tobacco policy. However, few of these examined the policy process, instead focusing on a range of outcomes, such as the re-spatialisation of smoking, denormalisation and stigma, or the reaction of smokers to increased restrictions in the home, at work or in public spaces. Except for one paper on China (Yang et al. 2015), no papers focused on the tobacco industry despite its continuing importance in influencing local political cultures of smoking and especially youth smoking.

Third, in terms of the places of research, the locations were, much as expected, very western-centric. Excluding one paper from Mexico, the majority of studies originated from the USA (32%) and Canada (16%), followed by Britain (18%) and the rest of Europe (18%) and Australasia (5%). The rest focused mainly on China and the rest of Asia (10%), with no papers on Africa or Latin America.

While this review is necessarily selective, it indicates the western centric nature of smoking research, and possibly health research more broadly, along with a somewhat myopic focus on locality issues. In what Slater (2013) has termed a 'cottage industry of neighbourhood effects studies', the predominantly local focus has limited the development of health geography as a more critical part of the discipline. This view is also taken by Herrick (2016) with respect to critical studies of global health. Herrick sees geography as being "side stepped" by other disciplines in studies of global health, most notably anthropology and political science.

More recently Bambra et al. (2019, p. 36) have indicated that, while health geography has enhanced our understanding of the effects of local environments on health, 'it has done so at the expense of marginalising the influence of macroeconomic and political structures on both place and health'. They note that the undue emphasis on context-composition arguments has privileged horizontal influences over vertical ones relating to the effects of macro political and economic forces on both place and health. Arguments that 'place matters for health, but politics matters for place' (Bambra et al. 2019, p. 37), however, are not new and have been emphasised in the political economy literature in the past (Cooke 1989; Harloe et al. 1990).

Nevertheless Bambra et al.'s call to 'scale up' research to take greater account of the politics of health and place is a welcome development. Such 'scaling up' is clearly necessary. Geography has been overtaken by other disciplines, such as international political economy and business studies, which have started undertaking research that was once more common in Geography, during the 1980s and 1990s. Despite its concern with space and place, research in health geography has paid little attention to examining national or regional level factors influencing smoking and tobacco control. Global economic changes, such as deindustrialisation, remain one of the basic causes of persistence of smoking in more deprived areas of richer countries. These changes, which underlay the marked increase in national and regional income inequality in western countries, have received little attention, as have the influence of national political cultures on smoking prevalence. In both cases the most insightful work has come from European public health researchers such as Richard Wilkinson (e.g. Wilkinson and Pickett 2009) or Johan Mackenbach (e.g. Mackenbach 2014).

The same could be said for the neglect of work on the globalisation of the tobacco industry, the absence of research on smoking in low- and middle-income countries and the record of tobacco control. As was argued in an earlier monograph, 'geographers have led research on globalisation yet have remained remarkably silent about the activities of "Big Tobacco" and how it is changing global geographies of smoking. Conflicts between economic development and health concerns in these countries deserve increased attention' (Barnett et al. 2017, p.9). Lee et al. (2016), for example, provide a recent review work on the globalisation of the tobacco industry. They call for the development of an interdisciplinary approach to explore a range of issues including the types of business structures (e.g. state monopolies or companies partly state owned) which have evolved and the types of strategies used in pursuing globalisation, in how the industry has strategically framed tobacco control issues in ways favourable to its interests and how the progress of globalisation has varied over time across different geographies. Work on these lines has already begun, including recent research on the globalisation strategies of Asian tobacco companies (Lee and Eckhardt 2017a, b).

Similarly, with the exception of smoking ban legislation, which has a distinct spatial component, studies of tobacco control legislation have largely been absent from the geographical literature. This is unfortunate especially given national and regional variations in political cultures and policy enforcement which have accompanied the implementation of the World Health Organisation's Framework Convention on Tobacco Control (FCTC). To date such research has been largely conducted by political scientists and economists and those working in public health (e.g. Joossens and Raw 2006; Berg et al. 2018), with the result that, despite obvious national and local differences in policy implementation, these topics seldom feature in geographic research.

There has also been a reluctance to explore the relevance of western theoretical concepts (e.g. such as denormalisation and stigmatisation) to understanding smoking trends in these contexts. Much the same could be said about the relevance of western methods of tobacco control to non-western contexts where different cultural values or methods of policy implementation may affect the success of any policy initiatives. As we shall see in the case of tobacco taxation in China, what was relatively effective in western countries was not effective there. In summary, while geographical research on tobacco control has recently begun to emerge, its approach has been partial and left many important questions unaddressed. With these issues in mind, we shift our attention to smoking research in China and to the key questions posed in this book.

1.1.4 Smoking and Tobacco Control in China

1.1.4.1 Why China?

This book focuses on environmental interpretations of the smoking epidemic in China. China, with its 44% share of the global cigarette market in 2018, is the world's largest producer (Fig. 1.5) and consumer of tobacco products, and smoking prevalence remains high as does consumption per capita (Hoffman et al. 2019).

Among LMICs, China is relatively unusual in that its tobacco industry remains in the hands of a dominant state producer, the China National Tobacco Corporation (CNTC), rather than being partly state-owned, like Japan Tobacco International, or controlled by global multinational companies, often as a result of the takeover of former state monopolies, which have stimulated cigarette consumption in much of the rest of Asia. Although much has been written on western TTCs and their global activities, much less attention has been paid to state-owned tobacco companies (Hogg et al. 2016). China is also unusual because despite moves to implement different tobacco control initiatives, the full implementation of these has been modest due to the economic importance of tobacco in the Chinese economy. In the past, therefore, economic imperatives have tended to dominate health concerns, although, with the recent announcement of China 2030 health goals, there are indications that this may be changing (Tan et al. 2017).

The idea for the book arose from collaboration between the Department of Social Medicine at Zhejiang University and Ross Barnett of the Department of Geography at the University of Canterbury. The initial collaboration focused on exploring smoking and other public health issues highlighting the importance of different environmental factors in China's economic and social transition. Although there have been numerous academic articles exploring smoking and tobacco control in China, no book has examined the importance of the various contextual factors that influence smoking prevalence and the processes and outcomes of tobacco control efforts. To date, most books have either had an economic or policy focus but have said little about the importance of place and space in affecting smoking and policy outcomes. China is a large and diverse country and the significance of this diversity has to be appreciated in understanding smoking and other forms of health behaviour.

This book is important for a number of reasons:

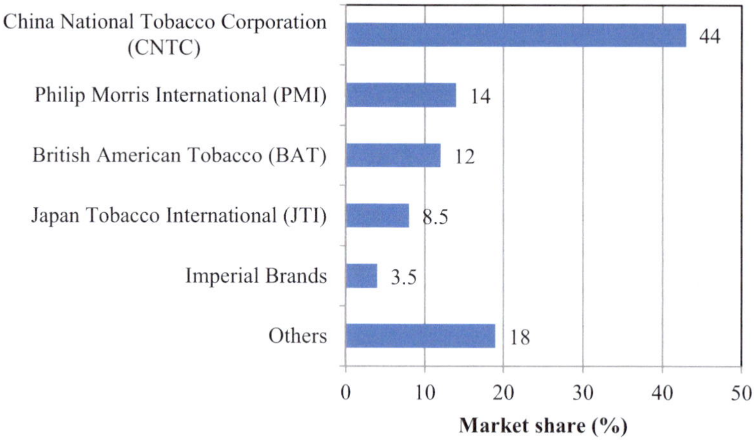

Fig. 1.5 Global cigarette market share by company, 2018. (Credit: Based on data presented in Campaign for Tobacco Free Kids, 2019 https://www.tobaccofreekids.org/assets/global/pdfs/en/Global_Cigarette_Industry_pdf.pdf)

- China is the world's largest cigarette consumer and smoking norms remain prevalent and still partly institutionalised.
- It will be the first major work to focus on environmental factors influencing smoking prevalence and tobacco control in China.
- It examines a particular cultural and institutional context where the aims of tobacco control conflict with economic policy.
- It highlights the importance of Chinese cultural perspectives to understanding smoking behaviour and the ineffectiveness of many tobacco control initiatives.
- It brings together Chinese and western academic scholars in its writing and synthesises Chinese and western research on the smoking epidemic.

Of particular importance is our attempt to synthesise the influence of cultural, economic and social attributes of place and to explore the interconnections between these factors.

1.1.4.2 Past Smoking and Tobacco Control Research in China

Over the last few decades, a large body of research has explored the smoking epidemic in China. Despite this, few studies have explored the significance of geographical factors and the role they have played not only in affecting the prevalence of smoking but also in tobacco control responses to the smoking epidemic. Given the size and economic, cultural and social diversity of China, this is surprising. We argue that the smoking epidemic needs to be more broadly situated in the context of the rapid economic and social changes that have occurred in Chinese society since 1979 and the way these have played out in different parts of the country. Similarly, policy responses for tobacco control need to be considered in such terms.

This is not to say that environmental factors have been neglected in Chinese academic research. Recently much research has been devoted to public health problems such as the increasing obesity epidemic in China (e.g. Jia et al. 2017; Wang et al. 2018), issues of aging (e.g. Zhang and Wu 2017; Liu et al. 2017b) and the negative effects of migration (e.g. Li and Rose 2017; Song and Smith 2019) and urbanisation upon health and social outcomes (e.g. Miao and Wu 2016; Liu et al. 2019). Much of this work focuses on issues of place and space and the impact of different kinds of environmental factors, including the effects of climate change (Wang et al. 2018).

With respect to smoking, similar trends have occurred (Chen et al. 2018). Since 2007, Professor Tingzhong Yang and his colleagues at Zhejiang University, Hangzhou, in particular, have examined the importance of geographic context in a large number of papers concerned with smoking and tobacco control in China. Many of the research outputs of the Center for Tobacco Control, established by Professor Yang in the Department of Social Medicine at Zhejiang University, reflect such concerns. These papers highlight the importance of place effects, at different geographical scales, as being important explanatory factors in understanding smoking and other forms of health behaviour. The work of the Centre has not been confined to smoking, but has also considered the role of environmental factors in influencing obesity, problem alcohol use, sleep deprivation and mental health status, along with a range of other health issues.

As noted in *Smoking Geographies* (Barnett et al. 2017, p. 206), smoking, like so many other health behaviours, is a product not only of individual factors but also of the environment. For example, while models of smoking transitions help us understand the broad shifts that have occurred over time, these models often fail to acknowledge the nuanced, but important, differences in smoking patterns both between and within nations. However, it is not enough to tritely claim that 'geography matters', but to demonstrate how it does so in complex and multifaceted ways. In this book we will unravel the different aspects of geography (economic, institutional, cultural or social) that allow for a deeper understanding of the national and industry determinants of smoking and the particular ways the Chinese state has responded to the need for tobacco control in China.

Although a substantial literature has evolved on different aspects of smoking in China, there are few books on the topic and, apart from Wang (2013), few deal with geographical considerations. With few exceptions (e.g. Yang 2017b), research to date has paid little attention to the dynamic societal changes that have taken place in China in recent decades or to locate smoking and other forms of health behaviour in the context of these changes. Since 2010 six important books have appeared on smoking and tobacco control in China:

- Yang, Gonghuan (ed), *Tobacco Control in China* (Springer, Singapore, 2018). This book focuses largely on the policy process of tobacco control and its successes and failures. It provides a valuable contribution to understanding health policy but pays less attention to the environmental factors influencing smoking and also how these have helped shape differences in the implementation of tobacco control policy in China. The book is rich on detail and provides a valuable resource for scholars interested in the evolution of tobacco control.
- Kohrman, M. et al. (eds), *Poisonous Pandas: Chinese Cigarette Manufacturing in Critical Historical Perspectives* (Stanford University Press, Stanford, 2018). This book provides a historical account of the rise of the Chinese tobacco industry and its political, economic and cultural influences. Important sections in the book are its focus on 'money and malfeasance' and tobacco governance, which takes account of subnational stakeholders and the tobacco industry's role in obstructing tobacco control.
- Hu, Teh-wei, *Economics of Tobacco Control in China: From Policy Research to Practice* (World Scientific Publishing Company, Singapore, 2016). This book largely deals with the economic costs of smoking, the demand for cigarette consumption and tobacco taxation and makes only minor reference (one chapter) to tobacco control. It builds on Hu's earlier book, *Tobacco Control Policy Analysis in China* (World Scientific Publishing Company, Singapore, 2008).
- Wang J, *State-Market Interactions in China's Reform Era: Local State Competition and Global Market Building* (Routledge, Abington, Oxford, 2013). The main aim of the book, as the title suggests, is to examine the regional decentralisation of the tobacco industry in the 1980s and 1990s and the importance of

local governments in fostering protectionism as a result of inter-regional competition. The book is one of the few works that adopts a geographic focus in the regional development of the Chinese tobacco industry and the contradictions that this produced.

- Li, Cheng, *Mapping China's Tobacco Industry and Anti-Smoking Campaign Politics and Policies* (Brookings Institution Press, Wash. DC, 2012). Like Kohrman's more recent work, this book primarily focuses on the political map of China's tobacco industry and its main stakeholders. In doing so geographic factors are considered in the analysis but are not the main focus of the book.
- Benedict, C., *Golden Silk Smoke. A History of Tobacco in China* (University of California Press, 2011). The book provides a detailed history of smoking in China from circa 1550 to recent times. It focuses on two well-defined themes. First it compares China's historical and cultural experience of adopting tobacco to that of other societies at the time undergoing social transformations. Second it looks at differences within China of changes in consumption patterns and thus provides an insightful analysis of how tobacco has become such a central part of Chinese culture today.

Together these books provide a valuable resource on smoking and tobacco control in China. Some, but not all, examine the geography of smoking and tobacco control, but this was not their main focus. In order to provide a more wide ranging analysis of smoking research in China, this book takes account of the key geographical themes, such as urbanisation, migration and socio-economic transition, which have been important in China over the last few decades. In examining such themes, we undertook a comprehensive literature search of Chinese literature primarily in western academic journals using a variety of databases. We were particularly interested in the types of topics researched, the nature of the analyses undertaken, and the extent to which different environmental factors were identified as being important.

1.1.4.3 Critique of Chinese Research

The literature search was instructive in highlighting the key themes of recent research on smoking and tobacco control in China, including the lack of engagement with geographic factors. This provided an initial impetus for writing this book. In reviewing recent research, we came to the following conclusions.

First, many studies focused on personal attributes of smokers such as socio-economic status, gender, age or ethnicity and the factors which resulted in group differences in smoking prevalence (e.g. Yang 2017a). For example, while the gendered nature of smoking has received much attention (e.g. Hermalin and Lowry 2012), there have been few detailed analyses of the role of gender in China or in other Asian or LMIC countries. Few studies place gender within a wider context of socio-economic change or, for example, examine how gender differences in smoking prevalence vary by geographic region or by socio-economic status. While gender and socio-economic status are key attributes of the western smoking epidemic model (Lopez et al. 1994), the relevance of this model needs more detailed examination, especially for countries, like China, which have strong social constraints against women smoking.

Second, many studies rely on large national surveys but often fail to capitalise on the diversity of their samples by exploring place effects. Contrasts between different regions or localities are often ignored, but these are important in a country as large and diverse as China. If geographical differences are explored, as by some of the present authors, they tend to be coarse regional ones or defined at the citywide scale. Also the range of place effects that have been examined is quite modest and has tended to focus on urban or regional GDP (as a measure of economic health) and sometimes measures of urban inequality. Surprisingly the debates on the negative effects of income inequality on health (Inoue et al. 2019; Du et al. 2019) and especially smoking (Li and Zhu 2006) that have characterised western health research have barely touched China.

Third, as Yang (2017b) has indicated, there is a lack of understanding of how communal socio-economic developments, as a result of China's transition from a socialist to a capitalist economy, have affected smoking trends and substance use more generally. While models of the smoking epidemic highlight the importance of changes in smoking prevalence by gender and socio-economic status, they pay no attention to community socio-economic developments and how these shape smoking outcomes for different social groups. Thus it is important to understand the extent to which individual changes in smoking prevalence are shaped by different socio-economic contexts.

Fourth, unlike western contexts, there has been very little research at the intra-urban scale. Given the rapid urbanisation of China in recent decades, this omission is surprising. While migration status and its effects on health behaviours has received increased attention (e.g., Liu et al. 2016), few researchers have considered the types of localities that migrants move to and the stresses encountered there. There is, surprisingly, little research on the local retail sector and little on how the structure of tobacco retailing affects the availability and pricing of tobacco products (Gong et al. 2013). There have also been few studies of the clustering of tobacco outlets, especially around schools or of the effects of such clustering on the persistence of smoking norms. How such trends have affected smoking initiation among adolescents has barely been considered.

Fifth, there has been very little qualitative research. The cultural turn which affected geography in western academic departments is not very visible in China other than through the writings of some western-based Chinese academics (Zhao et al. 2018). Most Chinese studies tend to be large-scale quantitative projects, often very repetitive in their methodological approach and more akin to the positivist viewpoint of western scholars in the 1970s. On the positive side, many Chinese studies have employed multilevel modelling and, in doing so, have considered a range of contextual effects in their analyses. However, there is room for other research methods in exploring smoking and tobacco control, along the lines of recent work by Aimee Mao who explored smoking control in the home and the relative power of young women compared to their husbands and grandparents (Mao 2013).

Sixth, there is a need to more fully explore the political economy of smoking policy implementation in China and how it varies across the country. Although much policy is 'top-down', regional variations exist (e.g. Li et al. 2013) as they do in other areas of policy, such as the production of green space, for example (Chen and Hu 2015). Apart from the books previously mentioned, few research articles have explored variation in the policy development. We found only one article which examined geographic differences in policy implementation across China (Astell-Burt et al. 2017). There is clearly a need for more nuanced research along the lines of Zhou (2000) and more recently Jin (2014), which looks at local economic and social conditions and how these have influenced the policy process and policy outcomes. Particularly important here is the influence of the tobacco industry on local political elites and their decision-making (Kohrman et al. 2018; Yang 2018).

Finally, we found relatively few articles linking smoking to other forms of health behaviour (e.g. Wang 2015; Sun et al. 2019). Given the growing obesity epidemic in China, it is important to make such connections since reductions in smoking may increase levels of obesity. Already it is clear that different social gradients are emerging for these two health behaviours, so research which considers the wider impacts of tobacco control on different groups in different regional contexts will be valuable.

1.1.5 Key Objectives of This Book

In the light of the previous critique, and building on our own research, this book has the following objectives:

- To identify and better understand the influence of economic, cultural and social environmental factors on the smoking epidemic in China.
- To locate smoking trends within the wider context of urbanisation and socio-economic transition.
- To compare smoking trends in China with those in other transition economies that have been subject to globalisation of the tobacco industry.
- To understand the regional and local geography of tobacco control and how and why this has varied across the country.
- To examine particular cultural and institutional contexts where the aims of tobacco control have conflicted with economic policy.
- To highlight the importance of Chinese cultural perspectives in understanding why many tobacco control initiatives have proved ineffective.

In highlighting these objectives, we seek to add a geographic perspective to the growing body of health research on China. As was emphasised in *Smoking Geographies* (Barnett et al. 2017), an in-depth understanding of the relationship between place and smoking behaviour requires appreciating that smoking represents one of many social practices that are accepted in particular contexts. Bourdieu's concept of 'habitus' (Poland et al. 2006) is significant here to understanding smoking in China. Bourdieu extended Marx's idea of capital beyond the economic and into the more symbolic realm of culture. Both are important in China: the first in terms of the influence of political and economic power that has legitimised smoking through patterns of institutionalised norms and practices and the second through collective lifestyles which, in turn, have influenced individual behaviour. As we shall see, however, dominant social practices do not apply across all social groups. Other philosophies, for example, the resurgence of traditional Confucian values, also may have affected rates of smoking especially among higher socio-economic status groups and women. Understanding the interaction and conflicts between dominant societal norms and social practices and individual/community agency (resistance to the dominant norms) in particular places is one of the challenges faced in understanding smoking in China. Such ideas are useful because they simultaneously consider the role of, and emphasise the interaction between, wider structural and societal forces and mediating local group or individual factors, such as resources, social capital and social practices, which also influence the decision to smoke.

1.1.6 Structure of the Book

1.1.6.1 Theoretical Perspectives

While the concept of 'habitus' is helpful in understanding the context of smoking in China, in this book we were guided by three other important theoretical perspectives. First, the political economy of place (Hanlon 2018) provided an important first step in understanding geographies of smoking and tobacco control. Macro level policy decisions have resulted in the continued importance of tobacco consumption within the broader economy of China. This is just not a national issue, however, as protectionist policies of provincial and local governments have also been important (Zhou 2000). As in the USA, there are a range of regional and local elites that have resisted moves towards greater regulation of tobacco products and encouraged their consumption (Fallin and Glantz 2015). In the light of such trends, a political economy approach helps us understand where power lies and in whose interests that power is exercised (Schrecker and Bambra 2015). Thus, as Bambra et al. (2019) have emphasised, explanations of health inequalities need to go beyond the neighbourhood and consider those central engines in society that generate and distribute power, wealth and risks. Political choices can thereby be seen as one the fundamental causes of inequalities in health and variations in health behaviour.

A political economy perspective also helps us to understand cultural practices in smoking and the impact of socio-economic factors. Although, as Benedict (2011) has indicated, 'tobacco has many social, psychological, and somatic attractions that transcend culture', cultural practices of smoking in China have been strongly influenced by class and power. This is seen particularly in the gendered nature of smoking, influenced by patriarchal power, but also evident in the ways in which economic and political interests have moulded cultural patterns, such as cigarette gifting, which helped legitimise smoking as a positive social practice.

Similarly, a political economy perspective can also provide insights into the links between economic transitions and smoking. In common with many other places, including western countries, such as New Zealand, and former Communist countries in Eastern Europe, since 1978 the Chinese economy has undergone a marked socio-economic transition. As in the West, China's adoption of neoliberal economic and social policies has had far-reaching consequences in terms of a promotion of consumption and the subsequent emergence of lifestyle disorders such as the growing obesity epidemic and smoking-related illnesses. Similarly, the increased focus on individual, rather than state, responsibility meant that smoking became an individual rather than a societal problem, a philosophy that has persisted until quite recently. Neoliberal political reforms also produced increased urbanisation and a host of other socio-economic changes in Chinese society that increased pressures to smoke. These pressures were not uniform, across places or population groups but must be understood as contributing factors in the Chinese smoking epidemic.

While we highlight some of the political determinants of health and links to culture and patterns of societal change, we also consider the local geographic processes emphasised in traditional context-composition studies of health behaviour. It is one thing to clarify the importance of higher level processes but another to examine how these have played out in different spatial contexts, whether provinces, cities, urban neighbourhoods or rural towns and villages. For example, decisions regarding the tobacco industry are made at national level, but its effects on health play out at regional and local levels. At the regional (meso) level, the presence of cigarette manufacturing will be an important contextual factor that will influence individual smoking behaviour through a variety of pathways, the effects of which will be felt at the local scale. Understanding these links goes beyond the traditional confines of context/composition research and broadens its emphasis and significance. In order to understand these links, we adopt a political ecology approach (Chitewere et al. 2017) which highlights the historical and structural approaches that produce and maintain patterns of behaviour that affect health and well-being. This approach is particularly useful in understanding how historical patterns of power shape community relations and attitudes and shed light on why communities exhibit certain patterns of health behaviour.

Finally, in examining issues associated with the emerging globalisation of the Chinese tobacco industry and the implementation of tobacco control, we engage with theories of global health governance (GHG) (Ng and Ruger 2011). GHG stresses the importance of non-state actors ranging from local to international civil organisations who are engaged in global health issues including the implementation of the provisions of the FCTC. In the case of the latter, it is especially important in examining 'agreed rules for decision making and accountability' (Lee and Kamradt-Scott 2014) which become important when assessing China's compliance with the FCTC as well as the tension between state and individual responsibility in the implementation of smoking cessation programmes (Glasgow and Schrecker 2016).

1.1.6.2 Chinese Smoking Geographies

Our goal in writing this book is to develop a greater understanding of smoking in China by not only considering the range of environmental factors related to smoking but also locating smoking as a health behaviour in the context of the wider socio-economic changes that have affected Chinese society since the late 1970s. While our focus is mainly a contemporary one, we also explore longer-term cultural factors that have affected smoking in Chinese society and to look at these through a modern lens. We also compare western and Chinese perspectives on smoking and pay particular attention to the relevance of the former in understanding the significance of smoking in Chinese society and how different cultural factors have affected the practice of tobacco control. This approach is strengthened by our past co-operation and discussions about the necessity of blending both western and Chinese perspectives.

Like the approach taken in *Smoking Geographies*, this monograph consolidates and extends our joint research from the last five years. The Center for Tobacco Control has deliberately engaged in international co-operation with scholars from a variety of countries strengthening their past work and ability to engage with different theoretical perspectives. This monograph not only considers contemporary work in health geography but also assesses other theoretical perspectives, concerned with smoking environments, in medical sociology, public health, health economics and epidemiology. In developing our ideas, we worked collectively on different chapters and aimed to produce a text to which we have all contributed.

As part of our approach, we set out a critical review of different theoretical approaches and their relevance to smoking in China. Concepts of neighbourhood, for example, so important in western research, are interrogated as to their relevance in Chinese cities. Similarly, we question the extent to which western ideas, such as denormalisation and stigmatisation, are relevant to China. In each chapter, we critically evaluate current research in China, in terms of the nature of the research and the methodologies used, and use this critique as a basis for suggesting future research priorities. One of the challenges of applying western theories and methods to other contexts is assessing their relevance and considering approaches might be more fruitful in explaining smoking or other trends in health behaviour.

The monograph is structured to reflect our main interests and concerns. The first major section of the book (Sect. 1, Chap. 2) places the smoking epidemic in China in a broader global context of the smoking transition. Changes in smoking prevalence by socio-economic status, age, gender and ethnicity, and the ways in which these contrast with the western experience, are discussed. The goal here is to provide a deeper understanding of the trends among different population groups and to assess the data on which such trends are based. We are particularly interested in the degree to which socio-economic trends in smoking mirror those in western countries and the contexts in which such trends have been most evident. The same is true for gender differences in smoking and the degree to which such differences have a particular geography. Smoking transition models have normally been discussed in a global context, but can also be applied to larger countries with regionally diverse populations and differences in economic development.

Part 2 of the book examines the importance of various environmental influences on smoking. Chapters 3 and 4 examine the economic context of the tobacco industry. Chapter 3 outlines the evolution and structure of the CNTC monopoly, the reasons for its restructuring, particularly in response to the regional protectionism which characterised the industry in the 1980s and 1990s and to China's entry into the WTO in 2001. Changes in the regional geography of tobacco growing and cigarette manufacturing are considered, along with the recent globalisation of the CNTC and the illicit tobacco market. Chapter 4 is concerned with consumption, how the latter has been influenced by CNTC marketing, and the extent to which an industry presence is associated with regional differences in smoking. Throughout the chapter, different aspects of the supply chain are discussed, including tobacco growers, cigarette manufacturers and tobacco retailers.

Chapter 5 pays particular attention to the role of geographical context and cultural practices affecting smoking. Chapter 5 identifies cultural factors that have affected smoking prevalence. We pay particular attention to the role of Confucian values and how these have shaped gender patterns of smoking and traditional social practices, such as cigarette sharing and gifting. Other cultural practices, such as the discriminatory *hukou* system and its effect on urban migrants, also receive attention.

Chapter 6 locates smoking in the context of neoliberal patterns of urban development. We address three issues: the growth of income inequality in China, patterns of urban development and their significance for smoking and wider health inequalities and how urbanisation has affected the social distribution of smoking. In many respects the social differences in smoking that we see between richer and poorer countries are replicated within urban and rural areas in China. Thus it is important to consider the geography of urbanisation in models of the smoking transition.

Part 3 of the book considers the institutional context of tobacco control. In some respects, because of the lack of research in China, this was the most challenging part to write. In order to understand the complexity of policy making in China, Chapter 7 examines the different policy environments for tobacco control. The chapter outlines the evolution of the national policy environment in China over the last forty years (1979–2019) and the social, institutional and political barriers to tobacco control during that time. Particularly important has been the conflict between national economic and public health goals and how this conflict has shaped China's approach to implementing the goals of the WHO FCTC. The chapter sets out two critical events that led to change: China's ratification of the Framework Convention on Tobacco Control in 2005 and the move by Xi Jinping since 2013 to provide 'top level design' for social as well as political and economic issues. As an introduction to Chaps. 8 and 9, there is a brief narrative of tobacco control activity since 1979 that includes the important measures of the last decade.

In the light of the discussion in Chap. 7, Chapter 8 specifically outlines the range of tobacco control instruments that have been applied in China. The main aim of the chapter is to assess the extent to which China's tobacco control efforts have been implemented nationally, regionally and locally and to evaluate the effectiveness of different tobacco control policies. In doing so we compare China internationally in the extent to which it is achieving the goals set out in the WHO MPOWER strategy.

Chapter 9 rounds out this section by providing a case study of smoke-free policies. Like so much policy development in China, these have only been partially developed and have a strong local emphasis. Consequently there is much geographical variation in policy implementation which has been restricted largely to larger cities and particularly those in provinces less dependent upon the tobacco industry. The chapter concludes by examining tobacco control in the private space of the home and the extent to which such developments have been stimulated by wider public smoke-free policies.

Chapter 10 concludes the book. We draw the book together by identifying common themes that emerged as well as placing the work in the wider context of global trends in smoking. We return to the role of environmental factors, including how a knowledge of these is important to understanding both smoking as a major health issue in China and formulating more effective tobacco control policies. We conclude the book by identifying important methodological and theoretical issues which still need to be addressed to provide a deeper understanding of the smoking epidemic in China.

References

Albaek E, Green-Petersen C, Nielsen LB (2007) Making tobacco consumption a political issue in the United States and Denmark: the dynamics of issue expansion in comparative perspective. Journal of Comparative Policy Analysis 9(1):1–20

Anlin TMJ, Annechino RA, Hunt G, Lipperman S, Young M (2017) The gendered experience of smoking stigma: implications for tobacco control. Crit Public Health 27(4):443–454

Astell-Burt T, Zhang M, Feng X, Wang L, Li Y, Page A, Zhou M, Wang L (2017) Geographical inequality in tobacco control in China: Multilevel evidence from 98,058 participants. Nicotine and Tobacco Research 20(6):755–765

Ayuka F, Barnett R (2015) Place effects on alcohol consumption: A literature review. Journal of Addiction Research & Therapy 6:1000207

Bambra C, Smith KE, Pearce J (2019) Scaling up: The politics of health and place. Soc Sci Med 232(July):36–42

Barnett R, Moon G, Pearce J, Thompson L, Twigg L (2017) Smoking Geographies: Space, Place and Tobacco. Wiley Blackwell, Oxford, 301p

Benedict C (2011) Golden Silk Smoke. A History of Tobacco in China 1550–2010. University of California Press, Berkeley, 349p

Berg CJ, Fong GT, Thrasher JF, Cohen JE, Maziak W, Lando H, Drope J, Mejia R, Barnoya J, Nakkash R, Salloum RG, Paracandola M (2018) The impact and relevance of tobacco control research in low- and middle-income countries globally and to the US. Addict Behav 87:162–168

Brown-Johnson CG, England LJ, Glantz SA, Ling PM (2014) Tobacco industry marketing to low socioeconomic status women in the USA. Tob Control 23(e2):e139–e146

Campaign for Tobacco-Free Kids (2019) The Global Cigarette Industry. Tobacco-free Kids, December: https://www.tobaccofreekids.org/assets/global/pdfs/en/Global_Cigarette_Industry_pdf.pdf

Cancer Research UK (2014) Smoking Prevalence, Adults aged 16 and Over, Great Britain, 1948–2010. Available at: http://www.cancerresearchuk.org/cancer-info/cancerstats/causes/tobacco-statistics#trends

Chen WY, Hu FZY (2015) Producing nature for public: Land-based urbanization and provision of public green spaces in China. Appl Geogr 58:32–40

Chen Y, Chen H, Li Z (2018) Effect of characteristics and life in cities in China on residents' smoking behaviour. J Int Med Res 446(10):4226–4234

CATC (Chinese Association on Tobacco Control) (2018a) Smoking violators to get 180-day train travel ban. CATC Domestic News 22 March, 2018. CATC, Beijing

CATC (Chinese Association on Tobacco Control) (2018b) Railway no-smoking. CATC Domestic News 27 July, 2018. CATC, Beijing

CATC (Chinese Association on Tobacco Control) (2019) China's train experience would only improve with total smoking ban. CATC Domestic News 22 January, 2019. CATC, Beijing

Chitewere T, Shim JK, Barker JC, Yen IH (2017) How neighbourhoods influence health: Lessons to be learned from the application of political ecology. Health Place 45(May):117–123

Cooke P (ed) (1989) Localities: The Changing Face of Urban Britain. Unwin Hyman, London, 321p

Crooks VA, Andrews GJ, Pearce J, Snyder M (2018) Chapter 1: Introducing the Routledge Handbook of Health Geography. In: Crooks VA, Andrews GJ, Pearce J (eds) (2018) Routledge Handbook of Health Geography. Routledge, Abington, pp. 1–7

Cummins S, Curtis S, Diez-Roux AV, Macintyre S (2007) Understanding and representing 'place' in health research: A relational approach. Soc Sci Med 65(9):1825–1838

Doll R, Hill AB (1954) The mortality of doctors in relation to their smoking habits. Br Med J 1(4877):1451

Du H, King RB, Chi P (2019) Income inequality is detrimental to long-term well-being: A large-scale longitudinal investigation in China. Soc Sci Med 232(C):120–128

Duncan C, Jones K, Moon G (1996) Health-related behaviour in context: A multi-level modelling approach. Soc Sci Med 42(6):817–830

Duncan C, Jones K, Moon G (1999) Smoking and deprivation: are there neighbourhood effects? Soc Sci Med 48(4):497–505

Dwyer-Lindgren L, Mokdad AH, Srebotnjak T, Flaxman AD, Hansen GM, Murray CJL (2014) Cigarette smoking prevalence in US counties: 1996-2012. Popul Health Metrics 12(1):5

Fallin A, Glantz SA (2015) Tobacco-control policies in tobacco-growing states: where tobacco was king. Milbank Q 93(2):319–358

Finian LJ, Lipperman-Kreda S, Abadi M, Grube JW, Kaner E, Balassone A, Gaidus A (2019) Tobacco outlet density and adolescents' cigarette smoking: a meta analysis. Tob Control 28(1):27–33

GBD 2015 Tobacco Collaborators (2017) Smoking prevalence and attributable disease burden in 195 countries and territories, 1990-2015: A systematic analysis from the Global Burden of Disease Study 2015. Lancet 386(10082):1885–1906

Glasgow S, Schrecker T (2016) The double burden of neoliberalism? Noncommunicable disease policies and the global political economy of risk. Health Place 39(April):279–286

Gong T, Lv J, Liu Q, Ren Y, Li L, Kawachi I (2013) Audit of tobacco retail outlets in Hangzhou, China. Tob Control 22(4):245–249

Hanlon N (2018) Chapter 13: Environments of health and care. In: Crooks VA, Andrews GJ, Pearce J (eds) Routledge Handbook of Health Geography. Routledge, Abington, pp 88–93

Harloe M, Pickvance C, Urry J (eds) (1990) Place, Policy and Politics. Do Localities Matter? Unwin Hyman, London, 224p

Hermalin AI, Lowry DS (2012) The decline of smoking among female birth cohorts in China in the 20th century: A case of arrested diffusion? Population Research &Policy Review 31(4):545–570

Herrick C (2016) Global health, geographical contingency, and contingent geographies. Ann Am Assoc Geogr 106(3):672–687

Hoffman SJ, Mammone J, Rogers Van Katwyk S, Sritharan L, Tran M, Al-Khateeb S, Grjibovski A, Gunn E et al (2019) Cigarette consumption estimates for 71 countries from 1970 to 2015:

systematic collection of comparable data to facilitate quasi-experimental evaluations of national and global tobacco control interventions. Br Med J 365:I2231
Hogg SL, Hill SE, Collin J (2016) State-ownership of tobacco industry: a 'fundamental conflict of interest' or a tremendous opportunity for tobacco control? Tob Control 25(4):367–372
Holden C, Lee K (2009) Corporate power and social policy: The political economy of transnational tobacco companies. Global Social Policy 9(3):328–354
Hu T-W (2016) Economics of Tobacco Control in China: From Policy Research to Practice. World Scientific Publishing Company, Singapore, 302p
Inoue Y, Howard AG, Yazawa A, Kondo N, Gordon-Larsen P (2019) Relative deprivation of assets defined at multiple scales, perceived stress and self-rated health in China. Health Place 58(July):102117
Jia P, Xue H, Zhang J, Wang Y (2017) Time trend and demographic and geographic disparities in childhood obesity prevalence in China – Evidence from twenty years of longitudinal data. Int J Environ Res Public Health 14(4):369
Jin J (2014) Why FCTC policies have not been implemented in China: Domestic dynamics and tobacco governance. Journal of Health Politics, Policy & Law 39(3):633–666
Joossens L, Raw M (2006) The Tobacco Control Scale: a new scale to measure country activity. Tob Control 15(3):247–253
Kasemets K (2018) Re-habituating the landscape of smoking. Emot Space Soc 29(September):55–61
Kohrman M, Quan G, Wennan L, Proctor RN (eds) (2018) Poisonous Pandas: Chinese Cigarette Manufacturing in Critical Historical Perspectives. Stanford University Press, Stanford, 309p
Lee K, Eckhart J, Holden C (2016) Tobacco industry globalization and global health governance: towards an interdisciplinary research agenda. Palgrave CommunicationsVol. 2: Article number: 16037
Lee K, Eckhardt J (2017a) The globalisation strategies of five Asian tobacco companies: An analytical framework. Glob Public Health 12(3):269–280
Lee K, Eckhardt J (2017b) The globalisation strategies of five Asian tobacco companies: a comparative analysis and implications for global health governance. Glob Public Health 12(3):367–379
Lee K, Kamradt-Scott A (2014) The multiple meanings of global health governance: a call for conceptual clarity. Glob Health 10:28
Li C (2012) Mapping China's Tobacco Industry and Anti-Smoking Campaign Politics and Policies. John L. Thornton China Center Monograph Series Number 5 Brookings Institution Press, Washington DC, 213p
Li H, Zhu Y (2006) Income, income inequality and health: Evidence from China. J Comp Econ 34(4):668–693
Li J, Rose N (2017) Urban social exclusion and mental health of China's rural-urban migrants: A review and call for research. Health Place 48(November):20–30
Li X, Gao J, Zhang Z, Wei M, Zheng P, Nehl EJ, Wong FY, Berg CJ (2013) Lessons from an evaluation of a provincial-level smoking control policy in Shanghai, China. PLoS One 8(9):e74306
Liu S, Zhang M, Yang L, Li Y, Wang L, Huang Z, Wang L, Chen Z, Zhou M (2017a) Prevalence and patterns of tobacco smoking among Chinese adult men and women: findings of the 2010 national smoking survey. J Epidemiol Community Health 71(2):154–161
Liu Y, Gao J, Shou J, Xia H, Shen Y, Zhu S, Pan Z (2016) The prevalence of cigarette smoking among rural-to-urban migrants in China: A systematic review and meta-analysis. Subst Use Misuse 51(2):206–215
Liu Y, Dijst M, Faber J, Geertman S, Cui C (2017b) Healthy urban living: Residential environment and health of older adults in Shanghai. Health Place 47(September):80–89
Liu Y, Wang R, Xiao Y, Huang B, Chen H, Li Z (2019) Exploring the linkage between greenness exposure and depression among Chinese people: Mediating roles of physical activity, stress and social cohesion and moderating role of urbanicity. Health Place 58(July):102168
Lopez AD, Collishaw NE, Piha T (1994) A descriptive model of the cigarette epidemic in developed countries. Tob Control 3(3):242–247

Macintrye S, Ellaway A, Sooman A (1993) Area, class and health: should we be focusing on places or people? J Soc Policy 22(2):213–234
Macintyre S, Ellaway A, Cummins S (2002) Place effects of health: how can we conceptualise, operationalise and measure them? Soc Sci Med 55(1):125–139
Mackenbach JP (2014) Cultural values and population health: a quantitative analysis of variations in cultural values, health behaviours and health outcomes among 42 countries. Health Place 28(July):116–132
Mao A (2013) Space and power: Young mothers' management of smoking in extended families in China. Health Place 21(May):102–109
Miao J, Wu X (2016) Urbanization, socioeconomic status and health disparity in China. Health Place 42(November):87–95
Mistry R, Pednekar M, Pimple S, Gupta S, McCarthy WJ, Raute LJ, Patel M, Shastri SS (2015) Banning tobacco sales and advertisements near educational institutions may reduce students' tobacco use risk: evidence from Mumbai, India. Tob Control 24(e1):e100–e107
Moon G, Smith D (2018) Un/healthy behaviour A bibliometric assessment of geographers' contributions to understanding the association between environment and health-related behaviour. In: Crooks VA, Andrews GJ, Pearce J (eds) Routledge Handbook of Health Geography. Routledge, Abington, pp 74–81
Neely AH, Nading AM (2017) Global health from the outside: The promise of place-based research. Health Place 45(May):55–63
Ng M, Freeman MK, Fleming TD, Robinson M, Dwyer-Lindgren L, Thomson B, Wollum A, Sanman E, Wulf S, Lopez AD, Murray CJL, Gakidou E (2014) Smoking prevalence and cigarette consumption in 187 countries, 1980-2012. J Am Med Assoc 311(2):183–192
Ng NY, Ruger JP (2011) Global Health Governance at a Crossroads. Global Health Governance: The Scholarly Journal for the New Health Security Paradigm 3(2):1–37
Pearce J (2018) Complexity and uncertainty in geography of health research: Incorporating life course perspectives. Ann Assoc Am Geogr 108(6):1491–1498
Pearce J, Barnett R, Moon G (2012) Sociospatial inequalities in health related behaviours: pathways lining place and smoking. Prog Hum Geogr 36(1):3–24
Peto R, Chen ZM, Boreham J (2009) Tobacco: the growing epidemic in China. Prevention & Control 4(1):61–70
Poland B, Frohlich K, Haines RJ, Mykhalovskiy E, Rock M, Sparks R (2006) The social context of smoking: the next frontier in tobacco control? Tob Control 15(1):59–63
Schrecker T, Bambra C (2015) How Politics Makes Us Sick - Neoliberal Epidemics. Palgrave Macmillan, London, 178p
Slater T (2013) Your life chances affect where you live: a critique of the 'cottage industry' of neighbourhood effects research. International Journal of Urban & Regional Research 37(2):367–387
Song Q, Smith JP (2019) Hukou system, mechanisms, and health stratification across the life course in rural and urban China. Health Place 58(July):102150
State Tobacco Monopoly Bureau (STMB) (various years) China Tobacco Yearbook. China Economic Publishing House, Beijing
Stefler D, Murphy M, Irdam D, Horvat P, Jarvis M, King L, McKee M, Bobak M (2018) Smoking and mortality in Eastern Europe from the PrivMort Retrospective Cohort Study of 177,376 individuals. Nicotine Tob Res 20(6):749–754
Studlar DT, Christensen K, Sitasari A (2011) Tobacco control in the EU-15: the role of member states and the European Union. J Eur Publ Policy 18(5):728–745
Sun M, Jiang Y, Sun C, Li J, Guo X, Lv Y, Yu Y, Yao Y, Jin L (2019) The associations between smoking and obesity in northeast China: a quantile regression analysis. Sci Rep 9:3732
Tan X, Liu X, Shao H (2017) Healthy China 2030: A vision for health care. Value in Health Regional Issues 12C:112–114
Taylor A, Chaloupka F, Guindon E, Corbett M (2000) The impact of trade liberalization on tobacco consumption. In: Jha P, Chalpouka F (eds) Tobacco Control in Developing Countries. Oxford University Press, Oxford, pp 353–364

Thompson L, Pearce J, Barnett JR (2007) Moralising geographies: stigma, smoking islands and responsible subjects. Area 39(4):508–517
Twigg L, Cooper L (2009) Healthy behaviour. In: Brown T, McLafferty S, Moon G (eds) A Companion to Health and Medical Geography. Wiley-Blackwell, Oxford, pp 460–476
Wang C, Zhang Z, Zhou M, Wang P, Yin P, Ye W, Zhang L (2018) Different response of human mortality to extreme temperatures (MoET) between rural and urban areas: A multi-scale study across China. Health Place 50(March):119–129
Wang J (2013) State-Market Interactions in China's Reform Era: Local State Competition and Global Market Building. Routledge, Abington, Oxford, 176p
Wang Q (2015) Smoking and body weight: evidence from China health and nutrition survey. BMC Public Health 15(1):1238
Whitehead M, Townsend P, Davidson M (1992) Inequalities in Health: the Black Report, the Health Divide. Penguin, London, 464p
Wilkinson R, Pickett K (2009) The Spirit Level: Why More Equal Societies Almost Always Do Better. Allen Lane, London, 337p
Wipfli H, Samet JM (2016) One hundred years in the making: The global tobacco epidemic. Annu Rev Public Health 37:149–166
Witten K, Pearce J (2010) Geographies of Obesity: Environmental Understandings of the Obesity Epidemic. Ashgate, Farnham, 364p
WHO (World Health Organisation) (2018) WHO Global Report on Trends in Prevalence of Tobacco Smoking 2000–2025, 2nd edn. World Health Organization, Geneva
Yach D, Wipfli H, Hammond R, Glantz S (2007) Chapter 3: Globalization and Tobacco. In: Kawachi I, Wamala S (eds) Globalization and Health. Oxford University Press, New York, pp 39–67
Yang G (ed) (2018) Tobacco Control in China. Springer, Singapore, 323p
Yang T (2017a) Health Research: Social and Behavioral Theory and Methods. People's Medical Publishing House, Beijing, 197 p
Yang T, Barnett R, Rockett IRH, Yang XY, Wu D, Zheng W, Li L (2015) The impact of regional economic reliance on the tobacco industry on current smoking in China. Health Place 33(May):159–171
Yang T, Zhu Z, Barnett R, Zhang W, Peng S, Jiang S (2019) Tobacco advertising, anti-tobacco information exposure, environmental smoking restrictions, and unassisted smoking cessation among Chinese male smokers: a population-based study. Am J Mens Health 13(3):1557988319856152
Yang XY (2017b) How community-level social and economic developments have changed the patterns of substance use in a transition economy. Health Place 46(July):91–100
Zatoński W (2003) A nation's recovery.Case study of Poland's experience in Tobacco Control. Health Promotion Foundation, Warsaw (In Polish)
Zhang W, Wu Y-Y (2017) Individual educational attainment, neighbourhood-socioeconomic contexts, and self-rated health of middle-aged and elderly Chinese: Exploring the mediating role of social engagement. Health Place 44(March): 8–17
Zhao X, Young RM, White KM (2018) 'I'm not a smoker … yet': A qualitative study on perceptions of tobacco control in Chinese high schools. BMJ Open 8:e019483
Zhou H (2000) Fiscal decentralization and the development of the tobacco industry in China. China Econ Rev 11(2):114–133

Chapter 2
Epidemiological Transition of Smoking in China

Xiazhao Y. Yang, Sihui Peng, and Ross Barnett

Abstract This chapter focuses on trends in smoking prevalence and examines classic models of the smoking transition and their relevance to China. Traditionally these models have been used to explain major shifts in smoking prevalence in richer nations and variations between different population groups. The chapter describes current smoking prevalence in China and how this compares to global patterns, especially those of other transition economies which have undergone substantial political and socio-economic change. This is followed by an outline of the smoking transition model and its main components. In keeping with western research, three main dimensions of prevalence, age/gender, socio-economic status (SES) and ethnicity, are discussed. The final section examines the issue of geographical variations in smoking transitions. Understanding such differences is particularly important in a country as large and diverse as China.

Keywords Cohort effects · Diffusion · Ethnicity · Gender differences · Smoking intensity · Smoking prevalence · Smoking transition model · Socio-economic status · Transition economies

X. Y. Yang (✉)
Sun Yat-sen University, School of Sociology and Anthropology, Guangzhou, China
e-mail: yangxzh27@mail.sysu.edu.cn

S. Peng
Jinan University, School of Medicine, Guangzhou, China

R. Barnett
School of Earth and Environment, University of Canterbury, Christchurch, New Zealand
e-mail: ross.barnett@canterbury.ac.nz

R. Barnett et al. (eds.), *Smoking Environments in China*,
Global Perspectives on Health Geography,
https://doi.org/10.1007/978-3-030-76143-1_2

2.1 Introduction

Any study of smoking in China must locate the country in the context of the global tobacco epidemic. This chapter focuses on trends in smoking prevalence and examines classic models of the smoking transition and their relevance to China. Traditionally these models have been used to explain major shifts in smoking prevalence in richer nations and the differences that have occurred among different population groups. There has been less attention to changes in the smoking population in low- and middle-income and non-western countries, and there remains some uncertainty over the relevance of the western experience to China and of the role of different cultural, political, economic and social values. It is important, therefore, to situate China within this broader context and to show how current trends in smoking differ from the western experience in important respects.

With this in mind, the chapter is organised as follows. We begin by focusing on current smoking prevalence in China and how this compares to global patterns, especially those of other transition economies which have undergone substantial political and socio-economic change. In the second part of the chapter, we briefly outline the smoking transition model and its main components. In keeping with western research, we focus on three main dimensions of prevalence: age and gender effects, socio-economic status (SES) and variations between ethnic groups. For each section we provide a brief synopsis of the key findings and how they relate to the smoking transition. In the final part of the chapter, we specifically examine the issue of geographical variations in smoking transitions, paying particular attention to gender and, to a lesser extent, SES differences in smoking prevalence, the latter being dealt with in detail in Chap. 6. Understanding such differences is particularly important in a country as large and diverse as China.

2.2 Smoking in China: Data Scorecard, 2015

China, in common with many other transition economies, continues to have a high prevalence of smoking. This peaked at 35.3% in 1996, with only a very modest decline occurring since then to 27.7% in 2015 (China CDC 2015). However, since smoking prevalence varies markedly by gender and other demographic factors, this is a relatively meaningless statistic especially in contexts which such group differences are most pronounced. For example, in 2015 current adult prevalence in China was 52.1% among men but only 2.7% for women. Because of such gender differences, therefore, overall smoking prevalence in China is relatively modest when compared to some other major consumers, such as Russia (40.9%) or Indonesia (39.9%).

When male smoking prevalence is considered separately, a different picture emerges. Males account for the majority of smokers in China; of the 318 million adults who smoked in 2010, 95.6% were men (Liu et al. 2017). In 2015, China, with an age-adjusted male prevalence rate of 47.6%, was ranked 20th internationally in terms of male prevalence rates (WHO 2016). Although lagging far behind some other large consumers, such as Indonesia (76.2%) and Russia (59.0%), China was similar to many other countries in Eastern Europe, the former Soviet Union and the Middle East that also recorded high smoking rates (Fig. 2.1). These figures are very similar to estimates obtained from other surveys, such as National Health Surveys conducted in China (2003–2013), which recorded an adult (15 years and over) male age standardised prevalence rate of 47.2% in 2013 (Wang et al. 2019a, b). This aligns with Zhang et al.'s (2019) estimate of 46.3% for regular male smokers (18 and over) in 2013–2014, but lower than the 54% prevalence rate for current smokers in 2010 recorded by Liu et al. (2017) for adults 18 years and over.

However, cigarette consumption per capita tells a different story. Figure 2.2, drawing on data from Hoffman et al. (2019), shows that after plateauing in the 1990s, after 2000 consumption rates increased again, a trend also evident only for Indonesia. While an even more dramatic increase had occurred in the Russian Federation after the break up of the Soviet Union, by the early 2000s this was largely over with per capita consumption declining after 2010. By contrast, other large tobacco consumers, such as the USA, Japan and Germany, all witnessed declining patterns of per capita consumption typical of countries in advanced stages of the smoking epidemic. In China Liu et al. (2017) reported that between 1996 and 2010 the number of cigarettes smoked per current smoker rose from 15.2 to 17.9 cigarettes per day, with the highest per capita rates occurring in middle age groups (40–59). The slight decline in smoking prevalence thus has been matched by a rise in consumption per capita, reflecting the impact of the industry in countering declining domestic demand among certain sections of the population (see Chap. 4).

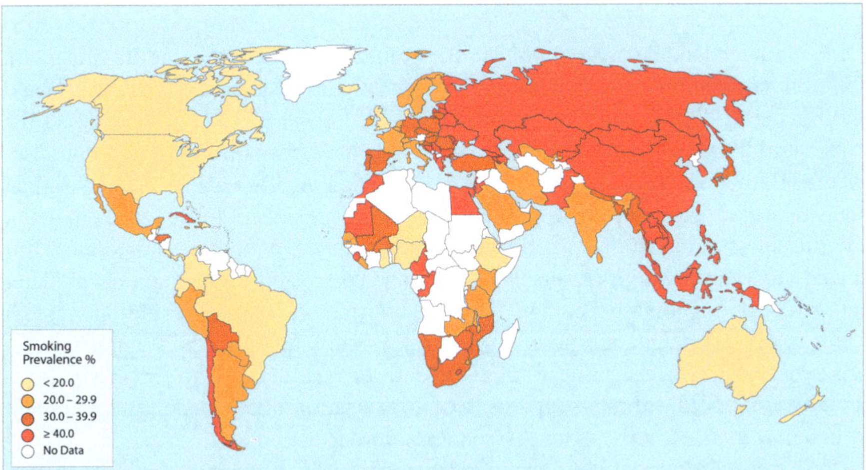

Fig. 2.1 Age-standardised current smoking prevalence among males aged 15 years and older (%), 2015. (Credit: WHO (2016) Global Health Observatory (GHO) data. https://www.who.int/gho/tobacco/use/en/ public domain)

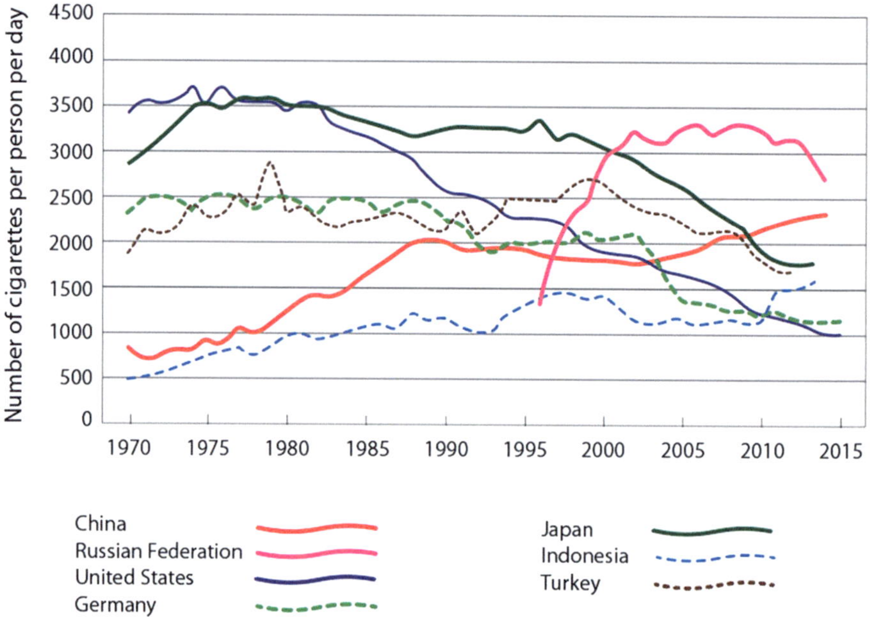

Fig. 2.2 Trends in cigarette consumption per capita 1970–2015 for countries which consumed the greatest volume of cigarettes in 2017*. (Credit: Reproduced courtesy of the British Medical Journal. Modified from Hoffman et al. (2019), p. 6; This file is licensed under the Creative Commons Attribution (CC BY 4.0) Licence; *Egypt, India and Vietnam are not included in Hoffman et al.'s dataset because of data quality issues)

2.3 Smoking Transition Models

To help understand changes in the smoking epidemic over time, and its impact on different population groups, western scholars, in particular Lopez et al. (1994) and more recently Dixon and Banwell (2009) and Thun et al. (2012), developed the concept of the smoking transition. These models have been discussed in detail elsewhere (Barnett et al. 2017) so only a brief overview is provided here. The original transition model developed by Lopez et al. (1994) consisted of four stages to detail the shift in smoking from a practice initially dominated by higher-income men, followed soon after by high-income women to one where gender differentials are later much reduced. Although Thun et al. (2012) consider their model useful as a rough guide to describing changes in male and female smoking (and death) rates in western countries, it suffers because it fails to consider the many of other factors, such as individual SES, ethnicity or levels of urbanisation and female empowerment which may also affect the extent of gender differences.

Perhaps more important than gender differences has been the relationship between socio-economic status (SES) and smoking. Here the work of Pampel (2001, 2005, 2006) has been significant in emphasising the importance of status differences in the initial uptake of smoking and its later diffusion from high to low status groups. The development of mass production and cheaper cigarettes initially led to a narrowing of the social gradient in smoking prevalence, but this soon reversed once the health costs of smoking became apparent to more educated sections of the population. As discussed by writers such as Dixon and Banwell (2009), the transition model highlights the concentration of smoking among more deprived populations. Again such trends hide important interactions between SES and gender and particularly ethnicity and how these have shaped temporal trends in prevalence.

A third important component of the transition model, but one which has been least explored, has been the links between ethnicity and smoking. Writers such as Factor et al. (2011) consider both macro and micro processes to explain such differences. They argue that ethnic differentials in smoking prevalence occur not only as a result of discrimination and alienation but also because of social resistance to dominant (white) public health narratives. To this can be added high levels of residential segregation which may help perpetuate social norms and high levels of smoking, particularly among lower-income women.

The three components (gender, SES and ethnicity) have all been well-rehearsed themes in research on smoking transitions in western countries. While they provide a general guide to demographic differences in smoking prevalence between countries, they have been less explored in non-western contexts, such as China. There has also been little attention to geographical differences in the pace of transitions within particular countries, a theme taken up later on in this chapter.

2.4 Demographic Differences in Smoking Prevalence in Post-Reform China

2.4.1 Age and Gender Differences

Without doubt, the contrast between males and females constitutes the major demographic difference in smoking prevalence in contemporary China. Much of the literature on gender differences in smoking has stressed differences in traditional sex roles and social norms which, for women, have emphasised a strong disapproval of smoking (Waldron 1991; Pathania 2011). This was not always so, and during the early twentieth century in China and some other Asian countries, female smoking rates were much higher than they are now (Hermalin & Lowry 2012). In China the initial reaction against female smoking began in the 1930s and reflected strong feelings of nationalism and resentment against western commercial domination and a desire to avoid decadent western practices. Trendy fashion conscious women (Fig. 2.3) began to be seen as immoral and a threat to traditional Confucian values and national well-being (Edwards 2000; Yen 2005). Other countervailing forces such as Madame Chang Kai-shek's New Life Movement in 1934 emphasised Confucian values and activities appropriate for each gender, which excluded smoking. Hermalin and Lowry's (2012) analysis of smoking rates of six 4-year birth

Fig. 2.3 Traditional and foreign cigarette posters targeting female consumers with visuals suggestive of femininity and modern lifestyle. (Credits: (left) Alamy Image ID: MHC64E Stock photo (Vintage cigarette advertising poster for Nanyang Brothers Tobacco Co, Shanghai, dated 1920s). (Right) Alamy Image ID: 2B0319F (Advertisement characteristic of 'Old Shanghai' (1920–1940s) started by American Carl Crow who established the first western advertising agency in the city featuring modern 'China Girl' posters))

cohorts between 1908–1912 and 1933–1937 reveals a sharp drop off in female smoking over this period.

These trends continued after the formation of the People's Republic of China in 1949. During the Maoist period, campaigns against unhealthy behaviours started in response to Mao's call to eradicate 'backward feudalistic traditions', which swept aside a wide range of traditional ideas and behaviours. It was officially proclaimed, and dubiously in the eyes of academic observers, that into the 1960s, prostitution had been eradicated from China along with opium poppy smoking – a hallmark symbol of the subjugation of the Chinese nation in the wake of the Opium Wars 1839–1860 (British-China Wars).

Smoking and drinking, however, was recalcitrant in the face of the campaign against substance use. Thus tobacco control during this period focused on education and health promotion often included in a range of moralistic teachings without concrete measures. By the time of the first major China national health survey in the 1980s, the gender gap in smoking had become substantial (26.9%), increasing slightly to 31.5% in 1996, but later decreasing to 25% in 2015, largely due to the greater decline in prevalence for men (Parascandola and Xiao 2019).

However, two identification problems in public health, which have been fully described by Bell (2020), thwart our full understanding of temporal trends in smoking in China and keep us from affirmatively answering how the decline in male smoking occurred. First, the decline may have been due to various trends that naturally advance as time passes: the level of education increases, the country becomes gradually modernised and knowledgeable of the harms of smoking and tobacco slowly loses its popularity. All of these occur as a structural trend over which individuals have little control. Second, perhaps no external trend in structural factors occurred, with the decline simply reflecting an ageing population and older people with more illness desisting from smoking.

The key difference between the two explanations is that the ageing effect is natural and inherent to individuals' health behaviour while the trend over time is exogenous. To complicate the matter further, there is also a separate explanation called the 'cohort effect': successive cohorts of people show different rates of smoking due to their unique life course experiences associated with their birth era. For example, we know that the 'famine cohort' tends to show worse health outcomes than their predecessors (i.e. age effect) (Chen and Zhou 2007; Luo et al. 2006) and their health outcomes still lag behind as the time passes (i.e. the period effect). These different mechanisms are methodologically and conceptually difficult to disentangle largely because they share collinearity. Thus, because of this, scholars have developed sophisticated statistical techniques to constrain or decompose the three effects (Rutherford et al. 2010; Yang and Land 2016).

2.4.2 *Cohort Analysis of Changes in Smoking Prevalence by Gender*

To analyse how age, period and cohort effects, respectively, affect the changes in smoking, the data need to at least contain a sufficient number of age points, cohort groups and survey periods (waves). Since the longest conducted survey, the China General Social Survey, does not include any consistent measure on smoking, we used data from the China Health and Nutrition Survey (CHNS). With ten survey waves spanning from 1989 to 2015 and including all adult respondents, this dataset should serve well to decompose the three separate effects. We used Stata add-on 'APCFIT' to conduct the analysis, separately for both genders since smoking in China is highly gendered, and plotted the growth curves of smoking.

2.4.2.1 Age and Cohort Effects

Usually in the tradition of demography, cohorts are constructed as 5- or 10-year age intervals for larger cohorts. Cohort refers to an aggregate of individuals sharing common characteristics in life trajectory and life events. Considering how people's behaviour is subjected to the significance of historical events and zeitgeist, the defining spirit or mood of a particular period of time (Mannheim 1952) and not simply as a product of numerical years, we decided to modify standard 10-year age intervals by setting the boundary of cohorts in accordance with important historical events in the history of modern China, such as the Sino-Japanese War and Cultural Revolution, among others. This produced seven cohorts.

The first cohort/generation named 'Silent Gen' comprises people born before 1937, the year of the Marco Polo Bridge Incident when the Sino-Japanese War fully exploded; the second cohort 'War Gen' comprises of people born between 1937 and 1948; the third cohort 'Boomers' are those born after the founding of the People's Republic of China (PRC) in 1949 until 1958; the fourth cohort 'Maoists' comprises those born between 1959 and 1965, a time which featured the Great Leap Forward and numerous other socialist projects; the fifth cohort 'Gen Red' are those born during the Cultural Revolution that lasted from 1966 to 1977; the sixth cohort 'Gen Market' are those born since the Open-up Reform in 1978 until 1989; the last cohort 'Millennials' combine all post-1990 births. We restricted the analysis to people older than 16 years of age and below 80 years old, so there are relatively few cases in the cohort of Millennials because the latest survey was conducted in 2015.

Figure 2.4 plots smoking prevalence rates on an age scale in order to compare smoking patterns of the seven different cohorts. There are diverging and converging trends in smoking as shown differentially by cohort groups. Overall, there is a

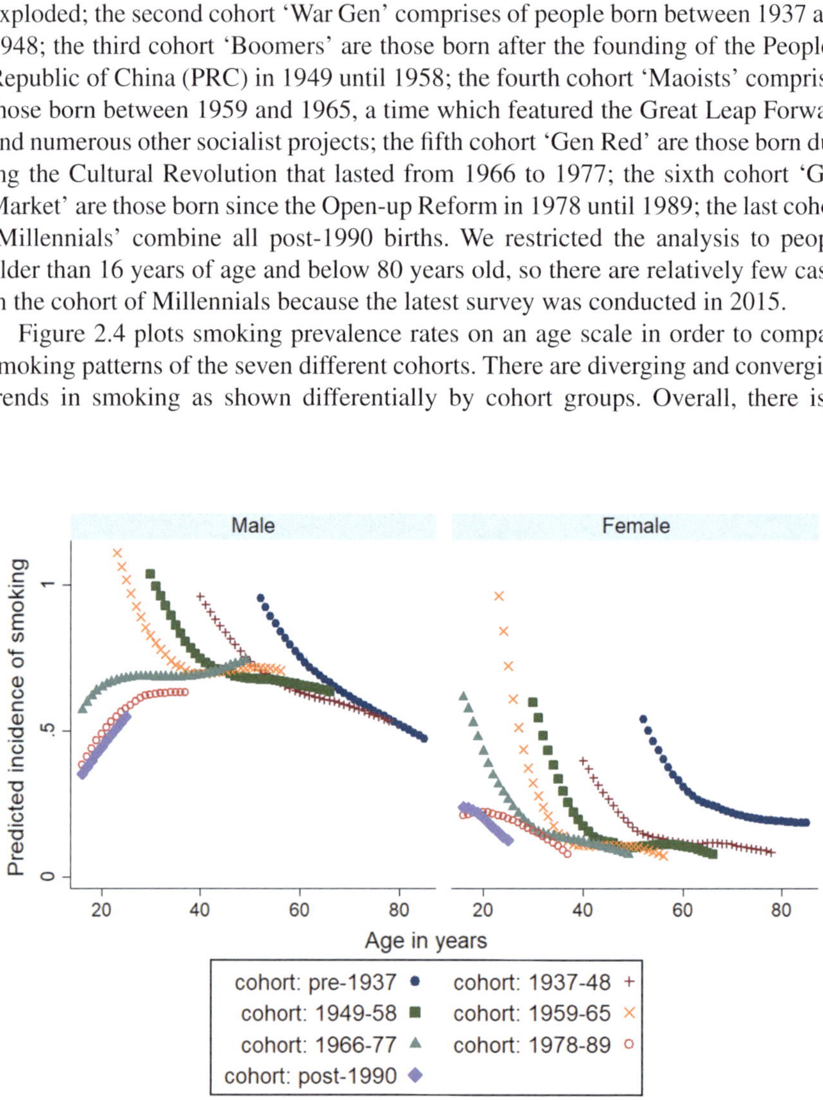

Fig. 2.4 Predicted smoking rate by gender and age for different birth cohorts. (Credit: Authors)

decline in smoking over one's life course. Older individuals tend to smoke less or be non-smokers. The shape that these trajectories merging together at the later ages in the life course resembles of the tail of a Chinese phoenix. A Chinese phoenix has a long and splendid tail made of feathers of different strings of colour, each representing a unique tribe of ancient China. At the end of the tail, each feather is unique and drags long and loosely from other feathers, but all feathers gradually join at the body of the tail. This 'phoenix tail' pattern holds particularly true for women, as all cohort groups show a falling trajectory of smoking with age. In contrast, younger male cohorts, Millennials, Gen Market and Gen Red, show upward trajectories of smoking as they age. Nevertheless, smoking prevalence among these groups is still lower than older cohorts at the same age, except for older members of the Gen Red generation (1966–1977) whose rates start to begin to exceed those of the War Gen (1937–1948), Boomers (1949–1958) and Maoist (1959–1965) generations once they reach 50 years of age. The three younger cohorts are still in their prime age for work and may be more obliged to conform to the smoking norm at workplaces and in the hierarchy of seniority. Women, on the other hand, are much less required to conform to such norms regarding drinking and smoking with the predicted smoking rate among Millennials being much less than other groups when they were the same age.

Contrary to the expectations of modernisation theories and some feminist scholarship, there is not a converging trend in smoking that merges the gender patterns of smoking over the course of a surging modernity in China. This expectation argues that women utilise the liberty afforded to them through equal rights and a cultural liberation of values and will slowly imitate men in all spheres of life and work. However, in the age, cohort and period trend of smoking, women of all cohort groups showed declining trend in smoking and still considerably distinguish themselves from men. In the Republic era (1911–1945), female smoking was promoted as a westernised, luxurious and labour-free lifestyle. Figure 2.4 suggests that possibility that older cohorts of Chinese women clung to smoking due to the particular historical context at the time of their birth and the gender roles and expectations they absorbed from a mixture of Confucian ethics and modernisation impulses. Almost all cohorts smoke more than their next-generation cohort. For example, female members, around 50 years of age, of the Silent Gen (pre-1937) cohort showed a higher probability of smoking compared to women of similar ages in later cohorts.

As Zhang et al. (2019) found, there is also evidence that older generations of men tend to smoke more than their younger counterparts. The Silent Gen born prior to 1937 have higher smoking likelihood at the age around 50, compared to Gen War, Boomers, Maoists and Gen Red, who were old enough to give us data at 50 years of age. Although Gen Market and Millennials were not senior enough at the time of the latest survey, we can reasonably expect they will not reach a level of smoking as high as the Silent Gen, given the lower starting level of smoking of these younger cohorts. If the female cohort pattern can be likened to a chorus, in which singers sing the same melody in synchrony, albeit at different pace, the male cohort pattern looks very similar to a polyphony, in which each singer has his own key. We can see in Fig. 2.4 that not only do the youngest three male cohorts show a surge in smoking during their early adulthood from 18 to 45, some cohorts have overall higher prevalence of smoking too, and these cohorts are not necessarily older. For example, 'Maoists' have higher smoking rate than Boomers after about 45, and Boomers have higher smoking rate than Gen War after 60. Although Boomers have higher smoking rate than 'Maoists' at the age of 30, their trajectories up to this time point almost are parallel to one another. This means that if we could magically ship Boomers back in the time and make them younger for 10 years, they would share the same smoking trajectory with 'Maoists'.

2.4.2.2 Period Effects

This retrospection leads to the possibility that some changes in smoking rates may not be a cohort effect. Instead, due to progress in public health knowledge, the socially defined popularity of smoking, intensifying modernisation and other structural-temporal factors that widely affect all populations rather than a few cohorts, people may show decreasing (or increasing) trends in smoking in later survey waves. Thus we should examine how the smoking rate for males and females changes across surveys, for different cohorts. The wave effect plotted in Fig. 2.5 largely confirms our suspicion: some older cohorts would not have smoked more

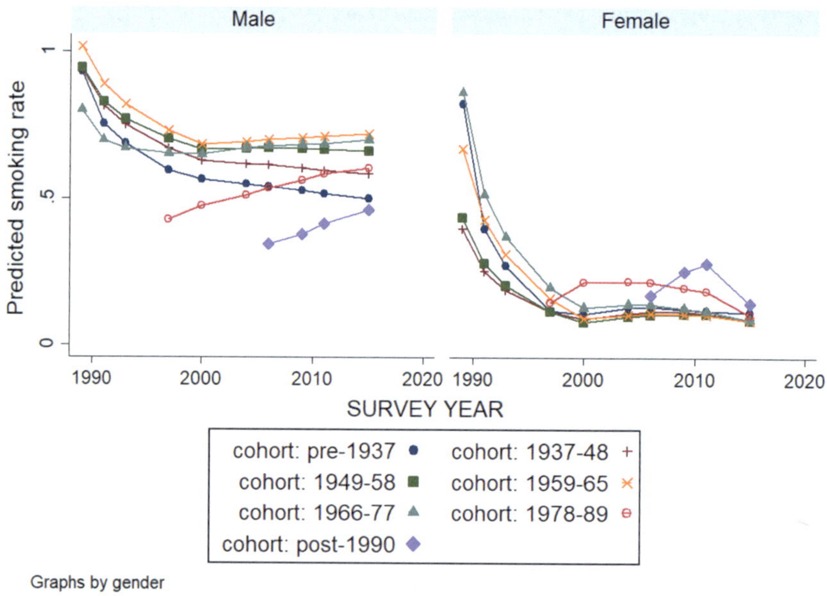

Fig. 2.5 The predicted smoking rate as time period passes, for different cohorts and separately by gender. (Credit: Authors)

were they were shifted temporally to an earlier time point, if age is not considered. When survey wave replaces age on the X-axis in Fig. 2.5, the male cohort that smoked the most is actually Maoists (1959–1965). Gen Red (1966–1977) and Boomers (1949–1958) smoked at similar levels after Maoists, followed by War Gen (1937–1948) and Silent Gen (<1937). Gen Market (1978–1989) and Millennials (1990–) still showed an increasing trend in smoking, and Gen Market had caught up with Silent Gen and War Gen at the most recent wave in 2015.

Interestingly, the overall period has featured a turning point of smoking. Before 1997, smoking was waning in all male cohorts, but rose after the year 2000. This trend is consistent with that of Zhang et al. (2019) who recorded that men born in recent decades were more likely to start smoking at younger ages and to smoke more cigarettes than those born in previous generations. For female cohorts, the initial gap in smoking between cohorts narrowed, with smoking largely declining over time. While Gen Market (1978–1989) and Millennials (1990–) had a higher level of smoking compared to older cohorts, by 2015 these differences had become minimal. This analysis suggests that younger female cohorts do not have a greater smoking rate. Instead, smoking prevalence among the youngest generations, with the partial exception of Millennials, saw no significant increase up to 2015, by which time all-female cohorts had similar smoking rates (Fig. 2.5). The main increase was seen among men. The youngest three generations of men, in particular Millennials and Gen Market and to a lesser extent Gen Red, reported an increasing smoking prevalence over time. Men may need to subscribe to the demanding smoking culture in China's job market and social hierarchy. The aspirant young men learn to smoke in order to join many smoking friendly social opportunities and use cigarettes for a variety of business and networking activities. The gender-divergent smoking pattern strongly reflects the context of Chinese society, with an extreme gap between male and female smoking rates. As far as the data of CHNS show, no foreseeable convergence of the gap can be expected in the near future.

2.4.3 Synopsis

Gender differences in prevalence are a marked feature of the smoking population in China. This was not always so, with older Chinese women being much more likely to smoke than their children. Despite economic development, much of China remains a male-dominated society, where smoking is seen as a symbol of men's status. Ethnographic studies have frequently noted that traditionally defined gender roles and expectations have prevented women from smoking (Mao et al. 2014). While socialist ideology liberated women by mobilising their labour force participation in the years after the 1949 Revolution, the pattern of greater equality in educational and employment opportunities did not persist following the economic reforms of the late 1970s (Gustafsson and Li 2000; Tu and Liao 2005). Thus female empowerment has not markedly improved with social and economic development and one indicator of this, female smoking prevalence, has only shown minor changes over this period.

Cross-national research suggests that women are much more likely to smoke at a similar rate to men in countries that have experienced a growth in political freedoms and where fertility rates are higher. On the other hand, women are less likely to smoke when a greater proportion of men work relative to women, where levels of urbanisation are lower and where gender differences in educational attainment are greatest (Lillard & Christopoulou 2015). China has many of these characteristics and thus until some, or all, of these factors change, smoking prevalence, at least among younger women is likely to remain low. Nevertheless a recent meta-analysis suggested a positive association between income and smoking among younger Chinese women, many of whom are migrants (Ding et al. 2014) (see Chap. 6). This indicates that women may take up smoking if migration and rapid social change brings alteration in traditional gender norms (Morrow et al. 2002). While little research on the interaction of individual female and areal social characteristics exists in China, the experience of Korean women indicates that manual work is associated with higher smoking prevalence, with this association being strongest among those living in affluent urban areas (Park et al. 2010). As Mackay (1996) indicated over two decades ago, there can be no complacency about the lower level of tobacco use among women which is bound to change as their independence and resources increase with urbanisation and as tobacco companies increasingly target this group.

2.5 Socioeconomic Status and Smoking

While there are clear gender differences in smoking in China, socio-economic inequalities in smoking are also important. These are yet to approach the differences found in western countries, but it is likely that as China continues to urbanise and institute stronger tobacco control measures SES smoking inequalities will increase (see Chaps. 5 and 8).

2.5.1 Research on SES Differences in China

Many studies have explored the effects of SES on smoking prevalence and smoking intensity, but the use of different measures of SES and different contexts of research have produced quite variable results. In simple terms, socio-economic inequalities in smoking are often defined as differences in smoking prevalence and intensity by income, occupation or education, but often these interact in complex ways and do not necessarily have a consistent effect on smoking. Wang et al. (2018), for instance, found among Chinese people 45 years and older that occupation and education had the most significant effect, whereas the effects of income were small. Similarly Si et al. (2018) found contradictory effects of household economic quantiles (annual household expenditure per capita) and individual educational attainment on tobacco consumption: those with a higher economic quantile, but who were less educated, consumed more tobacco than those with a lower economic quantile or with higher educational attainment. For these reasons, it is important to understand that SES may affect smoking through different mechanisms and via different causal pathways. Because of difficulties in defining SES, some studies have used composite measures of socio-economic position (e.g. Cai et al. 2019), a methodology common in western research. Depending on the variables used, such an approach may conflate the different effects of income and education on smoking prevalence.

In China, studies have shown that income and education have different effects upon smoking prevalence and consumption levels. Because China, as a middle-income country, is at a relatively early stage in the smoking transition, we might expect that both tobacco use prevalence and level of consumption would have a pro-rich distribution. This is evident in Kim et al. (2004) who showed that, compared to the USA, in China, as SES improved, lifestyles became less healthy with higher rates of smoking and drinking and diets which increased obesity. With respect to smoking, Table 2.1 provides some evidence of such trends, but shows that education has the most consistent effect in reducing both smoking prevalence and consumption levels, usually measured by cigarettes consumed per day. By contrast, income effects are more highly variable, even for the few studies which have examined SES differences in rural areas (Yang et al. 2008; Cai et al. 2019). The most consistent effect occurs for studies in urban settings (e.g. Yang et al. 2014; Chen et al. 2018), particularly those that focus on urban migrants (Chen et al. 2004; Yang et al. 2009; Liu et al. 2015).

These results need to be interpreted with caution but are strongly suggestive of the differential effects of urbanisation and migration on the social distribution of smoking, an issue discussed in detail in Chap. 6. Given the economic diversity of China, it is to be expected that the social distribution of smoking will not only reflect individual socio-economic factors but will also reflect community socio-economic development. However, only a few studies (Yang 2017; Zhang et al. 2019) have considered interactions between individual SES and community variables and how the SES distribution of smoking differs between different community contexts. There is also the problem of comparing studies that have used different methodologies, including controlling for different background factors.

From a theoretical point of view, we can also identify different explanatory frameworks such as the Marxian tradition which sees SES in terms of labour relations and a Weberian approach which pays greater attention to occupational groups' relative economic power in the market place as well as managerial concerns relating to differences in authority. Thus in broad terms there is an economic class effect that embodies way of life and collective opportunities and an income effect which relates to the purchasing power of labour. In this section we specifically focus on the issue of economic class. We define this in terms of control over the means of production and labour power. Class can also be closely linked to financial well-being, but by its very nature revolves around a range of valuable goods in addition to financial assets.

Table 2.1 Relationships between income and education and smoking prevalence and smoking intensity in China in studies published between 2004 and 2019

Study	Population[a]	Location	Smoking prevalence		Smoking intensity	
			Income	Education	Income	Education
Si et al. (2018)	China 2013 HRLS Adults 45+	China	**Positive***	*Negative*	**Positive***	*Negative*
Lee et al. (2019)	China 2011CHNS Adults >18	9 provinces/3 mega cities	*Negative*	*Negative*	ns	*Negative*
Yang (2017)	China 1989–2011 CHNS Adults	China			**Positive**	*Negative*
Liu et al. (2017)	China CCDRFS Adults >18	China	*Negative#*	*Negative***		
Wang et al. (2018)	2013 CHARLS Adults 45+	450 urban and rural communities	**Positive***	*Negative***	**Weak positive***	*Weak negative#*
Tian et al. (2013)	China CHNS 1993–2009 Adults	Urban and rural communities in 9 provinces	*Negative*	*Negative*	**Positive**	*Negative*
Chen et al. (2018)	China 2014 CLFDS Adults 35+	Urban China 22 cities	**Positive**	*Negative*		
Pan (2004)	Urban residents HIP	Urban China	ns	*Negative*	**Positive**	*Negative*
Mehta et al. (2014)	Government employees 18–61	6 cities		*Negative*		
Yang et al. (2014)	Survey 2011 Males >15	Hangzhou	**Positive**	*Negative*		
Xu et al. (2007)	Survey 2001 Adult males	Nanjing	*Negative*			
Wang et al. (2019a, b)	China 2013 NHSS Adults	Inner Mongolia	*Negative***	*Negative***		
Zhang et al. (2019)	China 2013–2014 CCRDFS Adults >18	China		*Negative***		
Yang et al. (2008)	Survey Rural males	Rural areas in 4 regions	**Positive**	*Negative*		
Cai et al. (2019)[b]	Survey Adults 35+	Rural areas in Yunnan	*Negative*	*Negative*	**Positive**	**Positive**

(continued)

Table 2.1 (continued)

Study	Population[a]	Location	Smoking prevalence		Smoking intensity	
			Income	Education	Income	Education
Chen et al. (2004)	Migrant workers 2002 MHBS	Beijing	**Positive****			
Yang et al. (2009)	Migrant workers >18	3 cities: Beijing, Shanghai, Chengdu	**Positive**	*Negative*		
Finch et al. (2010)	Survey female migrant workers	Beijing	Ns	*Negative*		
Liu et al. (2015)	Migrant workers	Shanghai	**Positive #**	Ns		

**Males and females; *Males only; Females only#

[a]*HRLS*, Health and Retirement Longitudinal Study; *NHSS*, National Health Service Survey; *CHNS*, China Health and Nutrition Survey; *CLFDS*, China Labour Force Dynamics Survey; *CCDRFS*, China Chronic Disease and Risk Factor Surveillance Survey; *HIP*, Household Income Project; *CHARLS*, China Health and Retirement Longitudinal Study; *MHBS*, Migrant Health Behaviour Survey. [b] The study used a measure of socio-economic position based on household assets, educational level and access to services

A higher class may be more educated, enjoy better prestige, jurisdictive power and cultural significance without necessarily being much richer. Wright (1997) has argued that in addition to the means of production, labour relationships must also take into account inequalities in authority of key managers within the private sector, state bureaucracies and state owned enterprises. With these caveats in mind, in the remainder of this section, we first provide a discussion of the importance of class in contemporary China. We then undertake a class-based analysis of smoking using data from 2012 to 2016 China Labor Dynamic Surveys (CLDS). Finally, in light of these results, we provide a brief synthesis of the importance of SES in smoking and the relevance of the smoking transition model to China.

2.5.2 *Class in China, if It Exists*

China is a country that uses authoritarian government and capitalist economic principles to guide economic development and fulfil national ambition. Even after the market reform in 1978, collective ownership or state ownership of enterprises and productive units comprised the bulk of the national economy, particularly in vital industrial sectors, such as aerospace and defence, education, and in natural resource extraction and processing. The Party Constitution of the Chinese Communist Party (CCP) repeatedly emphasises its leadership of the proletariat class, which was defined by orthodox Marxism as those do not privately own the means of production. At the same time, the history of the Chinese CCP and its strategies to secure control of China deviated from the orthodoxy of Marxist ideals as practiced by the USSR. Unlike the Russian October Revolution, which first occurred in major cities before proceeding more slowly against the White Army in rural areas, the early military victories by the CCP took place mainly in the countryside. This had implications for smoking which was first promoted in rural areas by the newly established Red Army as a way of obtaining the loyalty of the peasantry to the CCP in its base areas (see Chap. 3).

New issues in the class structure of China emerged in the market era. China in the twenty-first century has a 'steaming' capitalism engine boosting its economy and generating a class structure similar to that in many capitalist societies. Since 1978 the growing number of high-skilled workers and managerial occupants in the labour market poses the question regarding the appropriate place of managers and stock-owning employees in the class relationship. In spite of rising income inequality in Chinese society (see Chap. 6), socialist collective ownership of national property and resources does not allow an explicit admission of the existence of opposing classes. A democratic society's unwillingness to admit there are disenfranchised groups within it is reflected in China's reluctance to acknowledge that it, too, is increasingly a class society. This unwillingness has caused academic problems. Textbooks used for high school politics affirm that, since class is a product of private property ownership and the PRC practices collective ownership, then for this very reason, there cannot be class conflict in China. To circumvent the issues of obvious inequality and conflicts between social groups, the official narrative concedes that China does have social strata (阶层*Jieceng*) in the absence of real classes. We should take this as an implicit acquiescence for the analysis of class in China.

2.5.3 *An Economic Class Analysis of Smoking*

2.5.3.1 Defining Classes

To help understand class in contemporary China and how smoking varies across economic groups, we used the China Labour Force Dynamics Survey (CLDS) to get a glimpse of the class structure in twenty-first-century China. The CLDS is conducted and coordinated by Sun Yat-sen University under Prof. Cai He and is the most well-known authoritative population-wise representative survey on labour conditions and the labour market in China. The CLDS used a rotating panel method to ensure that, even in the face of attrition during follow-ups, rotated new cases make the entire sample random and representative of the national population. We combined the three waves from 2012, 2014, and 2016 and explored a wide range of questions to construct different classes. For example, questions related to whether people were working, their type of work, whether they worked for others and had particular skills, the nature of work contracts and whether they managed others or had started a business and/or hired others. Based upon these different questions, we derived nine different classes (Fig. 2.6) based upon the flow chart of questions adopted in the CLDS surveys.

CLDS first considers whether the person is currently working. If he/she is not working due to ageing (mostly for farmers without formal prior employment in a work unit), formal retirement (for those with a work unit prior to retirement) or health-related concerns, the person was defined as retired. This group makes up 13.9% of the sample. If the person is not currently formally employed for all other reasons (e.g. doing housework, giving birth, unemployed), he/she is in the status of non-employment (20.1%). This sector formed the reference group in our analyses.

If someone is undertaking agricultural work, defined in the survey as 'wunong' or literally transliterated as 'doing farms', they were coded as farmers. If the latter did not work the land but instead ran separate agricultural businesses, they were classified by interviewers as being self-employed or employers. Farmers, who actually tilled the land, were the second largest group comprising 28.8% sample.

CLDS also considers the question of whether the person works for others or for themselves. If they worked for others, but had no contract, they were defined as 'true proletariat', a class that is deprived of any means of production and likely to join the reserve army of the unemployed at any moment. Proletarians made up 9.8% of the sample. Next, the need for skill credentials in the current job sets apart skilled workers (7.2%) from unskilled workers (6.8%).

At the top of the pyramid were the self-employed, managers and bourgeoisie. Those who worked for themselves, but who did not own or start a business, were defined as self-employed. By contrast the bourgeoisie, which overall comprised 3.1% sample, referred to business owners who both employed (bourgeoisie) and did not employ (petite bourgeoisie) other workers. Managers who worked for others but, by virtue of their job, had responsibility for managing/employing other workers represented 4.1% of the sample.

Fig. 2.6 Classes based on the means of economic production and control over labour power. (Credit: Authors)

In addition to economic class, we also considered the effects of income on smoking. Detailed data on three different net income types (total income, salary income and revenue income generated from business profits, investment earnings and dividends) was obtained from the CLDS surveys. Since income is inextricably associated with social class, incorporating it is necessary to model both factors simultaneously in order to resolve their confounding relationship.

2.5.3.2 Class Gradients in Smoking?

Once the distribution of economic classes has been established, it is possible to analyse their links to smoking. Table 2.2 shows the age standardised odds ratios of smoking for males and females by different economic classes. Among men, compared to people in the non-employment category, who themselves have high rates of smoking (Wang et al. 2016), the highest odds of smoking are seen among farmers (2.67), the bourgeoisie who employed other workers (2.51) and proletarians (2.47), followed closely by the self-employed (2.36) and managers (2.34). Except for retirees, who may have given up smoking for health and ageing concerns, the lowest odds of smoking belong to unskilled (1.83) and skilled (1.79) contract workers and retired people (1.61).

Among females, the class gradient in smoking is less pronounced and the patterns slightly different. Compared to those without formal employment, the retired (2.20) and farmers (1.69) are the groups most likely to smoke, even after controlling for age. Compared to males and other female classes, proletarians are less likely to

Table 2.2 Logistic regression on smoking status and smoking intensity by social class

Smoking status					Smoking intensity[a]			
	Male		Female		Male		Female	
	O.R.	S.E.	O.R.	S.E.	Coeff.	S.E.	Coeff.	S.E.
Class (ref = non-employment)								
Retired	1.61***	0.10	2.20***	0.28	2.67***	0.32	0.28***	0.06
Proletarians	2.47***	0.15	0.62*	0.15	4.09***	0.32	−0.11***	0.03
Farmers	2.67***	0.15	1.69**	0.28	5.19***	0.32	0.09	0.05
Contract non-skilled	1.83***	0.12	0.66	0.15	1.80***	0.33	−0.07*	0.04
Contract skilled	1.79***	0.12	0.56	0.17	1.78***	0.32	−0.11**	0.03
Self employed	2.36***	0.16	0.96	0.24	4.80***	0.37	−0.01	0.05
Managers	2.34***	0.17	0.61	0.23	3.28***	0.39	−0.07	0.05
Petite bourgeoisie	2.17***	0.21	0.71	0.36	3.62***	0.55	−0.06	0.08
Bourgeoisie	2.51***	0.26	1.11	0.52	4.84***	0.63	−0.05	0.07
Age	1	0.00	0.99	0.00	0.00	0.00	0.00	0.00
Wave	0.93***	0.01	0.96	0.04	−0.24***	0.06	−0.01	0.01

Note: significance level ***$p < 0.001$, **$p < 0.01$, *$p < 0.05$
[a]Average number of cigarettes smoked per day

smoke (0.62). In terms of the magnitude of the odds ratio, the female bourgeoisie (1.11) shows higher risk of smoking, but unlike males the difference was not significant.

The class pattern also holds true for smoking intensity (number of cigarettes smoked). All classes smoke more than the not employed, with farmers (5.19) and the bourgeoisie (4.84) being the heaviest smokers of all. Skilled contracted workers only smoke 1.78 more cigarettes, on par with non-skilled contract workers (1.80). For women, the same class pattern can only be sporadically seen. Female farmers and bourgeoises do not smoke significantly more than non-employed people. Retired women smoke more cigarettes than the non-employed, while proletarians, contract non-skilled workers and skilled workers all smoke fewer cigarettes on average.

There is a suspicion that the distribution of smoking across the class gradient actually hides behind the fact that income determines smoking intensity with class correlated closely with income. The regressions in Table 2.3 allow rejection of this possibility. Although the addition of net income reduces the effects of some classes, most classes remain significantly different in terms of smoking intensity. After controlling for income, contract skilled workers smoke no more intensely than those who are not formally employed, and the effect size of virtually all classes has dropped. Income is also independently positively associated with smoking more cigarettes, offering evidence for the existence of a pure income effect. Controlling

Table 2.3 Regression on smoking intensity by class and income variables

	Model 1		Model 2		Model 3	
	Coeff.	S.E.	Coeff.	S.E.	Coeff.	S.E.
Age	−0.00	0.00	−0.00	0.00	−0.00	0.00
Wave	0.10	0.05	0.01	0.06	0.08	0.05
Gender	−8.0***	0.18	−7.9***	0.18	−7.9	0.17
Class (ref = non-employment)						
Retired	1.23***	0.13	1.32***	0.17	1.35***	0.13
Proletarians	1.83***	0.18	1.66***	0.20	1.65***	0.18
Farmers	2.40***	0.15	2.35***	0.17	2.42***	0.16
Contract unskilled	0.61***	0.18	0.38*	0.18	0.37*	0.19
Contract skilled	0.58***	0.18	0.35	0.19	0.34	0.18
Self employed	2.58***	0.25	2.41***	0.27	2.37***	0.26
Managers	1.50***	0.31	1.23**	0.32	1.21***	0.31
Petite bourgeoisie	1.86***	0.35	1.64***	0.35	1.55***	0.38
Bourgeoisie	2.31***	0.51	2.01***	0.51	1.91***	0.52
Net income (log)			0.10***	0.02	0.07**	0.02
Salary income (log)					0.02	0.02
Revenue income (log)					0.03	0.02

Note: Significance level ***$p < 0.001$, **$p < 0.01$, *$p < 0.05$

for social class, each 1000 Yuan increase in annual income led to men, but not women, smoking 0.03 more cigarettes and a one percent increase in the likelihood of heavy smoking (Yang 2020). The significant income effect notwithstanding, classes still differ from each other in terms of smoking intensity, further strengthening the explanation that labour relationships determine outcomes in lifestyle, including smoking and other risky health behaviours.

2.5.4 Synopsis

The CLDS data suggests that the current socio-economic distribution of smoking in China partly reflects a pattern that we might expect in the early stages of the smoking transition. Among men, smoking prevalence was high among the bourgeoisie, but also among other groups prominent in the economy, such as the self-employed and managers. But the pattern is more complex than this. Farmers and the proletariat also had elevated smoking rates compared to those who were not employed. The proletariat no doubt encompassed groups such as migrants with unstable employment conditions and this would also be true of many farmers. Farmers' long history of being a disadvantaged group in China encourages many to classify them as part of the proletariat. However, the complexity in their means to economic production (Wright 1997) fundamentally places this large group in a 'contradictory position' in class relations; they may exert some control over their labour inputs, but not over the means of production, that is, the communally owned land they farm. As we will see in the following chapter, often the latter is very tenuous, being determined by state rather than individual market decisions. Among females the patterns were different. For females the group with the highest prevalence and who smoked most heavily were the retired. The only other group to record high prevalence were farmers.

Overall the data suggest that contrary to the social gradient in smoking in western countries, higher class status and upward income mobility are both associated with more smoking among Chinese men, but not among women. The analysis adds to existing research in China which has often shown a strong relationship between smoking prevalence and SES variables. The effects, however, are inconsistent between urban and rural areas with the research generally failing to link SES effects on smoking to broader contextual processes such as urbanisation (see Chap. 6) and changes in labour relations. Understanding such processes and how they shape social mobility and SES has the potential to provide more insightful interpretations of smoking and other forms of substance use (Yang 2017, 2020).

2.6 Ethnicity and Smoking

Within western countries, ethnicity has emerged as an important component in the smoking transition. The marked declines in smoking prevalence in the general population has not been typical of certain ethnic groups, particularly indigenous ones which have suffered from a legacy of nineteenth-century colonialism, discrimination and economic marginalisation. There are also gender differences, with some of the highest recorded prevalence rates being recorded for indigenous women living in the most deprived communities (Barnett et al. 2017). The causes of ethnic variations in smoking are complex (Dressler et al. 2005) and arise from a combination of socio-economic, cultural and discriminatory factors (Factor et al. 2011).

2.6.1 Race and Ethnicity in China

In the western imagination, China is often thought to be an ethnically or racially homogeneous country, perhaps inspired by the Orientalist discourse, as somewhere exotic, but with little diversity. In fact, there are more than 50 minority groups in China, with the presence of ethnic groups being institutionalised following the adoption of the Stalinist model of ethnic identification in the 1950s. Prior to this, the modern concept of ethnic nationality was foreign to Imperial ideologies. Admittedly, Manchu rulers recognised different subject groups with distinct cultural customs and languages within the Empire, but their differences were not an important matter compared to the division that separated the royal Manchu-Mongolian banner nobles from the subjected groups, including Han Chinese and all other nations (Perdue 2009).

The identification of ethnic identities did not take place until the Communists came into power in 1949 (Zhang 2015). Ethnic identification in China during the 1950s, and later, strongly relied on the Stalinist definition of nation which emphasised a historically stable community of people with a common vernacular language occupying a single piece of territory (Stalin 1913). With this territorial definition, Stalin rejected the political nationalism that seeks self-administration in the name of some broad cultural identity. The Communist ethnic identification project, therefore, differs from the identity politics in contemporary western democracies, with ethnic identification born out of the political goal of regional and group-based administration rather than expressing the cultural autonomy of individuals holding common identities. For this reason, the number of ethnic groups under the Communist project of ethnic identification is limited and finite, for the benefits of governance.

The 56 officially defined nationalities of China possess very rich, and diverse, histories and cultures, many with their own languages and religious beliefs (Zhang 2015). At least three minority groups can be considered racially Caucasian, or 'white' in anthropological senses: the Tajiks, Russians and Tartars. Several are racially mixed by a sustained influx, throughout history, of Caucasian and Mongoloid makeups, such as Uyghurs, Salars, Kazakhs and Kirgiz. The Han majority make up 90% of the population. Due to wars, migration, and assimilation over the course of history, northern nationalities tend to share closer affinity in terms of language, with most ethnic minorities in northern China speaking some variants of the Altaic languages. The genetic and linguistic origins of the southern minorities are far more complicated, with groups of entirely different linguistic and genetic lineages overlapping and scattered among different ethnic groups. The southern ethnic groups comprise the Sino-Tibetan, Austroasiatic and Kra-Dai language families and Austronesian languages. Northern minorities tend to follow Islam and Tibetan Buddhism, whereas the southern ones tend to be Animists, Christians and Mahayana or Theravada Buddhists.

2.6.2 *Ethnicity and Smoking Behaviour in China*

As in the case of our exploration of gender differences, we used China Health and Nutrition Surveys between 1989 and 2015 to explore the association between ethnicity and smoking behaviour. The CHNS contains a broad range of health and biological information from a representatively chosen sample of Chinese residents from over ten provinces. Because of the attrition issue of panel surveys, we combined all respondents of the same age group from all survey waves into a pooled sample as a representative sample of a single age-group population, as the sampling at every single wave was representative. We retained the prime work-age group population from 25 to 54 years of age and tabulated their current smoking status by ethnic groups. Because racial and ethnic minorities are scattered and tend to have the 'small number problem' in large representative samples, we selectively merged with 'others' the ethnic groups with a sample size of fewer than 30 units in the currently defined prime work-age population sample and also merged together a few ethnic groups with proximate linguistic and cultural heritage. Thus, given the influence of Islam on smoking among Muslims (Ghouri et al. 2006), Uyghurs were merged with Hui (Dungan) to form a Muslim group (even though many may be only nominally Muslim today), Yao (Mien) were merged with Miao (Hmong) for their common linguistic family (known as the Hmong-Mien clan) and geographical proximity, and Dong (Kam) were merged with Zhuang as they both speak Kra-Dai languages and live in southern China (Buyi is another Kra-Dai-speaking ethnicity but retains its own group due to larger size in the sample), while the Sino-Tibetan-speaking Tibetans and Yi (Lolo) were collapsed into 'Others' due to small sample size.

This categorisation leaves us with ten major ethnic groups in the sample: Han, Mongolians (Altaic speakers), Muslims, Hmong-Mien, Kra-Dai, Buyi, Koreans, Manchu (Altaic speakers), Tujia (Sino-Tibetan speakers) and Others. In any classification, especially that involving 25 years of data, there will always be loss of information, typological discontent and ethnonymic errors, but the current classification does its best to retain the original volume of information, achieve computational efficacy that may be hampered by the small number problem and also make the most historical and cultural sense of grouping different ethnicities together.

As we can see in Table 2.4, the smoking rate is significantly higher among some ethnic minority groups. Hui, Buyi, Korean and Tujia tend to have a greater proportion of smokers than what would be expected by chance. However, compared to other groups, the Han majority do not display a unique smoking pattern more than what would be expected by randomness. In other words, Han ethnic status is not associated with a higher prevalence of smoking.

The ethnic distribution of smoking is also quite idiosyncratic, as the groups that have a significantly larger share of smokers do not share cultural and geographical similarities. Hui is a Chinese-speaking Muslim nationality distributed all over China but especially in the north and northwest; the Koreans native to China live along the Chinese-Korean border and traditionally speak Korean, whereas Buyi are a southern minority group speaking a branch of Kra-Dai language affiliated with Thai, while Tujia live in southern China and speak a Sino-Tibetan language closely related to Chinese. Despite their documented native tongues, most of these minorities de facto speak Chinese or local Chinese varieties in their daily life. Still, here we can find that the groups with a higher proportion of smokers are distributed everywhere

Table 2.4 Current smoking status by major ethnicities*

	Non-smoker	Smoker
Han	32,518 *32,385*	14,362 *14,494*
Mongolian	63 *68*	35 *30*
Muslims (Hui)	82 *95*	55* *42*
Hmong-Mien	1068 *1079*	497 *484*
Kra-Dai	300 *306*	143 *137*
Buyi	790 *856*	449* *383*
Korean	28 *40*	31* *18*
Manchu	1017 *998*	428 *446*
Tujia	439 *487*	267* *218*
Others	426 *413*	172 *185*

*Observed frequency on top and the expected frequency by chance in italics. The asterisk denotes the significant cell-wise difference between expected frequency and observed frequency, based on Pearson's chi-square with one degree of freedom (two categories)

without a unique and easily identifiable geographical pattern. Their higher rate of smoking may be explained by socioeconomic status.

However, a regression on smoking status by the inclusion of net household income, per capita household income, the individual's completed years of education and gender do not show any reduced significance of the ethnicities that reported the elevated proportion of smokers (Table 2.5). Although net household income, higher education and being female are all associated with lower risk of smoking, being ethnically Hui (0.54, $p < 0.05$), Buyi (0.19, $p < 0.05$), Korean (1.63, $p < 0.001$) and Tujia (0.38, $p < 0.001$) still constitute a risk factor of smoking. Differences in socioeconomic status, therefore, cannot explain the ethnic difference in smoking.

There is one problem preventing a valid conclusion that the ethnic variation in smoking is indeed caused by ethnic and racial identities, even after controlling for all what we can gather about demographic and SES measures. Ethnic and racial minority groups are not uniformly distributed geographically. Instead, a large number of minorities often form habitual clusters while a smaller number are scattered across the entire country. The official description of the ethnic distribution in China during the first ethnic identification project claimed that the peculiarity of our country's ethnic distribution is 'large mixed living, small clustering'. Of course, this characteristic is not particularly Chinese at all, as it applies to almost all large multiethnic countries where minority groups are geographically widespread, while also retaining their traditional localities. In China, while every province has a palpable presence of minorities, larger shares of minority populations still live in the northern and western borderlands. This clustering phenomenon undermines the use of conventional theoretical frameworks to explain the observed ethnic disparities in terms

Table 2.5 Regression of ethnicity and socioeconomic status on smoking

	Coefficient	Standard error
Ethnicity (ref = Han):		
Mongolians	0.29	0.29
Muslims	0.54*	0.25
Hmong-Mien	0.06	0.08
Kra-Dai	0.04	0.14
Buyi	0.19*	0.09
Korean	1.63***	0.43
Manchu	0.01	0.08
Tujia	0.38***	0.11
Others	−0.04	0.12
Net household income	−0.001*	0.00
Per capita household income	0.00	0.00
Education	−0.21***	0.01
Female	−4.38***	0.04
Age	0.01***	0.002
Wave	−0.03***	0.002

*$p < 0.05$, **$p < 0.01$, ***$p < 0.001$

of smoking. This is because any observed variation might be attributed to nothing more than the clustering of sampling units by geographical proximity.

The confluence of geographical clustering and group-level traits raises the difficulty of how to tease apart group-level stereotypes and individual-level factors. Fortunately, we may be able to detect the geographical clustering effect in racial/ethnic disparities of smoking through a random effects model that allows estimates of rates and coefficients to vary randomly across different geographical units and thus account for the extent of clustering. Conventional regression assumes the identical independent distribution of sampled units, but such an assumption is not met for ethnic groups who naturally occupy clusters of space and cannot be considered even remotely independently distributed. By allowing intercepts and coefficients to vary across all clusters randomly, a random effects model can adjust for the inflated interval estimates and account for the clustering phenomenon. In the preceding analysis, some ethnic groups reported significantly higher likelihood of smoking, but once we allowed communities to have random intercepts (the starting default level of smoking) and accounted for clustering, this was only partly true. Smoking rates no longer significantly vary between Han and Muslims, between Han and Kra-Dai people or between Han and Buyi. There is virtually no difference between Han and these minority people in terms of the likelihood of smoking.

The ephemeral significant association between smoking and some ethnicities, as suggested by random-effect model, comes as a result of the clustering effect in which some specific communities with a very large number of ethnic minorities have inflated the estimates of smoking rate that has nothing to do with ethnicity per se but emanates from the peculiarities of that community. For example, Muslims were more likely to report smoking, but we now know that the Muslim-smoking effect is unstable after such an effect was allowed to randomly vary across communities. Muslims are widespread across China, but with the exception of Yunnan province, most are located in northern China's major cities and villages along business routes. The notable cities featuring larger shares of Muslims in the CHNS samples include Zhengzhou, Beijing, Jinan and Shenyang, whereas the southern provinces have very few living there. The demographic and SES characteristics of Muslims in these northern cities may provide a partial image of the pattern of Muslim smoking. When the wider geographical distribution of Muslims is accounted for, the significant association between being a Muslim and smoking disappeared. As a noticeable exception, there is still a significantly elevated risk of smoking among Koreans (2.73, $p < 0.01$), indicating that this is not an artificial statistical product caused by geographical concentration or the clustering effect, but is consistent across the sampled communities.

2.6.3 Synopsis

The CHNS analysis suggests that ethnic differences in smoking are less important in China than in many western contexts where such differences have been magnified particularly at later stages of the smoking transition and where increasing concentration of smoking among particular ethnic groups has largely reflected high levels of socio-economic deprivation, rather than cultural or discriminatory factors. In China, while there is a suggestion that the latter may apply to Muslim communities, the results of the analysis do not totally support such an assertion. However, as Wang et al. (2019a, b) have noted, little information is available on the prevalence of smoking among minorities and the causes of the known variations are not well understood.

Given higher smoking prevalence among certain ethnic groups in China, there are calls for more in-depth studies of smoking in provinces, such as in Yunnan, with high concentrations of ethnic minorities. Wang et al. (2019a, b) showed that although there were significant ethnic variations in smoking prevalence in the province, there were also substantial variations within each group, largely reflecting differences in education rather than income. Unfortunately the authors did not examine between-group differences after controlling for SES and other factors.

There is also a need to pay closer attention to the relationship between the Han ethnic majority and ethnic minorities. Critical race theory (Delgato and Stefancic 1998) has long emphasised the effects of dominant ethnic majorities on maintaining their status through patterns of privilege in labour markets and politics. Because ethnic/racial minorities are comparatively evaluated and treated by the majority, how well minorities fare in terms of their health outcomes and health behaviours will depend on their relative power in the local demographic context (Smith 1985). Racial/ethnic minorities in China are found to have better self-reported health and mental stress outcomes when they reside in more diverse provinces, with a tendency for their health to deteriorate in Han-dominated areas (Yang et al., 2019a, b). Minorities also are subject to more severe emotional distress by feigning the appropriate emotional expressions and hiding their true feelings during interracial interaction, a phenomenon leading to the exposure of emotional racism by Bonilla-Silva (2019). For smoking, some ethnic and racial minority men and boys may resort to substance use as a conspicuous public display to regain respect for their masculinity denied them in a racist society (Bucerius 2014; Flores 2014). Others may use cigarettes to foster a sense of community in a sea of alienation (Yang and Yang 2018), the matter of communal belonging being particularly strong for minority smokers who live in deprived areas (Matheson et al. 2011).

2.7 Geographical Variations in Smoking Transitions

Smoking transitions have normally been conceptualised at the national scale, yet it is apparent that substantial differences can occur within countries in the timing and extent of smoking among different social groups. Such differences are likely to reflect a variety of contextual factors, at a range of geographical scales, all of which will affect the extent to which different individual factors are related to smoking prevalence. In this final section of the chapter, we briefly explore some of these in relation to gender. A discussion of contextual effects on socio-economic differences in smoking is left until the examination of the effects of urbanisation on health behaviour in Chap. 6.

The importance of considering gender differences in the impact of area effects is illustrated by a small number of studies that have examined non-western contexts at different stages in the smoking epidemic. Chuang et al. (2007), for instance, note that few studies of neighbourhood influences on health have occurred in Asia. Using data from the Taiwan Social Change Survey, they found that the relationship between area deprivation and smoking was different from that found in western countries. While lower-SES women tended to smoke more than higher-status women, this was particularly true when such women lived in higher-status neighbourhoods. Similar results have been found in Japan (Fukuda et al. 2005) and Korea (Park et al. 2010; Yun et al. 2015). In Japan, the relationship between regional per capita income, a measure of urbanisation, and female smoking, net of individual factors, was the strongest area effect reported for the four health behaviours examined. No significant relationship was evident for men. In Korea women employed in manual work were more likely to smoke when they lived, or worked, in more affluent areas (Park et al. 2010). Similar, but not significant, trends were also found for women employed in non-manual work while smoking rates among housewives, in traditional gender roles, were similar regardless of area deprivation. Taken together, these findings suggest that social attitudes to women's smoking varies by area of residence, although the reasons for this are unclear. It may be that lower-income women who are either living or working in affluent areas may experience more stress from relative deprivation when they compare their lives to those of other (more affluent) women around them. Alternatively residents of affluent areas may be more tolerant of women's smoking than residents in other neighbourhoods where smoking remains predominantly a male activity.

While much has been written on gender differences in smoking in China, there have been few studies of area effects and how these may accentuate or reduce the marked gender gap in smoking prevalence. Yang et al. (2016) examined the association between regional differences in urban sex ratios and smoking in China. The main finding is that, after controlling for individual and city level confounding factors, cities with a higher proportion of males tended to have higher male smoking rates and lower female rates. This suggests that, even though individually males smoke more than females, gender differences are accentuated in cities where males form a larger proportion of the population. Conversely, the fact that female smoking prevalence, other things being equal, is lower in cities with more males indicates that the factors inhibiting female smoking are much stronger there.

There are a number of possible interpretations of the association between city-level sex ratios and smoking. First, the gender ratio has long proved to be a crucial factor in influencing micro-level behaviours. Guttentag and Secord (1983) were among the first to propose the concept of dyadic power when they hypothesised that sex ratios affect the values and roles of both men and women. Dyadic power encourages men in societies with an undersupply of women to follow more traditional values and the pursuit of a macho culture. When high sex ratios occur (i.e. when women are relatively scarce), this may have a number of impacts. First, coping with marriage market pressure may place intense financial pressure upon men who endeavour to make their sons more attractive in places with a shortage of potential brides. Males, particularly of low SES, who fail to marry may suffer grave consequences such as not being cared for in old age. Chen et al. (2018) found that, in contrast to fathers with a daughter, those with a son smoked more when they lived in (rural) counties with higher sex ratios, that is, with a surplus of men, trends that were most marked among poorer families. Yang et al. (2019a, b) similarly found that the 'marriage squeeze' among rural men in Shaanxi Province reduced their quality of life and encouraged higher levels of smoking. Second, in such situations men may use their structural power to 'protect' (control) women, who are viewed as a precious good in short supply. Women are likely to be discouraged from pursuing educational or career goals, locking them into traditional domestic roles. Thus having a larger male population in a city or region is likely to be associated with the presence of traditional patriarchal norms (Song and Hird 2013) which further encourage the strict division of social practices between men and women. As in many other low- and middle-income countries, such norms include disapproval of women smoking, seeing the latter as a symbol of masculinity.

Higher male and lower female smoking rates in cities with more males may also reflect labour market conditions. As Attané (2014) has argued, Chinese women still do not have the same opportunities as men for social and professional advancement and, paradoxically, reduced fertility has not been accompanied by an increase in the number of working women. Rather the effects of fertility decline have largely been counteracted by the liberalisation of the labour market which has heightened competition between men and women, and limited female social empowerment, something that is likely to be even more restricted in 'male dominated' cities. Thus the Chinese situation is consistent with Hitchman and Fong's (2011) finding that females living in countries with lower levels of gender empowerment smoked less. But such an explanation fails to account for the fact that age and socio-economic differences in smoking were greatest in male dominated cities where Chinese women were least empowered. Perhaps because of this smoking rates were highest among the more disadvantaged younger and middle-aged groups of women.

One could also speculate that cities with larger male populations are more likely to be those with a pro-smoking culture which encourages cigarette gifting. This practice has strongly contributed to smoking initiation as well as failure to quit smoking among Chinese males (see Chap. 5). It has become an accepted practice in much of Chinese society and is particularly important in relationships between working males and their peers (Rich et al. 2012). Although no studies have assessed the geography of cigarette gifting in China, the practice is likely to be especially important in cities where male economic connections assume a more important role.

2.8 Conclusion

This chapter has explored the epidemiological transition of smoking in China with specific reference to gender, socio-economic status and ethnicity. It has also used new data to explore each of these issues. We draw three main conclusions. First, western transition models have little relevance to China. This is particularly so with respect to gender issues where, although female smoking prevalence has declined over time, the causes are different from the western case, as has been the timing with females preceding males. With increased urbanisation and social change, it is unclear how long the marked gender gap will continue, especially among younger women. Although currently smoking much less than men, compared to older female cohorts, this group nevertheless spends more years smoking with a younger initiation age (Liu 2015).

Second, with respect to socio-economic differences, the diffusion of smoking from high to low status groups and the increased association of smoking with community economic and social disadvantage commonly seen in western cities is not yet clearly evident in China. This is not unexpected, for as long as cigarettes form an important part of social currency among particular higher social classes then the familiar western social gradient in smoking is unlikely to become strongly apparent.

Third, regarding ethnicity, there is little evidence that higher smoking prevalence among minority groups is related to SES as in western countries, although there is some suggestion of the possible effects of discrimination, particularly with respect to Muslim minorities, in influencing smoking outcomes. Given the complexity of the political identification of ethnicity in China more work is needed on Han-minority ethnic relations and on how the political classification of ethnicity intersects with social and political issues such as ethnic inequality, the preservation and contribution of the rich traditions and customs of minority cultures, and their links to health behaviour.

While the epidemiological transition in China may bear little resemblance to the western experience at the national level, there is some evidence that it applies in more urbanised regions of the country, a topic we address in more detail in Chap. 6. It is also clear that epidemiological transitions reflect both individual and contextual factors with urbanisation and social change being just one component. The pace of change of the smoking epidemic and the groups most affected will also be influenced by the development of the tobacco industry and its influence on consumption trends, themes we explore in the following two chapters.

References

Attané I (2014) Being a woman in China today: a demography of gender. In: Attané I, Baochang G (eds) Analysing China's population: social change in a new demographic era. Springer Dordrecht, New York, pp 95–110

Barnett R, Moon G, Pearce J, Thompson L, Twigg L (2017) Smoking geographies space, place and tobacco. Wiley Blackwell, Oxford, 301p

Bell A (2020) Age period cohort analysis: a review of what we should and shouldn't do. Ann Hum Biol 47(2):208–217

Bonilla-Silva E (2019) Feeling race: theorizing the racial economy of emotions. Am Sociol Rev 84(1):1–25

Bucerius SM (2014) Unwanted: Muslim immigrants, dignity, and drug dealing. Oxford University Press, Oxford

Cai L, Wang XM, Fang LM, Cui WL, Golden AR (2019) Socioeconomic disparities in prevalence and behaviors of smoking in rural Southwest China. BMC Public Health 19(1):Article No. 1117

Chen X (2018) Do skewed sex ratios among children promote parental smoking? Longitudinal evidence from rural China. J Subst Abuse 23(4):366–370

Chen X, Li X, Stanton B, Fang X, Lin D, Cole M, Liu H, Yang H (2004) Cigarette smoking among rural-to-urban migrants in Beijing, China. Prev Med 39(4):666–673

Chen Y, Chen H, Li Z (2018) Effect of characteristics and life in cities in China on residents' smoking behavior. J Int Med Res 46(10):4226–4234

Chen Y, Zhou LA (2007) The long-term health and economic consequences of the 1959–1961 famine in China. J Health Econ 26(4):659–681

China Center for Disease Control and Prevention (CDC) (2015) China adult tobacco survey report. China CDC, Beijing. http://www.tcrc.org.cn/UploadFiles/2016-03/318/201603231215175500.pdf

Chuang YC, Li YS, Wu YH, Chao HJ (2007) A multilevel analysis of neighborhood and individual effects on individual smoking and drinking in Taiwan. BMC Public Health 7(1):151

Delgato R, Stefancic J (1998) Critical race theory: past, present, and future. Curr Leg Probl 51(1):467–491

Ding D, Gebel K, Oldenburg BF, Wan X, Zhong X, Novotny TE (2014) An early-stage epidemic: a systematic review of correlates of smoking among Chinese women. Int J Behav Med 21(4):653–661

Dixon J, Banwell C (2009) Theory driven research designs for explaining behavioural health risk transitions: the case of smoking. Soc Sci Med 68(12):2206–2214

Dressler WW, Oths KS, Gravlee CC (2005) Race and ethnicity in public health research: models to explain health disparities. Ann Rev Anthropol 34:231–252

Edwards L (2000) Policing the modern woman in republican China. Modern China 26(2):115–147

Factor R, Kawachi I, Williams DR (2011) Understanding high-risk behavior among non-dominant minorities: a social resistance framework. Soc Sci Med 73(9):1292–1301

Finch K, Novotny TE, Ma S, Qin D, Xia W, Xin G (2010) Smoking knowledge, attitudes and behaviors among rural-to-urban migrant women in Beijing, China. Asia Pac J Public Health 22(3):342–353

Flores EO (2014) God's gangs: barrio ministry, masculinity, and gang recovery. New York University Press, New York

Fukuda Y, Nakamura K, Takano T (2005) Accumulation of health risk behaviours is associated with lower socioeconomic status and women's urban residence: a multilevel analysis in Japan. BMC Public Health 5:Article No. 53

Ghouri N, Atcha M, Sheikh A (2006) Influence of Islam on smoking among Muslims. BMJ Clinic Res 332(7536):291–294

Gustafsson B, Li S (2000) Economic transformation and the gender earnings gap in urban China. J Popul Econ 13(2):305–329

Guttentag M, Secord PF (1983) Too many women? The Sex Ratio Question. Sage Publications, Beverley Hills
Hermalin AI, Lowry DS (2012) The decline of smoking among female birth cohorts in China in the 20th century: a case of arrested diffusion? Population Research & Policy Review 31(4):545–570
Hitchman SC, Fong GT (2011) Gender empowerment and female-to-male smoking prevalence ratios. Bull World Health Organ 89(3):195–202
Hoffman SJ, Mammone J, Rogers Van Katwyk S, Sritharan L, Tran M, Al-Khateeb S, Grjibovski A, Gunn E et al (2019) Cigarette consumption estimates for 71 countries from 1970 to 2015: systematic collection of comparable data to facilitate quasi-experimental evaluations of national and global tobacco control interventions. Br Med J 365:I2231
Kim S, Symons M, Popkin BM (2004) Contrasting socioeconomic profiles related to healthier lifestyles in China and the United States. Am J Epidemiol 159(2):184–191
Lee YH, Ang TFA, Lin HC, Chang YC (2019) Rural-urban disparities in smoking patterns among Chinese adults: a social-ecological approach. Journal of Ethnicity in Substance Abuse. Published Online July 4: 1–16. https://doi.org/10.1080/15332640.2019.1633980
Lillard DR, Christopoulou R (eds) (2015) Life course smoking behavior: patterns and national context in ten countries. Oxford University Press, New York, 304pp
Liu F (2015) Smoking in China. In: Lillard DR, Christopoulou R (eds). (2015). Life course smoking behavior: patterns and National Context in ten countries. Oxford University Press, New York, pp. 103–116
Liu S, Zhang M, Li Y, Wang L, Huang Z, Wang L, Chen Z, Zhou M (2017) Prevalence and patterns of tobacco smoking among Chinese adult men and women: findings from the 2010 national smoking survey. J Epidemiol Community Health 71(2):154–161
Liu Y, Song H, Wang T, Wang T, Yang H, Gong J, Shen Y, Dai W, Zhou J, Zhu S, Pan Z (2015) Determinants of tobacco smoking among rural-to-urban migrant workers: a cross-sectional survey in Shanghai. BMC Public Health 15:Article No. 131
Lopez AD, Collishaw NE, Piha T (1994) A descriptive model of the cigarette epidemic in developed countries. Tob Control 3(3):242–247
Luo Z, Mu R, Zhang X (2006) Famine and overweight in China. Appl Econ Perspect Policy 28(3):296–304
Mackay J (1996) Women and tobacco: international issues. J Am Med Womens Assoc 51(1&2):48–51
Mannheim K (1952) The problem of generations. In: Kecskemeti P (ed) Essays on the sociology of knowledge. Routledge & Kegan Paul, London, pp 276–320
Mao A, Yang T, Bottorff JL, Sarbit G (2014) Personal and social determinants sustaining smoking practices in rural China: a qualitative study. Int J Equity Health 13(1):12
Mehta N, Zhang C, Hua X, Redmon P, Eriksen M, Koplan J, Ali M (2014) Tobacco smoking among government employees in six cities in China. Heart Asia 6(1):179–183
Matheson FI, LaFreniere MC, White HL, Moineddin R, Dunn JR, Glazier RH (2011) Influence of neighborhood deprivation, gender and ethno-racial origin on smoking behavior of Canadian youth. Prev Med 52(5):376–380
Morrow M, Ngoc DH, Hoang TT, Trinh TH (2002) Smoking and young women in Vietnam: the influence of normative gender roles. Soc Sci Med 55(4):681–690
Pampel FC (2001) Cigarette diffusion and sex differences in smoking. J Health Soc Behav 42(4):388–404
Pampel FC (2005) Diffusion, cohort change and social patterns of smoking. Soc Sci Res 34(1):117–139
Pampel FC (2006) Global patterns and determinants of sex differences in smoking. Int J Comp Sociol 47(6):466–487
Pan Z (2004) Socioeconomic predictors of smoking and smoking frequency in urban China: evidence of smoking as a social function. Health Promot Int 19(3):309–315
Parascandola M, Xiao L (2019) Tobacco and the lung cancer epidemic in China. Trans Lung Cancer Res 8(Supp 1):s21–s30

Park EJ, Kim H, Kawachi I, Kim IH, Cho S (2010) Area deprivation, individual socioeconomic position and smoking among women in South Korea. Tob Control 19(5):383–390

Pathania VS (2011) Women and the smoking epidemic: turning the tide. Bull World Health Org 89(3):162

Perdue PC (2009) China marches west: the Qing conquest of Central Eurasia. Harvard University Press, Cambridge, MA

Rich ZC, Xiao S (2012) Tobacco as a social currency: cigarette gifting and sharing in China. Nicotine Tob Res 14(3):258–263

Rutherford MJ, Lambert PC, Thompson JR (2010) Age-period-cohort modeling. Stata J 10(4):606–627

Si Y, Zhou Z, Su M, Wang X, Li D, Wang D, He S, Hong Z, Chen X (2018) Socio-economic inequalities in tobacco consumption of older adults in China: a decomposition method. Int J Environ Res Public Health 15(7):1466

Smith EMJ (1985) Ethnic minorities: life stress, social support, and mental health issues. Couns Psychol 13(4):537–579

Song G, Hird D (2013) Men and masculinities in contemporary China. Brill, Leiden

Stalin (1913) Marxism and the National Queston. Retrieved from: https://www.marxists.org/reference/archive/stalin/works/1913/03a.htm

Thun M, Peto R, Boreham J, Lopez AD (2012) Stages of the cigarette epidemic on entering its second century. Tob Control 21(2):96–101

Tian G, Wang Z (2013) Socioeconomic determinants of smoking and drinking in China: urban-rural difference. J Cambrid Stud 8(2):1–16

Tu SH, Liao PS (2005) Gender differences and gender role attitudes: a comparative analysis of Taiwan and coastal China. J Comp Fam Stud 36(4):545–566

Waldron I (1991) Patterns and causes of gender differences in smoking. Soc Sci Med 32(9):989–1005

Wang M, Luo X, Xu S, Liu W, Ding F, Zhang X, Wang L, Liu J, Hu J, Wang W (2019a) Trends in smoking prevalence and implication for chronic diseases in China: serial national cross-sectional surveys from 2003 to 2013. Lancet Respir Med 7(1):35–45

Wang Q, Shen JJ, Cochran C (2016) Unemployment rate, smoking in China: are they related? Int J Environ Res Public Health 13(1):113

Wang Q, Shen JJ, Sotero M, Li CA, Hou Z (2018) Income, occupation and education: are they related to smoking behaviours in China? PLoS One 13(2):e0192571

Wang XM, Wu C, Golden AR, Le C (2019b) Ethnic disparities in prevalence and patterns of smoking and nicotine dependence in rural Southwest China: a cross-sectional study. BMJ Open 9(9):e028770

WHO (World Health Organisation) (2016) Global Health Observatory (GHO) data. WHO, Geneva. https://www.who.int/gho/tobacco/use/en/

Wright EO (1997) Class counts: comparative studies in class analysis. Cambridge University Press, Cambridge MA, 576pp

Xu F, Yin XM, Zhang M, Ware RS, Leslie E, Owen N (2007) Cigarette smoking is negatively associated with family average income among urban and rural men in regional mainland China. Int J Mental Health & Addic 5(1):17–23

Yang T, Barnett R, Jiang S, Yu L, Xian H, Ying J, Zheng W (2016) Gender balance and its impact on male and female smoking rates in Chinese cities. Soc Sci Med 154(April):9–17

Yang T, Li F, Yang X, Wu Z, Feng X, Wang Y, Wang X, Abdullah ASM (2008) Smoking patterns and sociodemographic factors associated with tobacco use among Chinese rural male residents: a descriptive analysis. BMC Public Health 8(1):Article No. 248

Yang T, Wu J, Rockett IRH, Abdullah AS, Beard J, Ye J (2009) Smoking patterns among Chinese rural-urban migrant workers. Public Health 123(11):743–749

Yang X, Jing W, Gao C, Attané I (2019a) Smoking behavior of "marriage squeezed" men and its impact on their quality of life: a survey study of China. Am J Mens Health 13(3):1557988319859733

Yang XY (2017) How community-level social and economic developments have changed the patterns of substance use in a transition economy. Health Place 46(July):91–100

Yang XY (2020) Class status and social mobility on tobacco smoking in post-reform China between 1991 and 2011. Nicotine Tob Res. https://doi.org/10.1093/ntr/ntaa103

Yang XY, Hu A, Schieman S (2019b) Relative deprivation in context: how contextual status homogeneity shapes the relationship between disadvantaged social status and health. Soc Sci Res 81(July):157–169

Yang XY, Kelly BC, Yang T (2014) The influence of self-exempting beliefs and social networks on daily smoking: a mediation relationship explored. Psychol Addict Behav 28(3):921–927

Yang XY, Yang F (2018) Acculturation versus cultural retention: the interactive impact of acculturation and co-ethnic ties on substance use among Chinese students in the United States. J Immigr Minor Health 20(3):546–560

Yang YY, Land KC (2016) Age-period-cohort analysis: new models, methods, and empirical applications. Chapman and Hall/CRC, Boca Raton, 353pp

Yen HP (2005) Body politics, modernity and national salvation: the modern girl and the new life movement. Asian Stud Rev 29(2):165–186

Yun WJ (2015) Household and area income levels are associated with smoking status in the Korean adult population. BMC Public Health 15(1):39

Zhang M, Liu S, Yang L, Jiang Y, Huang Z, Zhao Z, Deng Q, Li Y, Zhou M, Wang L, Chen Z, Wang L (2019) Prevalence of smoking and knowledge of smoking among 170000 Chinese adults 2013–2014. Nicotine Tob Res 21(12):1644–1651

Zhang X (2015) Ethnicity in China: a critical introduction. Polity Press, Cambridge UK, 236pp

Chapter 3
The Rise of China Tobacco: From Local to Global Player

Ross Barnett and Sihui Peng

Abstract This chapter sets out the economic context of smoking in China and explores the transformation of the Chinese tobacco industry from the late nineteenth to the early twenty-first century. The chapter adopts a historical approach which links political and economic changes to the evolution of the state tobacco monopoly, the China National Tobacco Company (CNTC). It highlights the stages in the development of this monopoly structure, the changing importance of central and local government stakeholders and how the tobacco industry restructured itself in response to domestic and global trends. There are four defining periods: the evolution of capitalist competitive market in the early twentieth century and its replacement by the state monopoly in 1982; the Chinese tobacco monopoly of the 1980s and 1990s, characterised by a decentralisation of control, local protectionism of tobacco manufacturers and tobacco governance; the restructuring of the industry in the late 1990s prior to China's membership of the World Trade Organisation (WTO) in 2001; and the recentralisation of political control of the industry since 2001 with changes in the geography of tobacco growing and cigarette manufacturing and the recent globalisation of the industry.

Keywords China Communist Party (CCP) · China National Tobacco Company (CNTC) · Cigarette manufacturing · Globalisation · Illicit cigarette market · Open door economic reforms · Protectionism · State restructuring · State Tobacco Monopoly · Tax system · Tobacco growing · World Trade Organisation

---

R. Barnett (✉)
School of Earth and Environment, University of Canterbury, Christchurch, New Zealand
e-mail: ross.barnett@canterbury.ac.nz

S. Peng
School of Medicine, Jinan University, Guangzhou, China

R. Barnett et al. (eds.), *Smoking Environments in China*,
Global Perspectives on Health Geography,
https://doi.org/10.1007/978-3-030-76143-1_3

3.1 Introduction

This chapter sets out the economic context of smoking in China and explores the transformation of the Chinese tobacco industry from the late nineteenth to the early twenty-first century. This period witnessed the transition of the tobacco industry (here defined as both tobacco growers and cigarette manufacturers) from a competitive capitalist market to the formation of a state monopoly and closed market under state leadership. Unlike the experience of many other lower- and middle-income countries (LMICs), this is still much the case. Despite changes since 1978, ushered in by Deng Xiao Peng to open up of the Chinese economy, political imperatives have meant that the tobacco industry has not followed these trends and the tobacco monopoly, originally established under Mao Tse Tung, remains relatively intact. This chapter adopts a historical approach which links political and economic changes to the evolution of the state tobacco monopoly, the China National Tobacco Company (CNTC). It highlights the stages in the development of this monopoly structure, the changing importance of central and local government stakeholders and how the tobacco industry has restructured itself in response to domestic and global trends.

The chapter is structured to reflect four defining periods. The second section (3.2) examines the emergence of the Chinese tobacco industry from the late nineteenth century, in particular the evolution of capitalist competitive market in the early twentieth century to its replacement by the state monopoly in 1982. The third section (3.3) briefly discusses the Chinese tobacco monopoly as it existed in the 1980s, with issues of decentralisation of control, local protectionism of tobacco manufacturers and tobacco governance. Attention is also directed at changes in tobacco growing and the illicit cigarette market. The fourth section (3.4) shifts attention to the restructuring of the industry prior to China's membership of the WTO in 2001. Here the recentralisation of political control of the industry and changes in the geography of tobacco growing and cigarette manufacturing are discussed. The final section (3.5) focuses on the recent globalisation of tobacco manufacturing and how this is reshaping the industry.

3.2 Emergence of the Chinese Tobacco Industry: 1902–1982

The first defining period of the Chinese tobacco industry since the nineteenth century loosely coincides with the time of the Republic of China (1902–1949) and the emergence of a state monopoly under Maoism (1949–1982). The first period is significant because it represented the early establishment of capitalist economic relations in China, and their eventual rejection and replacement by a state monopoly model which has persisted until the present. The tobacco industry continues to offer interesting insights into the interplay of capitalist and socialist models and the extent to which one has replaced the other at various points in the development of the industry.

3.2.1 Opening Up China: British American Tobacco and the Emergence of a Capitalist Monopoly (1902–1949)

A number of European scholars, in particular Carol Benedict (2011) and Sherman Cochran (1980), have documented the early development of the tobacco industry in China. More recently Chinese writers such as Sikang Song (2018), using a variety of Chinese archival sources, have added to this history bringing a particular Chinese perspective. The work of these writers provides a useful overview of the growth of the tobacco industry in China, social and spatial patterns of consumption and the main players in the tobacco market.

Traditional Chinese smoking largely involved pipe smoking and snuff, with cigars much less common. Song (2018, pp. 13–14) argues that Imperial China had already established tobacco agriculture by the time the Qing Empire had reached its zenith in the eighteenth century and a sophisticated smoking culture had developed which smoothed the way for the rapid spread and acceptance of the machine-rolled cigarette at the turn of the twentieth century. These ventures were aided by the political situation of the time. China, weakened by the effects of the Opium Wars (1840–1842), acquiesced to foreign influences through the establishment of 'unequal treaty ports' which enabled the importation of foreign goods, including tobacco. The first attempts in 1885 to market imported machine-rolled cigarettes, by the American Trading Company and British firm, W.D. & H.O. Wills, were relatively unsuccessful, but other companies soon followed. Most notable were James Duke's American Tobacco Company, which established itself in China in the 1890s and was largely active in Shanghai. A few small Chinese companies, such as the Maoda Rolled Tobacco Company, also existed but relied on labour-intensive methods and closed in 1900, lasting only two years (Benedict 2011).

The structure of the Chinese tobacco market was radically changed with the establishment of the British American Tobacco (BAT) Company in 1902. BAT, which became the leading multinational global tobacco business in the first half of the twentieth century, was formed when the UK's Imperial Tobacco Company and the US's American Tobacco Company agreed to form a joint venture in China. In 1902 BAT established itself in Shanghai and inherited a factory in the Pudong district from W.D. & H.O. Wills, one of the precursor companies of Imperial Tobacco (Fig. 3.1). Under the management of American James Augustus Thomas, by 1919, the Shanghai factory was producing more than 243 million cigarettes per week (Tinari 2005). Thomas worked closely with the local Wing Tai Vo Tobacco Company, which developed into BAT's principal Chinese partner after its success with the 'Ruby Queen' cigarette brand. The popularity of foreign brands was vividly highlighted in 1908 when the Empress Dowager Cixi died and was buried with a large supply of W.D. & H.O. Wills 'Passing Cloud' cigarettes which accompanied her into the afterlife (Fenby 2019, p. 113).

With its large production capacity and intensive marketing, BAT dominated the Chinese cigarette market until 1941 when World War II forced the closure of its

Fig. 3.1 British American Tobacco factory (right), Shanghai circa 1925. (Credit: Virtual Shanghai Project https://www.virtualshanghai.net/Data/Buildings?ID=1678 (Permission obtained))

Chinese factories. A brief revival occurred after the War until the Company's Chinese operations were nationalised by the Chinese Communist Party. Between 1924 and 1941, BAT's share of the Chinese market ranged from a high of 82% in 1924 to a low of 55% in 1934, but for most of the time hovered around 60% (Benedict 2011, p. 133). This success arose from the industrial nature of its operations compared to the traditional labour-intensive forms of cigarette manufacturing which had characterised the Chinese market in the past. BAT epitomised economies of scale, capital-intensive technology, vertical integration of production and distribution, low labour costs, diplomatic support from the United States and Britain, skilful advertising, aggressive pricing and a willingness to undercut all competitors, often using ethically questionable practices (Benedict 2011, p. 133). BAT was also aided during the Republican period (1912–1949) when a unified national government under Chang Kai-shek provided relative political stability and greater opportunities for the company to expand its business interests. BAT also introduced American bright tobacco into the agriculture of the North China Plain and established three tobacco-growing bases in Eastern Shandong, Central Henan and Northern Anhui (Song 2018). This, in turn, led to the formation of a commodity chain that reached from tobacco-growing areas to cigarette factories in the major cities. Developing a decentralised managerial system employing mostly Chinese nationals (Cochran 1980, p. 15) helped the company erase cultural and language barriers and extend the reach of its market. BAT was able to build on existing extensive sales networks that had been used to distribute exports and 'native' tobacco and was fortunate to be able to use Chinese sales agents who were already well connected with both regional and local markets. These strategies all helped BAT achieve and maintain its supremacy.

BAT had only a small amount of competition in China, drawn from two sources. First, during the early 1930s, BAT's main competition came from small mechanised rolling manufacturers who established themselves after 1925 in smaller cities and towns in provinces close to Shanghai. They were skilled at making fake counterfeit brands, including popular foreign ones using low-grade tobacco. By 1934 they occupied an estimated 25% share of the market (Cox 2000, p. 187). The presence of such operations, however, was short-lived. Following negotiations between BAT and the Ministry of Finance, which was concerned with the loss of state revenue, competition from such sources soon came to an end in 1934 when all hand-rollers had to be licensed and use only authorised cigarette papers in their products (Chi 1935).

Second, BAT's primary competition in the 1930s came not from other transnational tobacco companies (TTCs), but from the establishment of some indigenous Chinese cigarette companies using mass production methods. The most prominent of these were the Nanyang Brothers' Tobacco Company (Fig. 3.2) and the Huacheng Tobacco Company both of which operated in coastal treaty port locations. The establishment of such firms was closely related to periods of nationalism following the establishment of the Chinese Republic in 1912 and anti-foreign commodity boycotts (Rawski 1989). Anti-British protests and the deaths of demonstrators on Nanjing Road in Shanghai in 1925 ushered in a rapid expansion of local indigenous businesses (Yang 2002). Most of these were undercapitalised and soon disappeared, but others, such as the Nanyang Brothers' Tobacco Company, were more successful. Founded in 1904, and still in existence today in Hong Kong, the company managed to remain profitable by focusing on niche markets and by developing effective marketing systems that embodied both Chinese and western characteristics (Chen 2018). The founder of the company, Chien Chao-nan, had received training in Japan and recruited experienced Japanese technicians. Despite adopting kinship-based management networks, which limited the development of the company, and BAT attacks on the national legitimacy of the firm, Nanyang Brothers was one of the few Chinese tobacco companies to survive from the early period of the industry. Nevertheless, after 1928, the number of Chinese-owned firms declined rapidly from 182 in 1927 to only 60 in 1932 (Yang 2002, p. 69).

Cigarette consumption rose dramatically during the Republican era, BAT's sales alone soaring from 26 billion sticks in 1928 to 55 billion in 1937 (Cox 2000). However, the Chinese cigarette market during this time was highly stratified with significant regional and class differences. There had been a relatively rapid spatial diffusion of manufactured cigarettes from the coast to the interior and from cities to the countryside, effected by BAT and its agents. Although the data is highly fragmentary, estimates provided by writers such as Cochran (1980) and Gibbs (1938)

Fig. 3.2 Nanyang Brothers' Tobacco Company, Shanghai (circa 1920s). (Credit: Virtual Shanghai Project https://www.virtualshanghai.net/Photos/Images?ID=1655 (Permission obtained))

suggest that annual per capita consumption of factory-produced cigarettes was highest in coastal provinces containing major metropolitan areas and lower in the interior. Benedict (2011, p. 147) also suggests that consumption rates were high in more industrialised areas where cigarettes were produced, a pattern that has continued through to the present (Yang et al. 2015).

It should also be mentioned that, for a brief time during the Japanese invasion of China (1937–1945), Japanese cigarettes replaced Chinese brands in the occupied territories. A special 'Golden Bat' brand was marketed by the Mitsui Tobacco Company and is still produced by Japan Tobacco today (Roberts and Yamakawa 1973). Circulation of Golden Bat cigarettes was prohibited in Japan and caused controversy in the 1940s when it was revealed that small doses of opium or heroin were hidden in the mouthpieces of the cigarettes destined for the Chinese market (Seth 1974). Such actions, highlighted at the War Crimes trials in 1948, not only helped cause the breakdown of Chinese society but also generated substantial revenues for the Japanese wartime economy.

As predicted by the smoking transition model, the Chinese market was also socially stratified. Benedict (2011) cites a number of studies showing that wealthy tobacco users, who had previously used water pipe tobacco, foreign snuff or expensive premium regional hand-rolled tobacco, now preferred imported cigarettes, with consumption rates reflecting per capita income (Benedict 2011, pp. 151–152). This, however, depended upon local circumstance. In contrast to Shanghai whose mass market smoking culture was quite socially inclusive in the 1930s, in Beijing the tobacco trade remained much more segmented. In Shanghai manufactured cigarettes had largely replaced other forms of tobacco consumption by 1928, yet in Beijing native pipe tobacco remained significant, and the number of cigarettes smoked per person a day was only about half that of Shanghai (Benedict 2011, p. 158). Benedict cites three reasons for the difference in consumption patterns between the two cities. First, the supply of manufactured cigarettes was poor in Beijing. While the city produced cheap hand-rolled cigarettes, the Guomindang Nationalist Government, based in Nanjing, did not allow cigarette manufacturing in certain cities after 1934, including Beijing. Second, Beijing was a much poorer city than Shanghai and smokers tended to prefer cheaper pipe tobacco rather than machine-rolled cigarettes. Third, in Beijing the retail distribution of tobacco outlets tended to favour richer neighbourhoods in the inner city. This can still be seen today in the location of 'Tobacco Pipe' Street (Yandai Byway) in one of the city's hutongs near the Forbidden City (Fig. 3.3). While Beijing's wealthier consumers had access to the same brands as their counterparts in Shanghai, access to cheap machine-rolled cigarettes for poorer residents of the city was far more restricted. For these reasons Beijing did not develop a genuine mass market for manufactured cigarettes until after the Communist Party came to power in 1949.

While differences in consumption occurred between various cities, the largest differences occurred between cities and the countryside. Benedict (2011, pp. 165–168) claims that the introduction of the 'modern' cigarette had a very limited impact in rural China citing evidence from studies undertaken in the late 1920s in the county of Dingxian in Hebei Province. However, as Song (2018, p. 155) points out, even though Chinese cigarette consumption up to the 1930s did not overwhelmingly cover all regions and social classes, the scale of cigarette diffusion among the lower class in inland regions should not be underestimated. Looking at cigarette consumption regionally, Song notes that rural areas in North China and the Lower Yangzte River Valley had relatively high rates of cigarette smoking due to the rapid growth of the hand-rolling industry. Similarly for industrially produced cigarettes, it is likely that a spatial diffusion occurred from treaty port cities to their immediate hinterlands and that urban-rural differences in the consumption of new cigarette products would have been less in those areas.

3.2.2 *From Capitalist to State Monopoly (1949–1982)*

BAT dominated the tobacco industry in China until 1953. However, following the Chinese Communist Party (CCP) takeover in 1949 and the formation of the People's Republic of China (PRC) in 1953, when the tobacco industry was nationalised, BAT was forced to leave the country (Lee et al. 2004). The 1950s and 1960s were a time of economic and social crises in China. Nevertheless, events such as the Great Leap Forward (1958–1960) and ensuing famine (1959–1961) and the Cultural Revolution (1966–1976) with its associated social trauma had less impact on the tobacco industry than other sections of the economy. To a large extent this reflected the use of the

Fig. 3.3 Yandaixie ('Tobacco Pipe') Street, Beijing. (Credit: Authors. Renamed during the Qing Dynasty to Yandai, literally meaning 'Tobacco Pipe' Byway, Yandaixie Street is one of the oldest commercial thoroughfares in Beijing. Formed during the Yuan Dynasty (1279–1368), the street later rose to prominence during the Qing period as a place of tobacco pipe shopping. Its name also reflects in the curved shape of the street. In 2010 the street was awarded the title of 'China's Famous Historical and Cultural Street' by the Ministry of Culture and State Administration of Cultural Heritage)

tobacco market for political propaganda and to demonstrate the superiority of state ownership of production under the socialist economy.

The production of domestic cigarettes grew by 11% between 1949 and 1958, but during the politically turbulent years of the Great Leap Forward tobacco production slowed and by the time of the Cultural Revolution cigarettes were in short supply, especially in rural areas (Benedict 2011). The commitment to ensuring increased production had actually begun before the end of World War II when the breakdown of a brief and tenuous Communist-Nationalist anti-Japanese alliance led to the need for the CCP to promote self-reliance and support peasant livelihoods, especially since tobacco was at this time more lucrative than other cropping alternatives. The CCP was fully committed to building up the tobacco industry in its base areas in the Shaan-Gan-Ning Border Region of Western China. Wennan (2018, p. 48) reports that the 1944 ban of cigarette imports 'supercharged' the creation of factories, both small private ones and larger state-owned enterprises. By 1945 there were more than 200 factories in total producing over 700,000 cartons of cigarettes (500 cigarettes per carton). For a time 'there was something of the Wild West of cigarette manufacturing in China' (Wennan 2018, p. 48)

This did not last long. As early as 1952 China established the 'Tobacco and Alcohol Monopoly Agency' to control all aspects of production, distribution and marketing, a process that was largely complete by the 1960s (Chen 2018). However, as in any command economy model, inefficiencies arose particularly due to the lack of market signals from consumers and the need for manufacturers to be more responsive to the performance of their products in the market. These problems were accentuated by the lack of provincial control over the industry which was highly uncoordinated and lacked economics of scale.

As a response to such problems, the early 1960s saw further reorganisation of the industry as part of a broader effort to rationalise industrial production. In 1963 the China Tobacco Industrial Corporation was established. This aimed to achieve greater efficiencies through more centralised management processes (procurement, production, sales). It also made a number of more micro-adjustments to ameliorate the tensions between state bureaucracies and tobacco farmers. However, despite these innovations the corporation was dismantled in the wake of the Cultural Revolution and the industry reverted to its former fragmented structure (Fang et al. 2017).

While the first moves to improve the efficiency of the state tobacco monopoly produced some gains, perhaps the longer-lasting legacy of the Mao years was the promotion of a mass market of consumption, especially of cheap cigarettes among the peasantry. CCP political propaganda celebrated cigarette smoking by relating it to notions of masculinity and revered political leadership. In this way the Chinese state promoted a ready demand for cheap cigarettes among the masses and helped legitimatise the concentration of its political power. Qiushi and Kohrman (2018, p. 87) remarked that, in writing about the effects of the Great Leap Forward, not only did it have a profound influence on China's cigarette industry by eroding the hegemony of former capitalist strongholds such as Shanghai but 'it proletarised the cigarette, transforming it from an item largely enjoyed by the elite to something that 'the people' had the right to access, acquire and smoke. It furthered an institutionalisation … and helped to cement cigarette manufacturing as a bedrock of the normal order of things across China…'.

3.3 Structure of the Tobacco Industry (1982–2003)

This section details the structure of the Chinese tobacco industry following the enactment of the *State Monopoly Law* in 1982. This period is significant because it marked a new phase whereby central government attempted to formalise its control over the industry through the establishment of new institutional structures. Consequently it is important to outline the structure of the Chinese tobacco industry as it existed in the 1980s and early 1990s, its key stakeholders and their importance at national and provincial levels.

3.3.1 Institutional Structures

The rudimentary tobacco monopoly established during the 1960s and 1970s was not immediately altered by the 'Open Door' economic reforms introduced by Deng Xiaoping after 1978. The 'open door' marked a radical policy shift signalling the desire to integrate market economics with Chinese socialism, but, ironically rather than liberalising the tobacco market in China, it did exactly the opposite. If anything, the formalisation of the tobacco monopoly structure in 1982 solidified the localism that already characterised the industry. Although TTCs explored opportunities that this policy shift presented, in reality China remained closed to transnational tobacco companies. Nevertheless, despite an absence of foreign involvement, China's cigarette market did become more competitive, both as a result of the growth of entrepreneurial spirit unleashed by the national reforms and also by increased competition between local governments aiming to encourage and support the growth of tobacco production and cigarette manufacturing in their areas.

Before 1980 both tobacco growing and cigarette production were largely managed by provincial and local governments (Hu 2008). However, with the enactment of the State Monopoly Law in 1982, Central Government instituted a system of vertical leadership and monopolised operation dominated by the State Tobacco Monopoly Administration (STMA) and the China National Tobacco Corporation (CNTC) (Fig. 3.4). The STMA is the administrative arm responsible for the strategic direction and planning of the tobacco industry, but since 1991 it has delegated authority to the CNTC for most aspects of tobacco production, marketing, distribution, sales and pricing of cigarettes. The STMA is a complicated vertical bureaucracy that operates at the central, provincial, district and county levels through bureaus which manage the tobacco monopoly system (He et al. 2013). It sets overall policies for its local agencies which manage the allocation of provincial tobacco production quotas, tobacco leaf prices and cigarette production quotas in different localities. Each bureau interacts with government organisations at the appropriate level. The CNTC, the production arm of bureaucracy, allocates production quotas to the various tobacco factories, signs contracts with tobacco farmers and regulates prices charged by wholesalers and retailers in the supply chain. Provincial and local governments benefitted from the receipt of different taxes (see Sect. 3.4.2) but also traditionally played an independent role in influencing production quotas and prices received by growers.

The establishment of the state monopoly in the 1980s was a reflection of the importance Central Government placed on the tobacco industry and why, until recently, it has been ambivalent towards the WHO Framework Convention for Tobacco Control (FCTC) and the limitations that this could impose (Huang 2014). The STMA/CNTC has promoted smoking as an economic pillar of the country; it has framed tobacco control as a threat to social stability and maintained that the tobacco industry is a legitimate actor in tobacco control policy making (Jin 2014). Consequently, unlike its western counterparts, the tobacco industry enjoys a certain amount of respectability in China and is a source of national pride, and symbol of

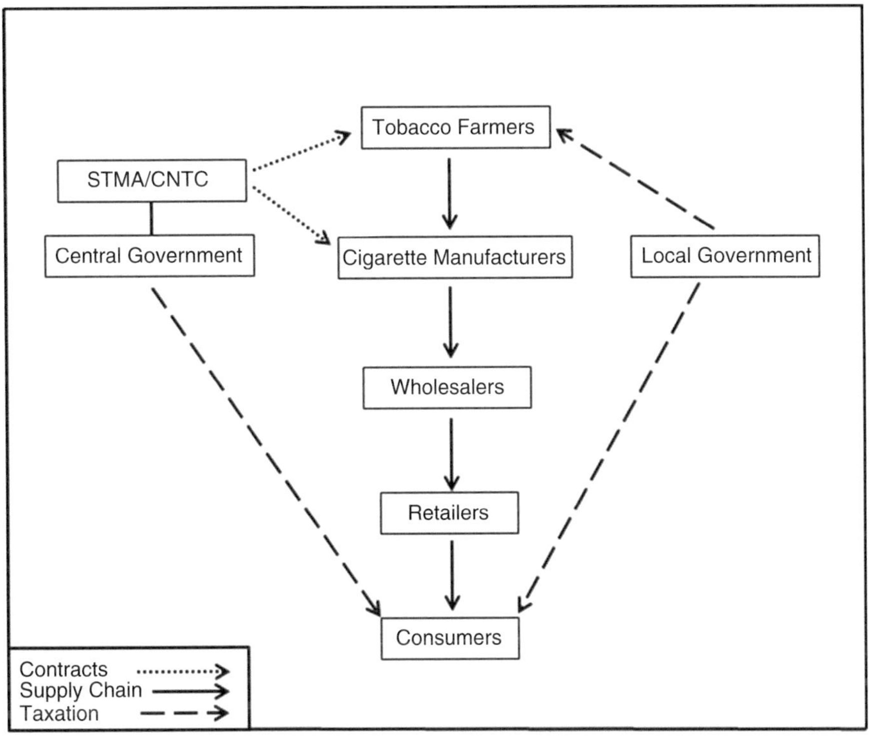

Fig. 3.4 Structure of the Chinese tobacco industry. (Credit: Authors)

resistance against foreign influence (Hu et al. 2013). Tobacco is a major source of income for the Chinese government and, although its share has recently declined slightly, over the last decade has constituted between 7% and 8% of all revenues (Fig. 3.5).

3.3.2 Subnational Stakeholders: Provincial Governments and the Industry

When central government established the STMA in the early 1980s, it had the clear intention of centralising and monopolising the tobacco sector. This occurred because provincial governments were now important stakeholders in the tobacco industry within their boundaries in what Li (2018, p. 180) has described as an 'incestuous relationship'. The enforcement of state monopoly by local STMA bureaus ensured a steady flow of revenue to local governments. It was in their interests to work closely with the provincial and local STMA bureaus and for the different organisations to support each other. This occurred, but not always in ways envisaged by central government. Provincial governments, along with their local

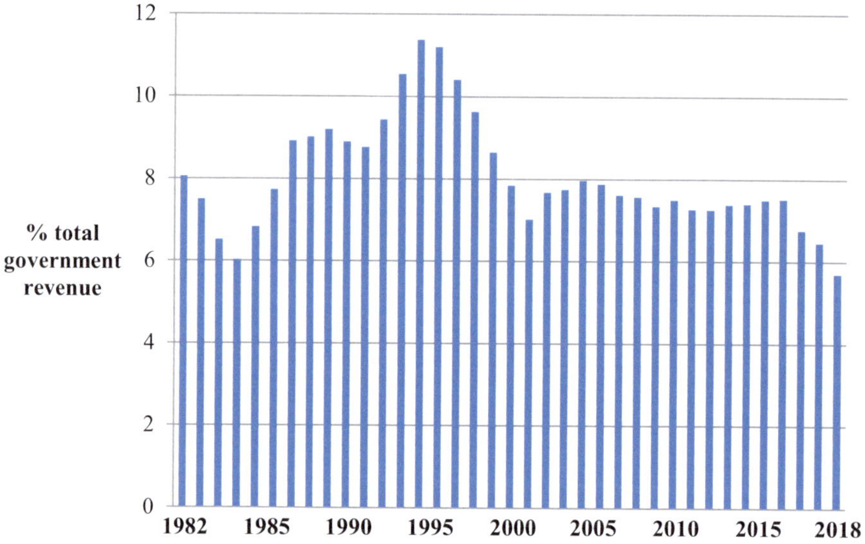

Fig. 3.5 Cigarette sales as % total government revenue (100m Yuan), 1982–2018. (Credit: State Tobacco Monopoly Administration (STMA) (various years) *China Tobacco Yearbook* (Science and Technology of China, Beijing); *China Tobacco Almanac* (The Economic Daily Press, Beijing) Public Domain)

STMA bureaus, encouraged levels of production, among tobacco farmers and cigarette manufacturers, beyond those set in centrally imposed quotas (Eng 1999). Such contracts granted local manufacturing enterprises a certain freedom in pricing and sales above quota production and, in so doing, provided them with an incentive to increase production especially since each branch company was allowed to retain its profits (Tong et al. 2008). Similarly, tobacco farmers increased their output to meet the demands of manufacturers. This, however, not only weakened central control over the production process, but inevitably created a glut in the market, heightening competition between cigarette manufacturers. This was doubly beneficial for local government which gained increased tax revenues from increased cigarette production, and, after the abolition of most agricultural taxes in 2005, the tobacco leaves tax remained provincial governments' only source of agricultural revenue (Jin 2014).

In 1994 a new central-local tax distribution system (*fenshuizhi*) was introduced. This was a complex tax sharing arrangement whereby certain taxes were independently raised by central and provincial governments but some were shared, most notably the value-added tax (75% central, 25% local), corporate income tax (60% central, 40% local for firms established before 2002) and the tobacco business income (sales) tax (60% central, 40% local). In 2009 these three taxes accounted for 89.5% tobacco-related tax revenue, with central government receiving approximately 81% total (Li 2018, p. 185). Some other minor taxes (e.g. tobacco leaves tax) went entirely to local government.

At the time, these new tax sharing arrangements were significant because they created a greater incentive for local governments to intervene in the local tobacco industry. The new tax policy came about because the fiscal power of central government had been weakened during the early days of market-oriented reforms when fiscal decentralisation had aimed to mobilise local economies. This weakening of the fiscal power of the central state was countered by the decision to recentralize fiscal revenue via a new tax sharing agreement (Cheng 2015). Under the new 1994 tax policy, the greater the sales tax (which varied according to the quality of cigarettes sold) and VAT collected by provincial governments, the higher was their tax refund from central government. These two taxes dominated central-local government transfers and were particularly important in leading to a surge in local protectionism especially after 1994 when the fiscal position of many provinces worsened (Cheng 2015). Protectionism was especially important for provinces with tax revenues highly dependent on the tobacco industry.

Local protectionism included a variety of administrative controls to protect cigarette manufacturers within their jurisdictions by making non-local products difficult to obtain. Protecting their tax revenues was especially important as consumption of non-local cigarettes eroded provincial revenues. Protectionism took a number of forms. For example, provincial governments often instructed tobacco companies within their jurisdictions to sell their production locally. Li (2012) reports that only three places (Shanghai, Yunnan and Guizhou) sold less than 40% locally made cigarettes, while Huang (2010) showed that 74% cigarettes manufactured in China were sold in their home province, cross-provincial sales being small. Some local governments conducted occasional inspections to check whether local retailers were selling cigarettes made outside their provinces, with fines being imposed, as in Jiangsu, on retailers selling non-local cigarettes (Cheng 2015). In some cases provincial governments ordered their public sector workers to smoke local brands in order to boost the local economy (Li 2012). Jin (2014) reports examples in Hubei Province where various counties within the province issued edicts for quotas of cigarettes to be consumed at official banquets and by teachers at local schools during the global financial crisis in 2009. These edicts also reflected the culture of cigarette gifting in China where the consumption of tobacco products is an important component of social interactions (see Chap. 5).

Provincial governments also had important relationships with tobacco manufacturers and local wholesalers. As in the case of the Hongta Group in Yunnan Province, top party and provincial leaders played a critical role in the rapid development of the industry at the enterprise level. This took the form of local financial investments in plant and equipment, political lobbying on behalf of particular factories, or encouraging the production of higher-grade cigarettes which in turn helped produce more government revenue (Li 2012, 2018). Also, in order to ensure favourable outcomes, provincial governments have played key roles in controlling appointments to tobacco companies operating in their regions. This has placed tobacco executives in an advantageous position when seeking government help to expand the industry. Li (2018), primarily using the example of Yunnan, shows how tobacco-related corruption penetrated provincial politics and how provincial and factory leaders conducted political lobbying in Beijing to influence policy outcomes. In this respect their activities have been no different to tobacco lobbyists in the United States who have blocked tobacco control initiatives, particularly in tobacco-growing states (Hoffer and Pellilo 2012; Morley and Pratte 2013).

In summary, the early period of the STMA/CNTC monopoly saw the emergence of intense market competition in the tobacco industry among provincial governments. Wang (2009, p. 181) notes that in the 1980s and 1990s these functioned as 'market competitors', working together with local tobacco companies in their jurisdictions to maintain their comparative advantage in the national market. This competition produced both 'winning' and 'losing' regions. Some 'winning' regions such as Yunnan, Hunan and Shanghai were able to expand their national market shares, but for many 'losing' regions, where local protectionist policies were strongest, local governments through a variety of policies were able to block imports from other provinces and protect their local markets (Wang 2013). As a result, because of local government protection, many cigarette factories with a low production capacity managed to survive, further weakening central government's role in the industry. With the exception of some high-quality brands, the huge increase in production (Table 3.1) resulted in a saturated market, leading to one-third of total production capacity being mothballed in the early 1990s (Zhou 2000, p. 131).

While decentralisation had allowed an enormous expansion of domestic cigarette production and sales, intense regional competition weakened the overall competitive capacity of the industry. Local governments exerted strong pressures in their regions for factories to be established, leading to the establishment of many marginal small enterprises. By the late 1990s, there were over 200 tobacco enterprises located in almost every province. Tong et al. (2008, p. 217) report that by the early 2000s non-profitable companies still existed because local governments were

Table 3.1 Rates of growth in cigarette production, 1949–2018

Period	100m pieces		Change	% change
1949–1955	800	1785	985	**123.1**
1955–1960	1785	2750	965	**54.1**
1960–1965	2750	2065	−685	**−24.9**
1965–1970	2065	3385	1320	**63.9**
1970–1975	3385	4360	975	**28.8**
1975–1980	4360	6515	2155	**49.4**
1980–1985	6515	11798	5283	**81.1**
1985–1990	11798	16302	4504	**38.2**
1990–1995	16302	17425	1123	**6.9**
1995–2000	17425	16682	−743	**−4.3**
2000–2005	16682	19389	2707	**16.2**
2005–2010	19389	23753	4364	**22.5**
2010–2015	23753	25886	2133	**9.0**
2015–2018	25886	23356	−2530	**−9.8**

Credit: State Tobacco Monopoly Administration (STMA), China Statistical Yearbook (various years)

willing to subsidise them for the sake of employment and tax revenue. Even by 2004, two-thirds (65.4%) of the 127 cigarette enterprises were small companies producing less than 0.3 million cases per year, and only 12.5% produced over 0.5 million cases (Cheng 2015, p. 122). Such trends, coupled with the large impact of the illicit cigarette market on production and prices, illustrated the fundamental weakness of the 1982 *State Monopoly Law* and the conflicts of interest between local governments and cigarette manufacturers over production targets. These issues prompted regulatory change to the tobacco monopoly system with the enactment of a new *Tobacco State Monopoly Law* promulgated in 1992. This was not a free market, but one that represented a mixture of market elements and state control under which manufacturers had autonomy over exchange and pricing, but could not determine their production targets which were now subject to higher level controls. The same applied to local government influence on the industry which was subsequently curtailed in the restructuring which followed. In essence the industry increasingly moved from a decentralised to a more centralised framework.

3.3.3 Changing Geographies of Production: The Case of Yunnan

Following the 1978 economic reforms, there was a remarkable surge of cigarette production compared to the years of the Great Leap Forward and Cultural Revolution when growth rates slowed and even became negative (Table 3.1). The rate of growth in cigarette production peaked between 1980 and 1985, a direct consequence of the reforms, only to slow substantially during the subsequent period of rationalisation that characterised the industry from the late 1990s. Most growth occurred within a cigarette manufacturing ‘belt’ of ten inland provinces stretching from Shandong in the east to Yunnan in the west (Fig. 3.6). These provinces were also important tobacco-growing areas; in 2000 the correlation between the provincial share of land devoted to tobacco farming and the provincial share of cigarette manufacturing was $r = 0.87$, $p < 0.001$. While the share of national cigarette production for the top ten provinces in 2000 was little different from that of 1986 (72.1% versus 72.6%), the ranking of the provinces was not, largely due to the spectacular growth of cigarette manufacturing in Yunnan. In 1986 Yunnan produced only 8.7% China’s cigarettes but this rose spectacularly to 19.0% by 2000.

Understandably the extraordinary growth of Yunnan’s tobacco economy has attracted a lot of attention about the dynamics of growth in a new socialist ‘market’ economy. Eng (1999) stressed the importance of external economies of agglomeration in tobacco production whereby tobacco firms gained a variety of benefits from locating close together, such as a pool of skilled labour, the sharing of information between different firms regarding markets and quality assurance, or production techniques. However, she noted that the success of the industry could not be viewed simply as an outcome of interaction between local producers as this neglected the

Fig. 3.6 Provincial distribution of cigarette production (100 million pieces), 2000. (Beijing, Tianjin, Shanghai and Chongqing are municipalities controlled directly by Central Government. Credit: Authors)

role of the local state in the production process. This is the view of Li (2018) who argues that the rise of Yunnan was not simply due to macro-economic factors. In his view 'the most important reason for the triumph of tobacco in Yunnan can be traced to certain personal and political connections of top provincial leaders, who managed to rise to political power through tobacco linked economic initiatives' (p. 186). Wang (2013), on the other hand, in an insightful analysis of the growth of the Yuxi Cigarette Factory (which became part of the Hongta (Tobacco) Group Ltd), places greater emphasis on the role of individual agency – that of tobacco entrepreneurs and the key roles they played in negotiating with local and central governments. Important here was Chu Shijian, nicknamed 'China's Cigarette King' (Tobacco China Online 2019a) who managed and promoted the Yuxi Cigarette Factory in the early years of reform. Taking advantage of the more liberal economic climate promoted by central government, he departed from the central planning system model by developing new sources of finance and management structures. For example, Chu mortgaged the factory in order to purchase new production equipment to produce filtered cigarettes. Other initiatives concerned providing more financial and technical assistance to peasants in order to improve leaf quality (de Oliveira 2016) and beginning informal trade relationships with other provinces wishing to preserve their production quotas (Wang 2013, p. 84).

Chu also played an important role in creating a new management structure, the '3 in 1' combined system, where leadership was shared between the Yuxi Tobacco Factory, the STMA Yuxi Bureau and the Yuxi Tobacco Trade Company. This somewhat radical arrangement in a centrally planned system gave greater flexibility to the manufacturing enterprise but posed no power threat to the STMA local bureaus as it was a compromise arrangement whereby all parties shared leadership. As Wang (2013, p. 90) concluded, 'as it turned out, the … enterprises gained economic autonomy from local governments and controlled the tobacco industry, and that was why Yunnan's tobacco firms eventually succeeded'. The success of this model, and its spread to other tobacco enterprises, helped Yunnan become China's pre-eminent cigarette manufacturing and tobacco producing region and enabled the province to outperform competing regions. It also embodied the new entrepreneurial spirit of the 1978 reforms and was seen by central government to be a positive thing as it stimulated tobacco production in a poor part of China with a high proportion of ethnic (non-Han) minorities.

3.3.4 The Illicit Cigarette Market

While the impact of the 1978 liberalisation produced a certain flexibility in the operation of CNTC monopoly, the continued dominance of the CNTC inevitably produced a black market in counterfeit cigarettes (Shen et al. 2010). von Lampe (2012) has suggested that the illicit cigarette market has gone through three stages, each involving many actors, state and otherwise, at different points in the supply chain. Initially an illegal cigarette market emerged in the 1980s, particularly in the southern provinces of China. This focused largely on the domestic market with the production of popular domestic brands. The STMA monopoly was undermined by the establishment of many unauthorised out-of-plan factories built by towns and villages. Farmers unable to sell their crops through official sources as well as former workers of state-owned cigarette factories were involved. Similar to the private hand-rolling workshops that had competed with BAT and other companies during the Republican period, these small plants captured a significant share, especially of the rural market, by offering low-income smokers very cheap cigarettes (Benedict 2011; Cheng 2015, p. 97). They also syphoned off raw materials needed by state-run factories and affected their profits and state fiscal revenues (Zhou 2000). It was not until 1994 that STMA local bureaus, as part of local protectionist measures, began to constrain the growth of illicit trade by better managing their wholesale and retail networks (Cheng 2015).

von Lampe (2012) notes that during the 1990s the illegal diversion of cigarettes from legal outlets into black market channels and the production of counterfeit domestic brands was less of a problem, but was replaced by the illegal importation of Western brand cigarettes. Many of these schemes were aided by TTCs, such as BAT, aiming to circumvent China's rigid import quotas and high tariffs (Lee et al. 2004; Lee and Collin 2006). von Lampe (2012, p. 22) suggests that, by supplying their cigarettes to illicit distribution channels, Western tobacco companies fostered a smuggling trade that became increasingly skilful and sophisticated, eventually shifting its direction from bringing genuine cigarettes into China to smuggling counterfeit cigarettes out of China.

This latter trend marks the third phase. China has become the leading global exporter of counterfeit cigarettes not only because of increased domestic anti-smuggling campaigns but also because of a ready demand in western markets for cheap western brand cigarettes (von Lampe et al. 2014; von Lampe 2019). Shen et al. (2010) estimated the Chinese production of counterfeit cigarettes to be 400 billion per year. Many of these are smuggled to Chinese enclaves in Europe, North America and Australasia (Von Lampe 2012). For example, Wang-Schweig et al. (2018) found that illicit Chinese brands were common in Oakland's Chinatown, pointing to the importance of Chinese ethnic enclaves in the importation and distribution of counterfeit products.

Trade in illicit cigarettes has received a great deal of attention (Campaign for Tobacco-Free Kids 2008), but is generally agreed to be carried out by organised criminal groups operating in a variety of different countries (Melzer and Martin 2016). However, the variety of smuggling routes and the role of different actors, governments, organised crime, terrorist organisations and tobacco companies, in smuggling and subsequent distribution remains unclear. Melzer and Martin (2016, pp. 145–146) have drawn attention to problems of ascertaining who is involved in the black market and their impact in different contexts. They noted that in the United States, Operations Smoking Dragon and Royal Charm, conducted by the FBI in 2005, linked the smuggling of Chinese and North Korean counterfeit cigarettes to organised crime rings that targeted US West Coast cities like Oakland.

Within China it is hard to estimate the size and distribution of the illicit tobacco industry. Data from Euromonitor suggests, that, compared to other world regions, between 2009 and 2014 China recorded a substantial decrease in the trading of illicit cigarettes (Melzer and Martin 2016, p. 128). Nevertheless despite increased state regulation to protect official CNTC output, the illicit tobacco market remains strong. Shen (2010, p. 243) reported that in 2005 almost 350,000 cases of counterfeit cigarettes were detected by authorities, and over 10,000 people were arrested for counterfeiting. Many of these people had bought low-quality tobacco from farmers and had bought used rolling machines from state-owned cigarette factories. Most counterfeit cigarettes were sold through traditional retailers but also informally at tourist spots and village marketplaces, with the industry 'tolerated' by local officials particularly when the brands were either foreign or non-local, thus not conflicting with local authorised manufacturers. It is also likely that officials have been less concerned with the illicit tobacco market when its impacts are felt outside China rather than within it.

In one of the few more recent studies, Wang et al. (2016) found that the illicit trade in tobacco products was strong in Changsha County in Hunan and that illicit trade hotspots were related not only to population density (a measure of demand) but also to proximity to the county border where enforcement policies may have been weakest. This also appears to have been the case in the border regions of southern China, where anti-counterfeiting and anti-smuggling efforts have intensified, but tobacco smuggling remains strong particularly in Guangxi Province and Yunnan (Tobacco China Online 2019b). However, while the STMA monopoly and strong controls remain over the growing and manufacturing of tobacco, and counterfeit sales no longer pose a major threat to state revenue, the illicit trading of tobacco products is likely to remain part of the Chinese tobacco economy.

3.3.5 Tobacco Growers

The decentralisation of control over tobacco production which occurred in the 1980s and 1990s not only influenced cigarette manufacturers but also had a dramatic effect on tobacco growers. During this period tobacco leaf production expanded spectacularly in China rising from less than 0.4 million tons in 1961 to a peak of 4.3 million tons in 1997 (Fig. 3.7). Only since 2013 has production declined slightly back to 1980s levels. Today China dominates the world production of tobacco leaf. Its 36.8% share of world production in 2017 (Factfish 2019), while less than its global share in 1992 (FAO 2003), is still greater than the 35.9% share of the next five tobacco-producing countries (Brazil, India, United States, Zimbabwe and Indonesia). The marked increase in production that occurred in the 1980s and 1990s largely reflects an increase in the area planted in tobacco rather than a marked increase in yields, which have been modest compared to other tobacco-growing countries (Factfish 2019). As in the case of manufacturing, tobacco growers were subject to the same forces of localism and protectionism that affected the efficiency of the industry.

Traditionally, the China Tobacco Leaf Production Procuring and Sale Corporation, a part of the STMA, signed procurement contracts with tobacco growers through its local agencies. These contracts specified planting areas and may have included production inputs such as seed and fertiliser provided at fixed prices, often lower than market rates. But these apparent subsidies were sometimes offset by low contract procurement prices for tobacco leaf, which also depended on leaf grade. Growers were required to sell all of their production to the state which, in theory, was required to purchase, at fixed prices, all the tobacco leaf produced by farmers from their allocated acreage (Hu 2008).

Most farmers traditionally have not been specialised tobacco growers. They tend to work small plots of land, estimated to be 0.3–0.4 ha in the main tobacco provinces (FAO 2003, p. 38). Tobacco is usually produced in conjunction with other crops. However, as in many other LMICs, economically tobacco is not a not a particularly attractive crop as its revenue is often lower than other alternatives and it is labour intensive. Hu (2008), in surveys of farmers in Guizhou and Sichuan Provinces, found that while tobacco was the second highest revenue producer (after fruit), the ratio of revenue to costs was higher for crops other than tobacco (4.3 for beans, 3.7 for fruit and also vegetable oil compared to 2.6 for tobacco). For all farms

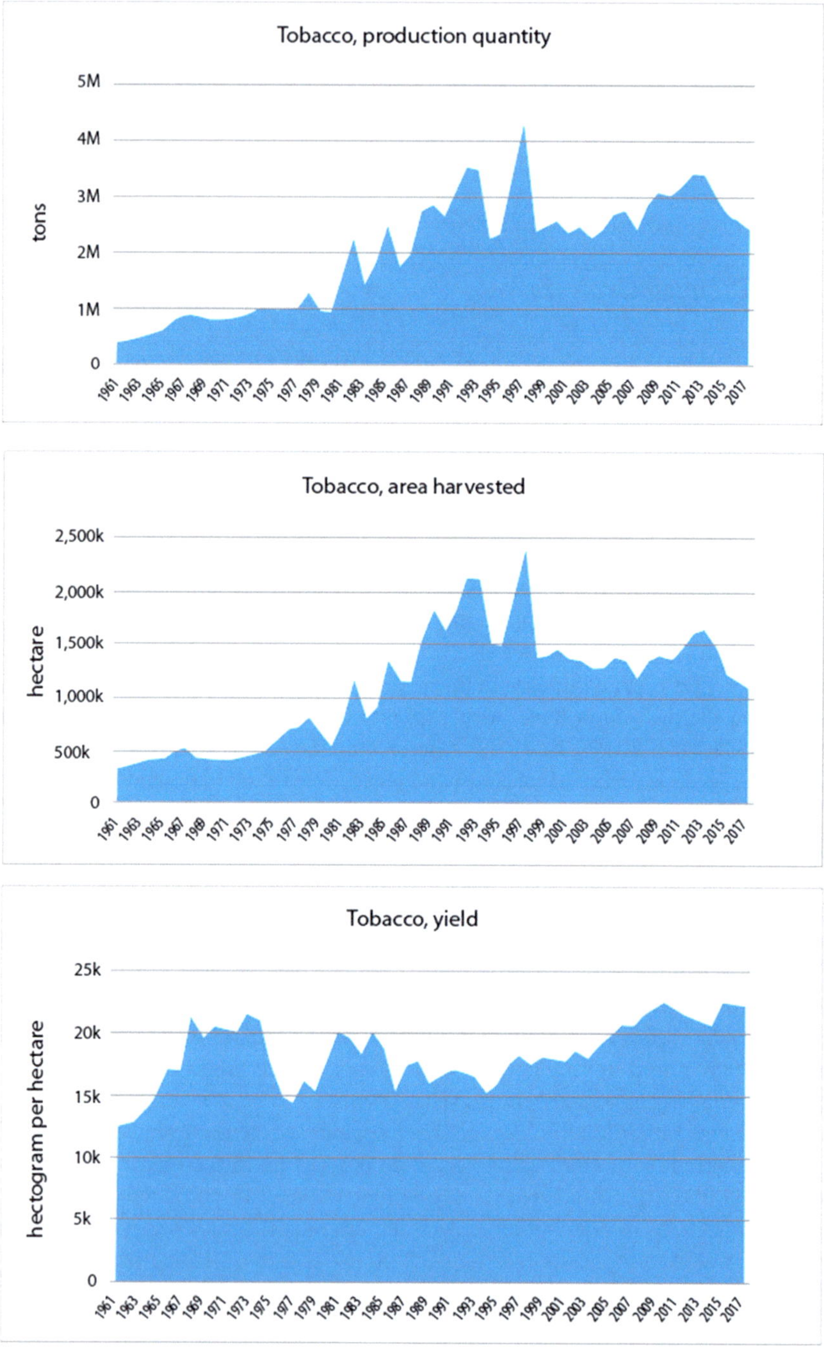

Fig. 3.7 Tobacco production, area harvested and tobacco yields, China 1961–2017. (Credit: Compiled from data in Factfish (2019) Public domain https://www.tilasto.com/en/country/world/geography-and-agriculture/tobacco-production-quantity)

only grain had a lower ratio (2.5). This was true for different sizes of farms except for small farms (≤0.5 ha) where tobacco had the lowest ratio.

Given the poorer economic returns from tobacco and the fact that it is more intensive than other cropping alternatives, Hu (2008) questioned why farmers produced it. Four reasons were suggested: (1) in the state-run industry, farmers were obligated to meet quotas assigned by local governments; (2) tobacco farming is a low-risk alternative. Although returns were low, there was a guaranteed revenue and farmers did not have to worry about storage or marketing problems; (3) soil and weather conditions are often ideal for tobacco, and (4) some farmers may not be aware of other cropping options. Of these four reasons the first was particularly pertinent in the 1980s and 1990s.

Since the formation of the STMA, local governments have been directly involved in tobacco production as tobacco taxes contributed an important part of their economic base. In 2003 an FAO report stated that 'Local governments … implement the production plan assigned by the state planning department and assist state tobacco companies to negotiate and sign production contracts with tobacco growers' (FAO 2003, p. 35). While this is correct, other evidence suggests that in the past the procurement system was far more coercive than this, particularly in poorer provinces where local governments had fewer financial options. Peng (1996) has argued that the political despotism that central government imposed on the peasantry during the collective farming era was replaced by a coercive system applied by local government to ensure its capacity to regulate tobacco cropping. In a study of two counties in the province of Guizhou, Peng found that farmers were pushed very hard to grow tobacco. Local county government agencies sent their employees to assigned villages to ensure that the peasants fulfilled a quota of tobacco planting. He found that many of the measures that local governments adopted were quite harsh. These included ordering that all land assigned to tobacco growing must be used for that purpose, that existing non-tobacco crops would have to be removed and that, in some cases, fines of 25 yuan were imposed on non-co-operative farmers. Farmers were often left in a difficult situation. This continued after harvesting particularly if there were disputes over the evaluation of the quality of tobacco leaves produced (Hu 2008).

Perhaps the biggest problem which characterised this period was the instability of production generated by the quota system, evident in Fig. 3.7. In response to low production years, especially in 1994–1995, local governments encouraged farmers to plant more tobacco. This, coupled with poorer grain prices in 1996, resulted in tobacco production reaching an all-time high of 4.3 million tons in 1997, only to almost halve to 2.4 million tons one year later. Local governments, in pressuring farmers to grow more tobacco, often created a situation where local cigarette factories were unable or unwilling to purchase tobacco above their quota levels and peasants were left with surplus production which they were generally unable to sell, either to private consumers or to out of province companies. As Peng (1996, p. 77) stated, for peasants 'tobacco for them is either cash or trash'. Because of the CNTC's monopoly, farmers were in a weak position. Production quotas and prices set by the CNTC did not always mean that planned production volumes would occur.

3.4 State Restructuring: Recentralisation of Control and Improving Productivity (2003–)

Despite some initial attempts to reform the tobacco monopoly, provincial governments, because of their institutional power and ability to influence decisions made by local STMA bureaucracies, remained important stakeholders in the tobacco industry, especially in provinces economically dependent upon tobacco. This system, heavily influenced by local government, lasted from 1985 to 2003 when central government began to institute a wider variety of reforms aimed at regaining control of the tobacco market and limiting provincial government protectionist policies. These reforms came about partly as a response to the inefficiencies created by local protectionism, but also reflected a reaction to the likely impact on the domestic market of China's membership of the World Trade Organisation in 2001. As Fang et al. (2017) describe, given potential erosion of the domestic market share, firmer plans were made to improve the competitiveness of local industry for it to be effectively positioned for overseas expansion.

3.4.1 Regaining Central Control: An End to Localism

In the late 1990s, two major reforms of the tobacco industry were instituted by central government. The first established a type of internal market of independent tobacco trade companies, creating cross-regional corporations or 'inter-provincial' tobacco groups, as they were officially called. The second reform in 2003 involved restructuring the industry at the provincial level to create a three-tiered system of new institutional entities. The previous CNTC now became independent of the STMA bureau system which had cooperated with local government and included two new bodies, the China National Tobacco Industrial Corporation (CNTIC) and the China National Tobacco Trade Corporation (CNTTC) to manage the newly merged tobacco enterprises. These changes to led to a new three-tiered vertical bureaucracy linking both organisations to new organisations at the provincial level; provincial *industrial companies* (of which municipal factories were subsidiaries) concerned with manufacturing, provincial *tobacco companies* concerned with sales and distribution of tobacco products both local and non-local and provincial *monopoly bureaus* to regulate and administer the industry (Zhou 2004; cited in Fang et al. 2017). Both industrial and tobacco companies were now able to operate and sell their products anywhere in China, thus reducing local influence and creating greater competition.

The 2003 reforms had two key objectives. The first was to break up the regional blocks of local government protectionism that had affected the productivity of the industry (Wang 2009). By eliminating regional restrictions on tobacco purchasing, local CNTC enterprises now had the opportunity of cross-area purchasing and the chance to obtain tobacco leaf from other provinces. A second, perhaps more important, objective was to merge existing tobacco enterprises to create more competitive large-scale tobacco corporations. This strategy, which officially became part of STMA policy in 2000, required the closure of small-scale tobacco factories (with an annual output of less than 100,000 cases) and the encouragement of mergers of middle size factories (100,000–300,000 cases) with larger ones. The first merger occurred in Anhui Province in 2003 (Fang et al. 2017), followed soon after by many others. Fang et al. (2017, p. 320) reports that between 2000 and 2009 the number of regional tobacco companies declined from 151 to 31, and by 2018 there were 19 major groups (Table 3.2).

A good example is the Hongta Group in Yunnan, formed from the Yuxi Hongta Tobacco (Group) Co., Ltd. in 2005 (Fig. 3.8). Since then it has developed into a trans-industry, trans-regional conglomerate. With annual sales representing 10% of the total sales volume, it is now the most important tobacco enterprise in China and is emerging as an important global player. The Hongta Group, well known for its trademark 'Yuxi', 'Hongtashan' and 'Hongmei' cigarette brands, owns four cigarette factories in Yunnan Province (Yuxi, Chuxiong, Dali and Zhaotong). Elsewhere in China the company website lists holdings in Hainan (Hongta Cigarette Co., Ltd.), in Liaoning (Hongta Liaoning Tobacco Co., Ltd), in Jilin (Jilin Tobacco Industrial Co., Ltd.), in Hong Kong (Hong Kong Hongta International Tobacco Company

Table 3.2 Corporation shares of cigarettes (% share of total retail volume), 2018

Company	Province	City	% share
Hongta Tobacco Group	Yunnan	Yuxi	10.0
Hongyun Honge Group	Yunnan	Kumming	9.9
China Tobacco Hunan Industrial Co Ltd	Hunan	Changsha	9.3
China Tobacco Zhejiang Industrial Co Ltd	Zhejiang	Hangzhou	6.7
China Tobacco Guandong Industrial Co Ltd	Guangdong	Guangzhou	6.4
China Tobacco Henan Industrial Co Ltd	Henan	Zhengzhou	6.3
China Tobacco Hubei Industrial Co Ltd	Hubei	Wuhan	5.9
China Tobacco Jiangsu Industrial Co Ltd	Jiangsu	Nanjing	5.7
Shanghai Tobacco Group		Shanghai	5.3
China Tobacco Shangdong Industrial Co Ltd	Shangdong	Jinan	4.8
China Tobacco Guizhou Industrial Co Ltd	Guizhou	Guiyang	4.6
China Tobacco Anhui Industrial Co Ltd	Anhui	Hefei	4.1
China Tobacco Fujian Industrial Co Ltd	Fujian	Xiamen	3.3
China Tobacco Sichuan Industrial Co Ltd	Sichuan	Chengdu	2.8
China Tobacco Shaanxi Industrial Co Ltd	Shaanxi	Xi'an	2.3
China Tobacco Hebei Industrial Co Ltd	Hebei	Shijiazhuang	2.0
China Tobacco Jiangxi Industrial Co Ltd	Jiangxi	Nanchang	1.9
China Tobacco Guangxi Industrial Co Ltd	Guangxi	Nanning	1.8
Gansu Tobacco Industrial Co Ltd	Gansu	Lanzhou	1.7
Others			5.1
Total			100.0

Source: Adapted from Passport, Cigarettes in China – Analysis (Euromonitor International 2019)

Fig. 3.8 Headquarters of Hongta Tobacco Group with its eight cigarette-shaped pillars, Yuxi, Yunnan. (Credit: National Public Radio, Washington D.C. (Permission obtained). https://www.npr.org/2011/02/18/133838124/china-dependent-on-tobacco-in-more-ways-than-one)

Limited) and in Macau (Asia Tobacco (Macau) Hongta Cigarette Factory) (China Tobacco 2018). Of these factories the largest is located at Yuxi which hosts the largest cigarette making machine in the world, the German PTOTOS2-2 capable of producing 16,000 pieces a minute! Although tobacco is its core business, the Hongta Group has diversified into a variety of other business activities including energy, transportation, chemicals, medicine and even tourism. It has won numerous awards and prides itself on the social duties it performs (China Tobacco 2018).

The Hongta Group epitomises the growth of large, sophisticated cigarette manufacturing enterprises that have characterised the industry in the last two decades. Many of the formal mergers after 2003 were built on informal mergers between different provincial tobacco enterprises that had been established in the 1990s. For a brief period between 1994 and 1997, regions with low production capacities were able to sell some or all of their production quotas (assigned by central government) to regions with high capacities. Wang (2013, p. 84) provides the example of Yunnan Province which had a particularly close relationship with Jiangsu Province. While the latter had a good productive capacity and, because of this, received good production quotas from the Central STMA, it lacked a favourable natural environment for tobacco production. This meant that it could not always meet its production quotas and was in danger of losing its centrally assigned production contracts and essential tax revenue. The informal trading arrangement helped resolve this issue, using a market solution to address a supply problem in the command economy.

This system meant that local governments that sold their quotas were still guaranteed their tobacco revenues and that more productive regions were able to expand their base without formally negotiating with the Central STMA. It also meant that local STMA bureaus could informally grow their production base and share the risks of doing so with the sellers of tobacco leaf without taking on the greater risks associated with new central STMA assigned contracts (Wang 2013). The Central STMA tolerated such practices because it meant that central production quotas could be more easily met, a particularly important consideration during the production shortages of the 1990s (Zhou 2001). At the time, the Central STMA also turned a 'blind eye' to manufacturers producing above their quota levels and selling their cigarettes locally. Such unplanned supplies were estimated to be approximately 10% national output (Zhou 2001, p. 161).

3.4.2 Changes in the Geography of Cigarette Production

The merger movement has also reshaped the geography of cigarette production. However, the presence of new regional cross-provincial corporations complicates any simple regional analysis of the geography of tobacco production, and it is unclear how much market dynamics will result in production eventually becoming concentrated into fewer provinces. Analysis of the changing provincial shares of national tobacco production between 2000 and 2017 reveals some interesting trends. Despite a 43.9% increase in absolute production, this was not shared evenly across provinces. Unlike predictions of a concentration of production in 'winning' regions (Song 2009), cigarette production has geographically decentralised rather than centralised. In 2000 the ten provinces/municipalities with the largest production base (over 600 hundred million pieces) (Fig. 3.6), that is, Yunnan, Henan, Hunan, Shandong, Hubei, Guizhou, Guangdong, Anhui, Sichuan and Shanghai, accounted for almost three quarters (72.1%) of China's total cigarette production. By 2011 this share had dropped to 63.4% with a further decline to 61.3% by 2017 (Table 3.3). Seven out of the ten provinces in this group recorded a declining share of national production. By contrast, provinces/municipalities with medium (200–600) and small levels (<200) of production increased their national share. The greatest gain in the share of national production occurred in smaller production provinces whose share increased from 5.8% to 11.9% between 2000 and 2017, while in medium-sized production provinces, their national share rose slightly from 22.1% to 26.8%.

Excluding provinces with no production, overall those with the largest cigarette production in 2000 had the greatest absolute loss ($r = -0.42, p < 0.05$) and declining share ($r = -0.78, p < 0.001$) of national production up to 2017 (Table 3.4).

The reasons for this trend are not completely clear especially since Chinese industry, as a whole, became more spatially concentrated at least in the period up to 2004 (Long and Zhang 2012). Nevertheless, they continue a trend partly evident in the 1990s when half of the top ten provincial cigarette producers in 2000 had already begun to experience relative declines in their share of national production (Wang 2013, p. 60). Geographically, decreases in production share were confined to a continuous belt of central provinces running from Yunnan in the southwest to

Table 3.3 Cigarette production (100 million pieces) in China by province 2000–2017 (provinces ranked by level of cigarette production, 2000)

Province	Cigarette production 2000	% share	Cigarette production 2011	% share	Cigarette production 2017	% share	Change % share 2000–2017
Yunnan	3061.2	**18.8**	3649.9	**14.9**	3589.4	**15.3**	**−3.5**
Henan	1471.2	**9.0**	1676.1	**6.8**	1517.0	**6.5**	**−2.6**
Hunan	1152.2	**7.1**	1816.2	**7.4**	1678.0	**7.2**	**0.1**
Shandong	1150.3	**7.1**	1361.3	**5.6**	1274.4	**5.4**	**−1.6**
Hubei	1036.3	**6.4**	1347.5	**5.5**	1265.9	**5.4**	**−1.0**
Guizhou	936.8	**5.8**	1226.2	**5.0**	1076.1	**4.6**	**−1.2**
Guandong	886.5	**5.4**	1345.6	**5.5**	1323.6	**5.6**	**0.2**
Anhui	789.0	**4.8**	1258.3	**5.1**	1026.8	**4.4**	**−0.5**
Sichuan	643.3	**3.9**	944.2	**3.9**	734.1	**3.1**	**−0.8**
Shanghai	612.1	**3.8**	896.1	**3.7**	896.7	**3.8**	**0.1**
	11738.6	*72.1*	***15521.4***	*63.4*	***14381.9***	*61.3*	***−10.8***
Hebei	539.9	**3.3**	807.5	**3.3**	751.5	**3.2**	**−0.1**
Jiangsu	516.0	**3.2**	991.1	**4.0**	1042.1	**4.4**	**1.3**
Zhejiang	496.0	**3.0**	885.2	**3.6**	957.0	**4.1**	**1.0**
Fujian	493.1	**3.0**	884.3	**3.6**	832.0	**3.5**	**0.5**
Guangxi	361.7	**2.2**	741.5	**3.0**	721.3	**3.1**	**0.9**
Chongqing	343.5	**2.1**	516.0	**2.1**	421.5	**1.8**	**−0.3**
Heilongjiang	320.0	**2.0**	436.1	**1.8**	364.3	**1.6**	**−0.4**
Jilin	280.0	**1.7**	430.0	**1.8**	530.1	**2.3**	**0.5**
Jiangxi	254.9	**1.6**	584.0	**2.4**	658.3	**2.8**	**1.2**
	3605.1	*22.1*	***6275.7***	*25.6*	***6278.0***	*26.8*	*4.7*
Gansu	147.6	**0.9**	410.0	**1.7**	482.5	**2.1**	**1.2**
Shanxi	144.7	**0.9**	155.0	**0.6**	150.0	**0.6**	**−0.2**
Inner Mongolia	140.0	**0.9**	287.5	**1.2**	314.5	**1.3**	**0.5**
Liaoning	115.0	**0.7**	274.5	**1.1**	269.1	**1.1**	**0.4**
Shaanxi	104.0	**0.6**	860.0	**3.5**	819.8	**3.5**	**2.9**
Beijing	85.4	**0.5**	208.9	**0.9**	166.4	**0.7**	**0.2**
Tianjin	85.0	**0.5**	226.0	**0.9**	213.1	**0.9**	**0.4**
Hainan	57.3	**0.4**	95.0	**0.4**	117.5	**0.5**	**0.1**
Xinjiang	52.5	**0.3**	160.0	**0.7**	175.5	**0.7**	**0.4**
Ningxia	15.6	**0.1**	0.0	**0.0**	80.0	**0.3**	**0.2**
Qinghai	0.0	**0.0**	0.0	**0.0**	0.0	**0.0**	**0.0**
Tibet	0.0	**0.0**	0.0	**0.0**	0.0	**0.0**	**0.0**
	946.9	*5.8*	***2676.9***	*11.0*	***2788.3***	*11.9*	*6.1*
Total	**16290.6**	**100.0**	**24474.0**	**100.0**	**23448.3**	**100.0**	

Source: China National Bureau of Statistics (various years). China Industry Economy Statistical Yearbook, China Statistics Press, Beijing

Table 3.4 Correlation coefficients between changes in cigarette manufacturing (2000–2017) and provincial characteristics[a]

Provincial characteristic	Change in provincial share of national cigarette production 2000–2017	Percent change in provincial cigarette production 2000–2017
GDP per capita (2011)	0.23	−0.06
Province population (2011)	−0.37*	−0.38*
Provincial cigarette production share (2000)	−0.78**	−0.42*
Province tobacco acreage (2000)*	−0.74**	−0.26
Change in tobacco acreage (2000–17)*	−0.09	−0.03

[a]Excluding provinces with no tobacco production $*p < 0.05$ $**p < 0.001$

Shandong and Hebei in the northeast, but also including the outlier of Heilongjiang in the far northeast. On the other hand, provinces which increased their share were geographically concentrated in eastern China.

There are a number of possible explanations for this pattern. First, given the existence of close linkage between the real estate industry and manufacturing inputs of different industrial sectors, including tobacco (Wu et al. 2020, p. 6), it could be argued that the real estate boom, beginning in 2003, had a negative effect on manufacturing investment. Not only did the boom 'crowd out' productive investment (Chen and Wen 2017) in favour of more profitable returns from real estate (Qian 2010), but it also had the effect of dramatically raising land prices. Gao et al. (2012), for example, found that rising real estate prices was an important factor, for manufacturing as a whole, in explaining why firms relocated from coastal to inland areas (Zheng et al. 2014). However, both arguments are not sufficient to explain locational changes in cigarette manufacturing which grew most, in both absolute and relative terms, particularly in eastern provinces most subject to the real estate boom. This regional pattern of growth, however, was unrelated to provincial GDP per capita (Table 3.4).

Second it is possible that cigarette manufacturing enterprises, while shifting some of their production base to north and western provinces, such as Shaanxi, also moved to lower cost small city locations in eastern provinces in order to maintain their competitive advantage. Given increased automation of the industry which was accompanied by a sharp drop in tobacco manufacturing employment between 2000 and 2013 (−3.1% in China compared to only −0.7% in India) (ILO 2014), the effect of higher land costs on labour availability would have been less significant as a determinant of firm location than for more labour-intensive industries. Examining these trends further is beyond the scope of this chapter, but analyses at smaller spatial scales is necessary for a fuller understanding of the locational dynamics of cigarette manufacturing and how it differs from other Chinese industrial sectors.

While changes in manufacturing production were greatest in tobacco-producing provinces, there was no evidence that the pattern of decentralisation reflected provincial changes in the area of land devoted to growing tobacco. Finally the changing geography of manufacturing might also reflect political interest in decentralising production to less favoured provinces given the importance of tobacco in their economies and of the Central Government's demonstrated commitment to reducing provincial inequalities in poverty (Wen 2004).

3.4.3 Changes in Tobacco Growing

The geography of tobacco farming has likewise undergone substantial changes. Today the industry is quite different from that of the 1980s and 1990s when a substantial variability in production characterised the years of local government control. Tobacco leaf production is now more consistent, leaf quality and tobacco yield have improved slightly and shortfalls have increasingly been met by imports. Since 2013, because of an excessive tobacco inventory, the industry has implemented total volume control, with tobacco purchases decreasing from 54.9 million loads in 2012 to 37.7 million in 2017 (Tobacco China Online 2018).

One outcome has been a decline in the area planted in tobacco, a trend typical of even some of the traditional tobacco-growing provinces (Table 3.5 and Fig. 3.9). Although official data needs to be viewed with caution, for China, as a whole, it appears that the proportion of cropland devoted to growing tobacco, while fluctuating between 8% and 9% between 2000 and 2013, dropped sharply afterwards from 9.9% in 2013 to 6.8% in 2017. When the provincial shares of China's total area planted in tobacco are examined (Table 3.5), some interesting trends emerge. Among the five main provincial producers in 2000 (Yunnan, Guizhou, Henan, Hunan and Sichuan), all but two increased their national share of cropland planted in tobacco between 2000 and 2017, representing a 14.6% gain over this period. By contrast, all but one (Fujian) of the medium-sized producers had declining shares amounting to an overall drop of −11.3%. The same trend occurred among minor tobacco-growing provinces but, because the amount of tobacco harvested was small so was their declining share (−3.3%). Some provinces, such as Jiangsu, Qinghai and Xinjiang, stopped producing tobacco entirely.

The changing geography of tobacco production to some extent reflects the implementation of official policy to reduce tobacco acreage and to substitute this with other more profitable cropping alternatives, particularly in impoverished counties (Tobacco China Online 2018). China's top leadership has increasingly recognised the problem of national food security, and in 2014 the STMA issued the Farmland Redline Policy on the reduction of tobacco acreage nationwide (Li and Tang 2018). A pilot tobacco crop substitution project in Yunnan has shown that a higher income can be derived from crops other than tobacco (Tang et al. 2018). The reductions in acreage are also likely to reflect declines in the production of lower-quality tobacco leaf, which is now increasingly being imported especially from Zimbabwe. Between 2000 and 2017 tobacco imports rose by 278% and in 2017 were equivalent to 6.0% of China's total production (National Bureau of Statistics

Table 3.5 Tobacco area harvested (1000 ha of tobacco crops) by province, 2000–2017[a]

Province	2000	2011	2017	Change 2000–2017	% share 2017	Change % share 2000–2017
Large producers 75 and above						
Yunnan	339.5	495.3	424.6	85.1	37.6	13.9
Guizhou	213.7	212.2	155.7	−58.0	13.8	−1.1
Henan	166.4	124.7	104.0	−62.4	9.2	−2.4
Hunan	88.9	104.8	94.8	5.9	8.4	2.2
Sichuan	82.2	117.1	86.3	4.1	7.6	1.9
	890.7	***1054.1***	***865.4***	***−25.3***	***76.6***	***14.6***
Medium producers (25.0–74.9)						
Hubei	74.6	67.2	39.7	−34.9	3.5	−1.7
Chongqing	70.8	46.2	34.9	−35.9	3.1	−1.8
Shandong	62.9	33.6	21.4	−41.5	1.9	−2.5
Fujian	55.3	67.9	52.7	−2.6	4.7	0.8
Shaanxi	52.0	36.2	17.1	−34.9	1.5	−2.1
Heilongiang	49.4	35.0	15.0	−34.4	1.3	−2.1
Guangdong	31.1	24.2	17.4	−13.7	1.5	−0.6
Jilin	26.6	22.3	6.2	−20.4	0.5	−1.3
	422.7	***332.6***	***204.4***	***−218.3***	***18.1***	***−11.3***
Small producers (<25.0)						
Guangxi	23.9	16.5	11.8	−12.1	1.0	−0.6
Gansu	21.0	3.7	1.5	−19.5	0.1	−1.3
Anhui	19.7	11.4	8.3	−11.4	0.7	−0.6
Liaoning	18.4	10.8	8.0	−10.4	0.7	−0.6
Jiangxi	14.1	20.0	25.7	11.6	2.3	1.3
Shanxi	8.0	3.2	1.5	−6.5	0.1	−0.4
Inner Mongolia	7.9	4.2	1.6	−6.3	0.1	−0.4
Hebei	5.5	2.9	1.3	−4.2	0.1	−0.3
Xinjiang	2.7	0.0	0.0	−2.7	0.0	−0.2
Zhejiang	1.4	1.2	0.6	−0.8	0.1	0.0
Ningxia	0.7	0.4	0.2	−0.5	0.0	0.0
Jiangsu	0.5	0.1	0.0	−0.5	0.0	0.0
Hainan	0.2	0.1	0.1	−0.1	0.0	0.0
Qinghai	0.1	0.2	0.0	−0.1	0.0	0.0
	124.1	***74.7***	***60.6***	***−63.5***	***5.4***	***−3.3***
Total	**1437.5**	**1461.4**	**1130.4**	**−307.1**	**100.0**	

Source: China National Bureau of Statistics (various years). China Rural Statistical Yearbook. China Statistics Press, Beijing

Places with no production (Beijing, Tianjin, Shanghai and Tibet) are excluded

[a]Provinces ranked by hectares of tobacco crops in 2000

Fig. 3.9 Area of tobacco harvested (1000 ha of tobacco crops) by Province, 2000. (Tibet and the municipalities of Beijing and Tianjin recorded no tobacco production in 2000. Credit: Authors)

2018). They could also reflect the consolidation of smaller less profitable farms into larger more productive enterprises.

The impact of institutional and economic reforms in the tobacco industry on tobacco farming practices and farmers' welfare has, so far, not received much consideration. Cheng and Ngo (2014) have argued that under the old decentralised monopoly system, where the enforcement of administrative accountability was weak and bureaucratic discipline poor, local governments were responsive to local needs and were able to manoeuvre the system to ensure direct subsidy to the peasants and to fuel local development. They contend that this stands in stark contrast to the new regulatory regime introduced since 2003, where power and influence has been tilted in favour of the line hierarchy at the expense of territorial authorities. Ironically, the enforcement of such an arrangement, with supposedly more transparency and better quality of bureaucratic rule, has placed state interests in direct conflict with peasants' interests and local welfare. This is evident in (1) local cadres compelling peasants to grow tobacco if their farmlands are located in tobacco zones as designated under local plans, (2) forcing farmers to rent out their lands to professional contractors, (3) loss of property rights evident in the enforced combining of individual farms for large-scale tobacco cultivation, (4) little allowance for intercropping (the growing of other crops in conjunction with tobacco) and (5) unfair tobacco leaf grading practices. All have produced unrest among tobacco farmers.

The authors conclude that given the reassertion of central state power and faced with the contradictory responsibilities of balancing their budget versus protecting the livelihood of peasants, local governments have turned their back on the latter. In such a situation the grievances of peasants are contained through an informal toleration of smuggling and illegal tobacco sales (Cheng and Ngo 2014, p. 237).

3.5 A Globalising Industry

3.5.1 International Competitiveness

The structural reforms of the 1990s and early 2000s were designed to make the Chinese tobacco industry more globally competitive in preparation for China's entry into the World Trade Organisation in 2001. Although entry into the WTO entailed some risks, it was also seen as an opportunity for export-led growth of key economic sectors, including tobacco. Accordingly the government began to formulate a 'Go Global' strategy, formally named in 2000 (Fang et al. 2017). Fang notes that the key factors underlying this strategy included (1) seeking high-quality tobacco leaf to feed growing industry needs; (2) given the increasing saturation of the domestic market and policy restrictions on smoking, developing a more focused export strategy; (3) developing tobacco growing and decentralising cigarette manufacturing to overseas markets through joint ventures; and (4) seeking wider business opportunities in such markets.

The major incentive for the structural reforms in China's domestic tobacco industry was the view that the CNTC could not compete with other TTCs internationally. Writers such as Tong (2008, pp. 221–222) have reported that the entire Chinese tobacco industry had income and profits lower than Philip Morris alone. For the CNTC, as a whole, its total revenue was a quarter (25.3%) of that of Philip Morris and about half of the revenue of BAT and Japan Tobacco (52.2% and 54.8 % respectively). No single Chinese brand dominated the Chinese market in a way that Marlboro does in the United States, and, until recently, the majority of cigarettes sold were low-grade high-tar products not suited for export.

While the CNTC is the largest tobacco company worldwide, its major subsidiaries are still small in comparison to most western TTCs. In 2018 the top four companies in China (Yuxi Honga Tobacco Group, Hongyun Tobacco Group, China Tobacco Hunan and China Tobacco Zhejiang) accounted for only 35.9% of the total retail volume and the top ten, 70.3% (Table 3.2). China's membership of the WTO in 2001 and the uncertainty which this produced was undoubtedly important in the STMA consolidating companies and brands, as well as acquiring new technology and improving product standards.

3.5.2 Trade Liberalisation and the Tobacco Industry

Entry into the WTO had a dramatic effect on the domestic tobacco industry, but this was mainly confined to tobacco growing rather than cigarette manufacturing. As a result of a lowering of tariff barriers, Chinese imports of tobacco slowly increased after 2001 but ten years later were still only equivalent to 4.3% of the total Chinese tobacco production (National Bureau of Statistics 2012). In 2016 most (47%) of the Chinese tobacco imports of US$1.2 billion came from Zimbabwe followed by Brazil (23%) and the United States (14%) (Observatory of Economic Complexity 2017a, b). Compared to trade in tobacco leaf, the Chinese cigarette market, unlike other markets in Asia (Lee et al. 2014), remained closed to foreign interests. In 2018 the CNTC remained the dominant player in the Chinese tobacco market, generating a 99.3% share of the market's volume (Marketline 2018).

Ironically, despite the overall liberalisation of China's economy since entry to the WTO, China's tobacco industry remains one of the last sectors to be solely state owned. Nevertheless over the past few decades, TTCs, in particular BAT, have attempted to increase their presence in China through three main strategies: (1) establishing a presence via tobacco leaf development, licensed manufacturing agreements or joint ventures; (2) complicity in the illegal cigarette trade, which has been hugely profitable and an integral part of BAT strategies in China (Lee et al. 2004; Shen et al. 2010); and (3) through attempts at trade liberalisation resulting from China's membership of the WTO (Holden et al. 2010). In a study of internal documents of BAT, released as a result of litigation in the United States, Holden et al. (2010) show that BAT worked closely with the Chinese Government in the WTO negotiations, no doubt fearful that its main competitors, notably Philip Morris and RJ Reynolds, would gain a foothold in China before them. Although the outcome of the WTO negotiations was to produce 'an open but heavily guarded door' (Holden et al. 2010, p. 423), this produced little benefit for TTCs wanting to access the highly lucrative Chinese market. However, for BAT the company's perseverance paid off. In 2004 BAT reached agreement with China Eastern Investments, becoming the first foreign company to legally manufacture foreign brands in China. While selling its most popular brands, Dunhill, LuckyStrike and State Express 555s in Asia for many years, high tariffs had previously limited BAT's penetration of the Chinese market (Watts 2004).

International disputes over China's compliance with the conditions of entry into the WTO continue and the current 'trade war' with the United States over trade imbalances and lack of access for western manufactured goods, including tobacco, is reshaping trade practices. The recent imposition of a new tariff on US imports has resulted in China turning elsewhere, particularly to India, for sources of raw tobacco (Weston 2019). The Chinese government, despite a more favourable stance at the time to trade liberalisation, has steadfastly resisted efforts of TTCs to set up domestic cigarette manufacturing factories and the few limited licensing agreements and joint ventures which have occurred pose little threat to China's cigarette manufacturing industry (Tong et al. 2008). For example, joint venture agreements, first established in the 1990s with TTCs such as RJ Reynolds, Rothmans, Philip Morris and Japan Tobacco (Lee et al. 2004), while giving foreign brands increased exposure to the Chinese market, did so on the CNTC's terms. For the CNTC joint ventures such as these gave access to new technology and knowhow, but for TTCs did not provide a much greater foothold in the Chinese market. This became very evident in 2006 when the Chinese government banned all new manufacturing facilities including foreign joint ventures (Fang et al. 2017, p. 324).

3.5.3 Globalisation of the Chinese Tobacco Industry

The globalisation of the CNTC represents a new stage in the development of the industry. Globalisation has involved a number of different strategies with both economic and political goals. On one level it can be seen as a consequence of saturated markets in China and the emergence and implementation of stronger tobacco control policies, which were the driving forces behind western TTCs shifting their emphasis from traditional domestic markets and expanding their operations to LMICs (Dunning and Lundan 2008). In this respect the activities of the CNTC are no different and reflect the economic imperatives of the industry to maintain its income and profitability and to replace lost domestic revenue with exports. In addition, political imperatives have contributed to the globalisation of CNTC operations with the expansion of the company being an important part of China's 'Belt and Road' initiative (Fang 2019). Both developments have resulted in considerable controversy regarding the potential negative health outcomes of such involvement, but also the degree to which 'Belt and Road' policies foster genuine cooperation, economic benefits and sustainable development in the countries involved (Gu et al. 2016). However, the globalisation of the CNTC's operations is not unique and is similar to the global business strategies of other Asian TTCs (Lee and Eckhardt 2017a, b).

Dunning and Lundan (2008) provide a useful fourfold framework for identifying why tobacco companies may pursue global business strategies. First, TTCs are likely to be 'market seekers' when faced with static or declining domestic markets. Second, they may also be 'natural resource seekers' to secure a steady supply of higher-quality tobacco leaf necessary for competing on the global market. Third, by attempting to set up offshore operations in key overseas locations, they may be 'efficiency seekers'. This strategy, involving direct foreign investment, mergers and acquisitions or joint ventures, helps to reduce operational costs, may break down cultural barriers through the employment of local labour and raises the market profile of the industry. It is also likely to involve the vertical integration of different parts of the production chain from leaf growing to manufacturing. Finally, firms may be 'strategic asset seekers' in that foreign acquisitions or joint ventures, not just in tobacco, help improve the economic viability of companies and consolidate their overseas base.

When compared to four other large Asian TTCs, the CNTC has pursued its global business strategies by engaging in all four strategies (Lee and Eckhardt 2017a, b). First, as a *market seeker*, it has stimulated the development of Chinese cigarette exports. In this respect it is similar to the Asian development model of strong government intervention in economic policy striving for export-oriented growth. Just prior to entry into the WTO, China developed a more strategic approach to exports with formation of the China Tobacco Import Export Group (CNTIEG). As the parent company, this was designed to handle all the CNTC's overseas operations and to coordinate exports from provincial companies. Renamed in 2008 into China Tobacco International (CTI), an explicit export policy was developed in 2011 focusing on gaining a foothold in overseas markets, product distribution and the establishment of local production facilities (Fang et al. 2017).

The effect of these initiatives was immediate. Between 2014 and 2018, the rate of growth in cigarette exports ranked fourth globally and by 2018 China, with 3.4% of world exports, ranked eleventh among tobacco exporting countries (Workman 2019). The majority of exports went to markets in Central and South East Asia with Indonesia being by far the biggest purchaser. Other developing markets included the Middle East, where Egypt was the largest purchaser, and Central Asia, along with Russia and Eastern Europe (Table 3.6). As of 2017 the CNTC had made little impact in Africa or Latin America. Despite the rapid rise in cigarette exports, the volume of these was relatively minor, with exports amounting to just 1.07% of China's total cigarette production in 2017 (National Bureau of Statistics 2018).

Over the last few decades, Asian markets were the first major target of the CNTC but recently its operations have become more global in scope (Fang et al. 2017). As an *efficiency* and *strategic asset seeker*, the CNTC has been proactive in forming

Table 3.6 Destinations of Chinese cigarette and cigar exports (2017)

Region	Value ($USm)	Percent of total exports	Key markets (2% & over)
Central and South East Asia	$317.91	49.61	Indonesia (34.6%), Philippines (3.5%), Vietnam (2.9%)
Middle East and Central Asia	$117.44	18.33	Egypt (11.4%), UAE (6.3%)
Western and Southern Europe	$92.15	14.38	Belgium (3.4%), Germany (3.2%), France (2.1%)
Russia, Eastern Europe	$85.05	13.27	Poland (4.7%), Russia (4.3%)
Africa	$15.07	2.35	
North America	$4.91	0.77	
Central and South America	$4.11	0.64	
Other	$4.20	0.66	
Total	$640.85	100.0	

Credit: Computed from data retrieved from the Observatory of Economic Complexity (2017a, b) https://oec.world/en/visualize/tree_map/hs92/export/chn/show/42401/2017/

joint ventures with other TTCs in a variety of contexts outside Asia. Lacking its own distribution channels, it has set up subsidiaries in countries in Africa (e.g. Namibia), Latin America (e.g. Argentina) and the Middle East (e.g. United Arab Emirates). Even in Europe the company signed a joint venture with Philip Morris International (PMI) to distribute Chinese cigarette brands in Switzerland (Workman 2019). Joint ventures with other TTCs, such as Japan Tobacco International (JTI) and Imperial Tobacco (IT), help facilitate expansion abroad, but with TTCs in turn hoping that such partnerships may help give them access to the Chinese market. For example, the joint venture agreement with PMI in 2005 allowed the latter to manufacture and sell Marlboro cigarettes in China. In exchange PMI agreed to a joint venture with the CNTC allowing greater access of the latter to the global market using PMI distribution channels. As a result, in 2008 Chinese brands were marketed in Central and Eastern Europe and Latin America (Argentina).

Although less common, the CNTC has established manufacturing facilities abroad. As of 2015 these included plants in Macau, Laos, Mongolia, North Korea, the UAE, Panama and Romania (Fang et al. 2017, p. 326). The Romanian plant, located at Pantelimon near Bucharest, was established in 2007 by the China Tobacco International Europe Company (CTIEC), of which the Hongta Group is a major shareholder. This is the only CNTC manufacturing plant in Europe, with the company holding a 97.5% share. Cigarette production, mainly using leaf tobacco sourced from Zimbabwe, Brazil and Argentina, has more than doubled since 2012 from 2.0 billion to 4.5 billion cigarettes in 2018 (CTIEC 2019). As the company's website explains:

> CTIEC is playing the leading role in implementing China National Tobacco Company's "going out" strategy, minded with world-class aspirations. CTIEC not only aims to become (an) outstanding representative of (the) cigarette industry in Romania, but also to become one of the world's leading name(s) … Our Company takes Romania's neighbouring countries, the Middle East, the Balkan countries and North African markets as expanded focus and market base and explores new markets in all areas by drawing upon the experience gained from key points to achieve sequential development. These expansion areas are also gateways for exporting an increasing amount of goods to remote overseas markets in South Eastern and Central Asia, Western Europe, and Southern Africa.

Finally the CNTC has been an active 'natural resource seeker' as it has attempted to source high-quality tobacco leaf from key producing countries such as Zimbabwe, Brazil and, until recently, the United States. Zimbabwe provides a good case study of CNTC activities. The Zimbabwe economy depends very heavily on tobacco exports which contribute around 14–18% of its national GDP, and it is China's main source of tobacco imports contributing 32% in 2015 (Gao 2017). Following the controversial 2000 land reforms, which redistributed land from a small number of European farmers to a large number of African smallholders, and the imposition of western economic sanctions against Zimbabwe, Chinese companies took advantage of the post-land reform context by developing contracting arrangements across the different resettlement areas (Fang et al. 2020). Important here was the establishment in 2005 of TianZe, a 100% CNTC owned subsidiary. Soon after China became the largest destination for Zimbabwe's tobacco (Gu et al. 2016). As of 2019 TianZe had contracts with 305 farmers covering 8720 ha (Masau 2019).

While such investment has provided technical advice and help to some local growers, it has been criticised on a number of counts. First demand for higher-quality tobacco leaf has not been met by many small holders resulting in disappointing prices (Scoones et al. 2018). Thus TianZe's contracting arrangements tend to favour more productive farmers capable of producing higher-quality crops. Second, Ojakorotu and Kamidza (2018) contend that Zimbabwe's 'Look East Policy' has primarily benefitted political elites, including large-scale tobacco growers, with a weak, corrupt government allowing the Chinese to operate in a tax-free environment not only in agriculture but also in mining and business sectors. In addition, China's willingness to support a 'pariah' political regime, which had little regard for human rights, was widely criticised, as were the effects of a tobacco focused agricultural policy on food insecurity, a serious issue in many African countries (Hu and Lee 2015). Nevertheless, as in China, the CNTC, through different corporate social responsibility initiatives, has attempted to cement social relations by encouraging perceptions of state friendship (Fang et al. 2020). Such involvement is just one part of China's promotion of 'soft power' in its wider Belt and Road strategy (Cai 2017; Huang 2016).

However, CNTC involvement in Zimbabwe changed in 2019 with the announcement of the first partnership between the CNTC and an African cigarette manufacturer (Masau 2019). In order to add value to the tobacco supply chain, it was announced that China Tobacco Shaanxi Industrial Co Ltd had formed a joint venture with the local Pacific Cigarette Co. (PCC). The aim was to develop a new Acacia cigarette brand using high-quality Zimbabwean tobacco and Chinese-preferred flavours to target Chinese living in Africa.

3.6 Conclusion

This chapter has examined the tobacco industry in China and how it has been shaped by major changes in the country's political and socio-economic environment. In summary, the following observations can be made. First, the development of the CNTC monopoly in China is a direct outcome of the establishment of the People's Republic of China in 1949. Despite economic liberalisation since the late 1970s, the CNTC has maintained its status as a monopoly producer of tobacco and cigarettes. Hogg et al. (2016) have noted that state-owned tobacco companies have been neglected in the literature on TTCs which is surprising given their wide influence both in the countries concerned and also on emerging trends of globalisation. In the case of the CNTC, its dominant position in China can be linked to broader issues of national pride, the development of Chinese influence in LMICs, the importance of tobacco in the economy as well as a continued reaction to the foreign control and influence that plagued the country in the past. Understanding these links is important because it will lead to greater understanding of the industry and its likely continuance as a monopolistic structure, despite this seemingly being at odds with the socio-economic development of China as a whole.

A second observation to be made is that the industry has been characterised by repeated restructuring and shifts in state power between central and provincial governments. Each round of restructuring also led to certain contradictions and the need for further restructuring. This is unsurprising given China's history following rejection of the capitalist market in 1949. Given the poverty and the economic uncertainties which characterised China during the years of the Great Leap Forward (1958–1962), it is not surprising that the first attempts to establish a centralised state tobacco monopoly were delayed until 1964, with the establishment of the Chinese Tobacco Industrial Corporation. This was soon disbanded during the Cultural Revolution (1966–1976), the industry being allowed to decentralise, weakening central power and revenue. In response, central government attempted to regain control of the decentralised regime by the establishment of the STMA/CNTC state monopoly in 1982 during the early years of market reform. Ironically this led to less, not more, central control as protectionism intensified competition between provinces resulting in a proliferation of tobacco factories encouraged by tax-dependent local governments. This was also true of tobacco farming where local government policy, aimed at avoiding peasant unrest, encouraged high levels of production. Under such circumstances the national plan for tobacco production was highly flawed. The result was a glut in tobacco production, aggravated by the illicit tobacco market, conflicts between local governments and production enterprises and a decentralised distribution system which was poorly controlled. As Cheng and Ngo (2014, p. 231) have suggested:

> (T)he national plan for tobacco production existed only in name. The STMA was unable to keep tobacco cultivation and procurement under control. Instead, blind production was encouraged by local governments under the fiscal responsibility system. In balancing the conflict between local interests and the national plan, most local governments chose to protect the former.

These deficiencies soon led to pressures for more economic restructuring and the enactment of a new tobacco monopoly law in 1992. These were, in part, generated by central government still wishing to regain control of the industry, but the major incentive for reforming this structure came from the decision of China to join the WTO. As in so many other global contexts, the assertion of neoliberal economic policies became evident in closure of many inefficient state enterprises (Prasad 2006). Although provincial protectionism was brought to an end, tobacco companies amalgamated, and firms given greater freedom to market their products, in reality the resurgence of state power in the tobacco industry did not produce the economic liberalisation hoped for by many foreign TTCs. The CNTC monopoly remained, and joint ventures with TTCs were limited. To some degree this was a blessing as market liberalisation, in other parts of Asia, led to a growth in market share by TTCs and an upward trend in cigarette consumption on the part of targeted population groups, notably youth and young women (Lee et al. 2014). By contrast, in China the growth in cigarette production after 2010 began to slow, prompting the

CNTC to streamline its domestic operations and to embark on a globalisation strategy. As yet, this has not led to further privatisation. However, the recent listing of China Tobacco International, a small CNTC Import/Export subsidiary, on the Hong Kong Stock Market may be a sign of change (Fioretti 2019).

A third major observation is that the restructurings of the industry have led to new geographies of tobacco growing and manufacturing, although these are poorly understood. For tobacco growers the period since 2000 has seen a certain centralisation of the industry with some, but not all, of the major tobacco-growing provinces, particularly Yunnan, becoming more important. However, the reverse has been true of smaller producing provinces which have lost market share. To some extent these trends are an outcome of reductions in tobacco planting in the name of food security and perhaps of shifts by tobacco farmers to more profitable crops. They also reflect an increased reliance on imports particularly of higher-quality crops from countries such as India which has made up the American deficit (Tobacco China Online 2019c). These trends deserve increased consideration as does the geography of cigarette manufacturing which has received little attention. Traditionally cigarette manufacturing, at the provincial level, has been one of the more decentralised of Chinese industries reflecting a tendency to locate close to sources of raw materials (He et al. 2007) and, apart from Yunnan, not exhibiting the industrial agglomeration trends evident in more export-oriented parts of the economy (Wen 2004). In contrast to tobacco growing, cigarette manufacturing became more decentralised on a provincial basis; however, the reasons for this are as yet unclear. More research is needed on the link between tobacco company amalgamations and the location of their production enterprises. The above trend is hardly a return to the localism of the past, but most likely reflects plant closures/relocations in order to improve economic efficiency.

A fourth dominant trend to emerge is that the CNTC, like TTCs elsewhere, has increasingly globalised. As yet this remains at a very early stage but a clear pattern of expansion involving joint ventures in both tobacco growing and cigarette manufacturing has occurred, particularly in Southeast Asia. However, given China's 'Belt and Road' strategy, this is likely to intensify in Africa and the Middle East. Such trends raise many questions regarding the effects of 'soft power' on trade issues involving tobacco as well as governance issues, related to the implementation of the FCTC, in countries where China's influence has grown the most. The restructuring of the domestic industry in the last two decades has provided the means to globalise, but the key reasons for doing so appear to have been the decline in domestic demand and the emergence of stronger policies of tobacco control. As in the case of western TTCs, the CNTC has responded to such trends by adopting new marketing strategies involving new products, forms of advertising and by targeting particular groups. These topics are the focus of the following chapter.

References

Benedict C (2011) Golden-silk smoke: a history of tobacco in China, 1550–2010. University of California Press, Los Angeles
Cai P (2017) Understanding China's belt and road initiative. Lowy Institute for International Policy. http://hdl.handle.net/11540/6810
Chen K, Wen Y (2017) The great housing boom of China. Am Econ J Macroecon 9(2):73–114
Chen Y (2018) From Lucky Strike to Chunghwa: the development of China's tobacco industry in the twentieth century. Senior Projects. Bard College, Annandale-on-Hudson, New York https://digitalcommons.bard.edu/senproj_s2018/140
Cheng Y-W (2015) State monopoly, Chinese style: a case study of the tobacco industry. PhD thesis, Leiden University, Leiden, Netherlands
Cheng Y-W, Ngo T-W (2014) The heterodoxy of governance under decentralisation: rent-seeking politics in China's tobacco growing areas. J Contemp Asia 44(2):221–240
CTIEC (China Tobacco International Europe Company) (2019) Who are we? Retrieved from: https://www.ctiec.cc/
China Tobacco (2018) About Hongta: creating a world-class brand. Hongta Group, Yuxi. http://www.hongta.com/language/en/aboutus/overview/
Chi CJ (1935) Cigarette industry in China. Chin Econ J 16(5):631
Cochran S (1980) Big business in China: sino-foreign rivalry in the cigarette industry, 1890–1930. Harvard University Press, Cambridge, MA
Cox H (2000) The global cigarette: origins and evolution of British-American tobacco, 1880–1945. Oxford University Press, Oxford
de Oliveira JAP (2016) Tobacco cluster in southern Yunnan. In: de Oliveira JAP (ed) Upgrading clusters and small enterprises in developing countries: environmental, labor, innovation and social issues. Routledge, Abington, pp 67–71
Dunning JH, Lundan SM (2008) Multinational enterprises and the global economy, 2nd edn. Edward Elgar, London
Eng I (1999) Agglomeration and the local state: the tobacco economy of Yunnan, China. Trans Inst Br Geogr 24(3):315–329
Euromonitor International (2019) Passport, cigarettes in China – analysis. Country report. https://www.euromonitor.com/cigarettes-in-china/report
Factfish (2019) Tobacco production quantity, tobacco area harvested, and tobacco yield for China, 2017. https://www.tilasto.com/en/country/world/geography-and-agriculture/tobacco-production-quantity
Fang J (2019) China: tobacco and belt and road initiative – the new 'Go Global'? Tobacco Control August 1. https://blogs.bmj.com/tc/2019/08/01/china-tobacco-and-belt-and-road-initiative-the-new-go-global/
Fang J, De Souza L, Smith J, Lee K (2020) "All weather friends": how China transformed Zimbabwe's tobacco sector. Int J Environ Res Public Health 17(3):723
Fang J, Lee K, Sejpal N (2017) The China National Tobacco Corporation: from domestic to global dragon. Global Public Health 12(3):315–334
Fenby J (2019) The Penguin history of modern China: the fall and rise of a great power 1850 to the present, 3rd edn. Penguin Books/Random House, New York
Fioretti J (2019) Overseas unit of China's cigarette monopoly files for Hong Kong IPO. Reuters, January 2. https://www.reuters.com/article/us-china-tobacco-ipo/overseas-unit-of-chinas-cigarette-monopoly-files-for-hong-kong-ipo-idUSKCN1OW0LB
FAO (Food and Agricultural Organisation) (2003) Issues in the global tobacco economy: selected case studies. FAO, Rome
Gao B, Chen J, Zou L (2012) Housing price's regional differences, labor mobility and industrial upgrading. Econ Res J 1:66–79. (in Chinese)
Gao X (2017) Can global health governance contend with transnational corporate activities? Geogr Honors Proj 49. http://digitalcommons.macalester.edu/geography_honors/49

Gibbs BJ (1938) Tobacco production and consumption in China. US Bureau of Agricultural Economics, Washington, DC

Gu J, Chuanhong Z, Vaz A, Mukwereza L (2016) Chinese state capitalism? Rethinking the role of the state and business in Chinese development co-operation in Africa. World Dev 81(C):24–34

He C, Wei YHD, Pan F (2007) Geographical concentration of manufacturing industries in China: the importance of spatial and industrial scales. Eurasian Geogr Econ 48(5):603–625

He P, Takeuchi T, Yano E (2013) An overview of the China National Tobacco Corporation and State Tobacco Monopoly Administration. Environ Health Prevent Med 18(1):85–90

Hoffer AJ, Pellilo A (2012) The political economy of tobacco control spending. Appl Econ Lett 19(18):1793–1797

Hogg SL, Hill SE, Collin J (2016) State-ownership of tobacco industry: a 'fundamental conflict of interest' or a tremendous opportunity for tobacco control? Tobacco Control 25(4):367–372

Holden C, Lee K, Gilmore A, Fooks G, Wander N (2010) Trade policy, health and corporate influence: British American Tobacco and China's accession to the World Trade Organization. Int J Health Serv 40(3):421–441

Hu TW (2008) The role of government in tobacco leaf production in China: national and local interventions. In: Hu T-W (ed) Tobacco control policy analysis in China. World Scientific Publishing Company, Singapore, pp 189–209

Hu TW, Lee AH (2015) Commentary: tobacco control and tobacco farming in African countries. J Public Health Policy 36(1):41–51

Hu TW, Lee AH, Mao Z (2013) WHO framework convention on tobacco control in China: barriers, challenges and recommendations. Global Health Promot 20(4):13–22

Huang H (2010) Zhongguo yancao pinpai dacaijun zhenxiang (The truth about the reduction of brand names of tobacco). Nanfang zhoumo (Southern Weekly) November 5, 2010 (Cited in Li, 2012, p 116)

Huang Y (2014) China's position in negotiating the Framework Convention on Tobacco Control and the revised International Health Regulations. Public Health 128(2):161–166

Huang Y (2016) Understanding China's belt & road initiative: motivation, framework and assessment. Chin Econ Rev 40(C):314–321

ILO (International Labour Organization) (2014) Tobacco sector: employment statistical update. International Labour Office, Geneva

Jin J (2014) Why FCTC policies have not been implemented in China: domestic dynamics and tobacco governance. J Health Polit Policy Law 39(3):633–666

Lee K, Collin J (2006) Key to the future: British American tobacco and cigarette smuggling in China. Plos Med 3(7):e228

Lee K, Eckhardt J (2017a) The globalisation strategies of five Asian tobacco companies: an analytical framework. Glob Pubic Health 12(3):269–280

Lee K, Eckhardt J (2017b) The globalisation strategies of five Asian tobacco companies: a comparative analysis and implications for global health governance. Glob Public Health 12(3):367–379

Lee K, Gilmore AB, Collin J (2004) Breaking and re-entering: British American tobacco in China 1979–2000. Tobacco Control 13(Supp II):ii88–ii95

Lee S, Lee K, Holden C (2014) Creating demand for foreign brands in a 'home run' market: tobacco company tactics in South Korea following market liberalisation. Tobacco Control 23(3):e8

Li C (2012) The political mapping of China's tobacco industry and anti-smoking campaign John L. Thornton China Centre Monograph Series Number 5. Brookings Institution, Washington, DC

Li C (2018) Tobacco governance. Elite politics, subnational stakeholders and historical context. In: Kohrman M, Quan G, Wennan L, Proctor RN (eds) Poisonous Pandas: Chinese cigarette manufacturing in critical historical perspectives. Stanford University Press, Stanford, pp 179–204, 308pp

Li VC, Tang S (2018) China's new road for tobacco control: tobacco crop substitution. Am J Public Health 108(10):1316–1317

Long C, Zhang X (2012) Patterns of China's industrialization: concentration, specialization, and clustering. China Econ Rev 23(3):593–612
Marketline (2018) Marketline industry profile. Tobacco in China. Marketline, August 2018. Retrieved from: https://store.marketline.com/report/ohme5864%2D%2Dtobacco-in-china/
Masau P (2019) Leaves of fortune. China lights up Zimbabwe's tobacco sector. ChinAfrica 11, June 14. Retrieved from: http://www.chinafrica.cn/Homepage/201906/t20190614_800170698.html
Melzer S, Martin C (2016) A brief overview of the illicit trade in tobacco products. In: OECD, illicit trade: converging criminal networks. OECD Publishing, Paris
Morley CP, Pratte MA (2013) State-level tobacco control and adult smoking rate in the United States: an ecological analysis of structural factors. J Public Health Manage Pract 19(6):e20–e27
National Bureau of Statistics (2018) 2018 China statistical yearbook. Statistics Press, Beijing
National Bureau of Statistics of China (2012) 2011 China statistical yearbook. Statistics Press, Beijing
Observatory of Economic Complexity (2017a) Import Origins Of Raw Tobacco (hs4) to China. Observatory of Economic Complexity, Cambridge MA. https://oec.world/en/visualize/tree_map/hs92/import/chn/show/42401/2017/
Observatory of Economic Complexity (2017b) Export destinations of tobacco (HS 92: cigars and cigarettes) from China. Observatory of Economic Complexity, Cambridge MA. https://oec.world/en/visualize/tree_map/hs92/export/chn/show/2401/2017/
Ojakorotu V, Kamidza R (2018) Look east policy: the case of Zimbabwe-China political and economic relations since 2000. India Q 74(1):17–41
Peng Y (1996) The politics of tobacco: relations between farmers and local governments in China's Southwest. China J 36(July):67–82
Prasad M (2006) The politics of free markets: the rise of neoliberal economic policies in Britain, France, Germany and the United States. University of Chicago Press, Chicago
Qian Z (2010) China tobacco enters real estate market. Global Times 15 January. http://www.globaltimes.cn/content/498646.shtml
Qiushi H, Kohrman M (2018) The Chinese cigarette industry during the "Great Leap Forward". In: Kohrman M, Quan G, Wennan L, Proctor RN (eds) Poisonous Pandas: Chinese cigarette manufacturing in critical historical perspectives. Stanford University Press, Stanford, pp 73–92, 308pp
Rawski TG (1989) Economic growth in pre-war China. University of California Press, Berkeley
Roberts JG, Yamakawa M (1973) Mitsui: three centuries of Japanese business. Weatherhill, Boston, 564pp
Scoones I, Mavedzenge B, Murimbarimba F, Sukume C (2018) Tobacco contract farming and agrarian change in Zimbabwe. J Agrarian Change 18:22–42
Seth RS (1974) Encyclopedia of Espionage. Doubleday, New York, 683pp
Shen A, Antonopoulos GA, von Lampe C (2010) 'The dragon breathes smoke' cigarette counterfeiting in the People's Republic of China. Br J Criminol 50(2):239–258
Song S (2018) Building a smoking society: culture and ecology of the tobacco industry in North China, 1902–1937. Phd dissertation, Department of History, Washington State University, Pullman
Tang S, Li VC, Cui Y, Sun J, Khow (2018) Successful experience with tobacco crop substitution in Yuxi, Yunnan, China. Tobacco Induced Dis 16(Suppl):A949
Tinari P (2005) China trade: the art and commerce of tobacco. Duke Mag 91(4):1–3
Tobacco China Online (2018) China aims to increase tobacco farmers' income. Tobacco Asia, May 18, 2018
Tobacco China Online (2019a) The legendary life of China's Cigarette King. Tobacco Asia, May 27, 2019
Tobacco China Online (2019b) China's anti-counterfeiting and anti-smuggling efforts. Tobacco Asia, July 22
Tobacco China Online (2019c) New Protocol to revive exports to China. Tobacco Asia, March 18

Tong E, Tao M, Xue Q, Hu T (2008) China's tobacco industry and the World Trade Organization. In: Hu T (ed) Tobacco control policy analysis in China: economics and health. World Scientific Publishing Company, Singapore, pp 211–244, 328pp
von Lampe K (2012) The changing role of China in the global illegal cigarette trade. Int Crim Justice Rev 22(1):43–67
von Lampe K (2019) The illegal cigarette trade. In: Natarajan M (ed) International and transnational crime and justice, 2nd edn. Cambridge University Press, Cambridge, pp 49–54
von Lampe K, Kurti M, Bae J (2014) Land of opportunities. The illicit trade in cigarettes in the United States. In: van Duyne P, Harvey J, Antonopoulos GA, von Lampe K, Maljević A, Markovska A (eds) Corruption, greed and crime-money. sleaze and shady economy in Europe and beyond. Wolf Legal Publishers, Oisterwijk, pp 267–290
Wang J (2009) Global market building as state building: China's entry into the WTO and market reforms of China's tobacco industry. Theory Soc 38(2):165–194
Wang J (2013) State-market interactions in China's reform era: local state competition and global market building. Routledge, Abington
Wang J, He S, Xu Y, Cao Z, Wang L, Zeng DD (2016) Spatial-temporal patterns and drivers of illicit tobacco trade in Changsha County, China. *2016* IEEE conference on Intelligence and Security Informatics (ISI), Tucson, AZ, pp. 127–132. https://doi.org/10.1109/ISI.2016.7745455
Wang-Schweig M, Lipperman-Kreda S, Lee JP (2018) Rapid assessment for establishing evidence of an underground cigarette market in Oakland Chinatown: a dual approach. J Health Care Poor Underserved 29(2):615–622
Watts J (2004) BAT breaks China barrier. The Guardian 17 July. https://www.theguardian.com/business/2004/jul/17/china.smoking
Wen M (2004) Relocation and agglomeration of Chinese industry. J Dev Econ 73(1):329–347
Wennan L (2018) Experimentation: cigarettes in the Communist base areas during World War Two. In: Kohrman M, Quan G, Wennan L, Proctor RN (eds) Poisonous Pandas: Chinese cigarette manufacturing in critical historical perspectives. Stanford University Press, Stanford, pp 37–55, 308pp
Weston D (2019) Outlook bleak for tobacco farmers. Lake Gaston Gazette Observer March 22, 2019. http://www.lakegastongazette-observer.com/news/article_81263a78-4c9b-11e9-8522-0f1db50a9e99.html
Workman D (2019) Tobacco cigarettes exports by country. Financial Times, May 31, 2019
Wu Y, Heerink N, Yu L (2020) Real estate boom and resource misallocation in manufacturing industries: evidence from China. China Econ Rev 60(C):101400
Yang G (ed) (2002) Zhongguo yanyeshi huidian (Documents on the history of Chinese tobacco). Guangming ribao chubanshe, Beijing. (In Chinese)
Yang T, Barnett R, Rockett IRH, Yang YY, Wu D, Zheng W, Li L (2015) The impact of regional economic reliance on the tobacco industry on current smoking in China. Health Place 33(May):159–171
Zheng S, Sun C, Qi Y (2014) The evolving geography of China's industrial production: Implications for pollution dynamics and urban quality of life. J Econ Surv 28(4):709–724
Zhou H (2000) Fiscal decentralisation and the development of the tobacco industry in China. China Econ Rev 11(2):114–133
Zhou H (2001) Implications of interjurisdictional competition in transition: the case of the Chinese tobacco industry. J Comp Econ 29(1):158–182
Zhou Y (2004) Industrial and commercial separation of the Chinese tobacco industry, its challenges and the way forward. Tobacco China. Cited in Fang (2017), p 334. http://www.tobaccochina.com/management/Industry/management/200410/2004108000_192491.shtml

Chapter 4
The Tobacco Industry: Marketing Strategies and Consumption

Ross Barnett, Sihui Peng, and Shuhan Jiang

Abstract This chapter examines industry marketing strategies and how these have influenced changes in consumption. The increased saturation of the domestic cigarette market in China has, since 2000, been an important reason for the increased globalisation of the China National Tobacco Company (CNTC). Exports alone, however, are not sufficient to maintain high levels of cigarette production. Changes in domestic product offerings are also necessary in order to cope with changing consumer tastes, foreign competition, new patterns of demand and the increased threat of tobacco control measures by the state. The chapter discusses some of the different marketing tactics adopted by the CNTC and the groups targeted to increase levels of consumption. Possible pathways which link regional economic reliance on tobacco growing and manufacturing to high rates of consumption are discussed. The chapter concludes by focusing on the geography of tobacco retailing and the influence of social media. Given the increased restrictions which have been imposed on tobacco advertising, both settings have recently assumed increased importance.

Keywords Advertising · Brand consolidation · Consumption · Marketing strategies · Point of sale (PoS) · Premiumisation · Price · Regional economic reliance · Social media · Tobacco retailing

R. Barnett (✉)
School of Earth and Environment, University of Canterbury, Christchurch, New Zealand
e-mail: ross.barnett@canterbury.ac.nz

S. Peng
Jinan University, School of Medicine, Guangzhou, China

S. Jiang
Zhejiang Chinese Medical University, School of Humanities and Management, Hangzhou, China

R. Barnett et al. (eds.), *Smoking Environments in China*, Global Perspectives on Health Geography,
https://doi.org/10.1007/978-3-030-76143-1_4

4.1 Introduction

This chapter focuses further on the economic context of smoking by examining industry marketing strategies and how these have influenced changes in consumption. The increased saturation of the domestic cigarette market in China has, since 2000, been an important reason for the increased globalisation of the CNTC. Exports alone, however, are not sufficient to maintain high levels of cigarette production. Changes in domestic product offerings are also necessary in order to cope with changing consumer tastes; foreign competition albeit limited, new patterns of demand; or the increased threat of tobacco control measures, such as gradual spread of smoking bans. Indeed the rise in per capita consumption after 2000 (Hoffman et al. 2019), after slowing during the 1990s, stands as testimony to the effectiveness of the product development and marketing policies put in to place by the CNTC.

The chapter comprises four sections. The first section briefly examines some of the different marketing tactics adopted by the CNTC and the groups who have been targeted to increase levels of consumption. This is followed by an analysis of whether individual smoking prevalence is higher in cigarette-producing regions. Based on an earlier study (Yang et al. 2015), we take a socio-ecological approach and focus on some of the possible pathways linking an industry presence to high rates of consumption. In the final two sections, we focus on the geography of tobacco retailing and the influence of social media. Given the increased restrictions which have been imposed on tobacco advertising, both settings have recently increased in importance.

4.2 Marketing Tactics

In its push to establish an increased global foothold for its products and to improve domestic sales, the CNTC has engaged in a number of marketing strategies. These include new product developments and sales innovations both at the point of sale and on social media (Gao et al. 2013; Wang et al. 2014a, b; Liang et al. 2015; Yao et al. 2016). With respect to product development, Lee and Eckhardt (2017) note that this has been central to the global business strategies of most Asian tobacco companies. This is also true of the CNTC where product development has involved two main strategies: (1) brand consolidation and (2) creating higher-value-added premium products which have been marketed as ‘less harmful, low tar’ alternatives to traditional cigarettes. Both of these strategies appeal strongly to foreign as well as local markets.

Brand consolidation has been important for achieving greater economies of scale and for creating greater product visibility. Also the development of higher-quality cigarette packaging, with anti-counterfeit features, has helped reduce the circulation of counterfeit brands (Euromonitor International 2012). Consequently, numerous small regional brands have been replaced by a smaller number of higher-quality brands. Between 2001 and 2013, the number of national cigarette brands was reduced from 1183 to 90 (Fang et al. 2017, p.322). Among the new styles of cigarette products are flavoured capsule cigarettes, slim cigarettes and short- and medium-sized categories (Tobacco China Online 2019a), many of these reflecting similar trends in western countries at an earlier point in time. For example, compared to traditional cigarettes, short-sized and slim cigarettes have been specifically aimed at the younger and female market and health-conscious consumers. Short-sized cigarettes, such as the *Guiyan (Xingzhe)* brand from Guizhou Province take less time to smoke and create the impression of being stylish, elegant and easy to carry. Similarly flavoured capsule cigarettes, also manufactured by Guiyan Industrial Tobacco Co., have proved popular because of their varied tastes such as dried tangerine or wine-flavoured capsules.

Product development has also involved a 'premiumisation' strategy to produce higher-quality cigarettes. Historically the majority of cigarettes consumed in China have been discount and mid-priced brand cigarettes and the lack of tax increases has kept prices low. Even as late as 2016, Yang et al. (2020a), in a study of responses to tobacco pricing in six cities, showed that just over one-quarter of smokers consumed cheaper brands costing ¥9 or less, compared to 38% in the mid-price category (¥10 to ¥14), while 36% purchased cigarettes costing ¥15 or more. Currently the CNTC classifies cigarettes into three groups based on the before-tax allocation price: discount brand cigarettes (classes IV (¥1.65 to ¥3 per pack) and V (less than ¥1.65 per pack)), medium brands (class III priced from ¥3 to ¥7 per pack) and premium brands (classes I (more than ¥10 per pack) and II (¥7 to ¥10 per pack)) (Xu et al. 2019a, b). Since 2007 there has been a marked shift in the price distribution of brands. Low-price brands (classes IV and V) share of total production dropped from 66.9% in 2007 to only 23.2% in 2016. By contrast, higher-cost brands increased their share from 13.1 to 33.5% (Fig. 4.1).

Three main factors caused the industry to move away from cheaper brands. On the demand side, rapid economic growth resulted in a surge of disposable income which meant that Chinese smokers could afford to spend more. On the supply side, the increased globalisation of the CNTC and a rise of tobacco exports prompted the need to raise product quality. Also, as rates of cigarette production slowed dramatically in the 1980s and 1990s, CNTC profits declined and the industry looked both to stimulate production and move away from cheaper brands with low profit margins.

As a result of such trends, the CNTC launched its 'premiumisation' strategy in 2009. This was designed to encourage smokers to buy more expensive higher-quality brands which have been promoted as less harmful (Xu et al. 2019a, b). Many of these are light 'low tar' cigarettes designed to persuade health-concerned smokers to switch cigarette brands rather than quit (Campaign for Tobacco-Free Kids 2012). In the 2010 *Global Adult Tobacco Survey* in China, there was a clear social gradient in the awareness of the health effects of low-tar cigarettes, but even among the most educated group, those with a college education and above, most (45%) agreed with industry claims that low-tar cigarettes were less harmful to health than regular cigarettes (Yang 2014). In view of such trends, it is not surprising that light

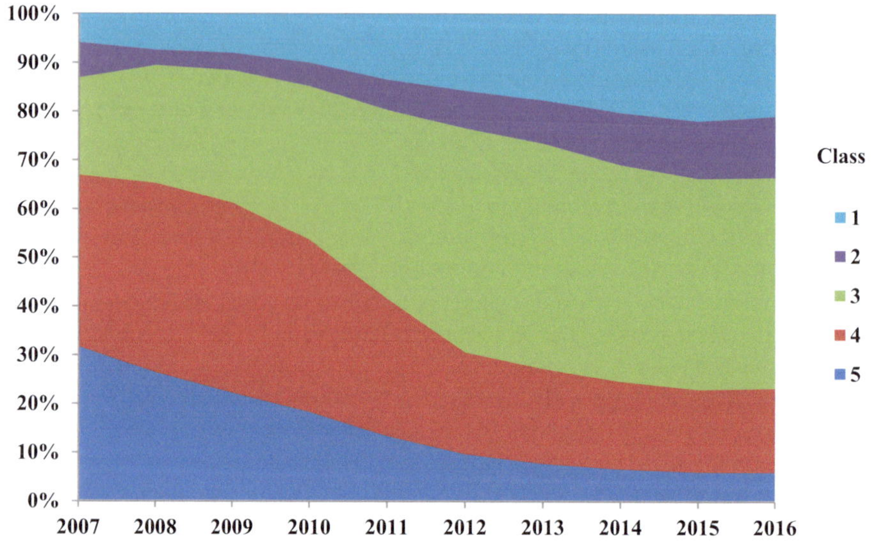

Fig. 4.1 Market share of cigarettes by price class, 2007–2016. (Credit: Data from the State Tobacco Monopoly Association (STMA), China Tobacco Yearbook. China Economic Publishing House, Beijing (various years). Public domain)

cigarettes have proven to be popular, particularly in places with environmental smoking restrictions, whether households, workplaces or public spaces (Yang et al., 2015a).

In a national evaluation of the impact of the 'premiumisation' strategy Xu, Gravely et al. (Xu et al. 2019a, b) found that between 2006 and 2015 most of the increase in consumption resulted from smokers switching to premium and, to a lesser extent, mid-priced brand cigarettes with the greatest decrease occurring among discount brands. Most did so because of their willingness to pay for better quality and what they perceived to be less harmful products. While premiumisation has spurred the production and sale of cigarettes in the Chinese domestic market, it has also been important globally in improving the perceived quality of Chinese products. The development of the new *Acacia* brand in Zimbabwe, primarily aimed at the Chinese market in Africa, for example, fits this trend. However, compared to other Asian TTCs, while the CNTC has dramatically reduced the number of Chinese brands to achieve economies of scale and improved their quality, as of the present, it has not yet managed to create substantial foreign demand for its products (Lee and Eckhardt 2017).

The changing mix of cigarette products reflects both domestic changes in China along with changes associated with the increased internationalisation of the tobacco industry and national pride associated with China's rise as a global power. The policy of 'premiumisation' has taken advantage of the rise of disposable household income and encouraged brand switching to higher-priced brands. Similarly, domestic brands, such as the Hongta Group's *Hongtasha*, the best-selling brand in China

(Campaign for Tobacco-Free Kids 2012), has exploited Chinese culture both in its packaging, often featuring classic Chinese temples or landscapes, and through sales promotions marketing *Hongtasha* as a gift for friends and relatives during Chinese holidays and on special occasions (Chu et al. 2011). A good example is the new range of products from China Tobacco Jiangxi Industrial Co. Ltd., the key operator of the industry in China's Jiangxi Province. The company has made great strides to be more competitive through product innovation but also to improve the attractiveness of its products through associating them with local heritage (Fig. 4.2). A description of one of its five new products of the Jinsheng brand, which includes the *Tengwangge* series, concluded:

> In the eyes of CT Jiangxi, the cultural and historical heritage that inspired the names of the products are yet another positive attribute of the brand. The development of the *Tengwangge* series cigarette products was a first step taken by the group in naming its products after famous cultural or historical heritage to back sales promotion. As one of the three great towers of southern China, Tengwangge is not only an ancient architecture, but is also part of the culture and history of the province, with much cultural potential yet to be tapped. Therefore, when consumers smoke an aromatic and tasteful Jinsheng *Tengwangge* cigarette, the cultural and historical significance of Tengwangge will come to mind. (Tobacco China Online 2016)

Besides efforts to promote sales by associating new products with China's cultural heritage, China Tobacco, like so many other companies, has made use of social media, by introducing marketing on the *WeChat*, to improve its corporate image.

The industry has used its marketing strategies to target specific groups, such as adolescents and women, but also to counter tobacco control policies (Yang 2014).

Fig. 4.2 Jinsheng brand cigarettes. (Credit: Tobacco Asia, September 7, 2016. https://www.tobaccoasia.com/features/profile-china-tobacco-jiangxi-industrial-striving-to-be-more/. October Multimedia Co Ltd. Permission obtained)

For example, the recent emphasis on point of sale (POS) advertising by tobacco companies, especially near schools, is aimed at encouraging smoking initiation among youth (see Sect. 4.4.3). Candy, aimed at even younger children, has also recently started to be marketed in packets, resembling those of cigarettes, often featuring well-known Chinese and foreign brands (Fig. 4.3). Once popular in China, such ‘candy cigarettes’ have again become common in many retail outlets reminding people of their childhood memories but also reinforcing the importance of cigarettes and stimulating consumption among adolescents. Parkour rallies aimed at teenagers have also been organised by different tobacco companies. For instance, in 2015 the Beijing Tobacco Corporation, which manufactures *Zhongnanhai* cigarettes targeted at fashionable young people, organised a series of parkour activities popular with teenagers. The activity background board featured the prominent Zhongnanhai tobacco brand logo (LeParkour 2015).

Similarly, while the proportion of women who smoke remains low, given trends in other Asian countries, this segment of the population remains an attractive potential market especially as women gain new freedoms and increased status and financial independence. Tobacco companies have increasingly shifted their focus to women as raising the already high smoking rate among men has proved difficult. Initially Chinese tobacco companies spent heavily on advertisements that associated tobacco use with beauty and liberation, but more recently have focused on new cigarette brands, such as light and fruit-flavoured cigarettes, a strategy which is most likely to affect low income and less educated women. Many of these brands are also slim cigarettes, developed by the CNTC to target young smokers pursuing fashion and elegance. They were also developed in conjunction with TTCs (Liao

Fig. 4.3 Candy cigarettes called ‘The Taste of Childhood’ in a snack shop, Hangzhou, 2020. (Credit: Authors)

2015a) as a strategic new product and one perceived to be less harmful than traditional cigarettes (Zhang et al. 2019).

Similarly, 'premiumisation' is a strategy targeting health-conscious consumers to switch cigarette brands rather than to quit smoking even though it has long been known that 'light' and 'low tar' cigarettes are no safer than regular cigarettes (Campaign for Tobacco Free Kids, 2012). In China, as a result of 'industry-induced misconceptions', most smokers believe the reverse, with the proportion of smokers doing so being the highest in over 20 ITC study countries (Xu et al., 2019a, b, p. s72). Premiumisation, especially the increased consumption of class I and II premium brands, has been especially attractive to younger male non-daily smokers and possibly to those of higher socio-economic status, although this was not considered in the 2006–2015 ITC surveys. By contrast, the general increase in affordability of cigarettes over the same period was significantly larger among younger and lower SES smokers, a trend reminiscent of the intense targeting of lower income groups by tobacco companies, especially in the United States (Cruz et al. 2019).

As in western countries, electronic cigarettes are increasingly being marketed in China as safer alternatives to conventional tobacco products. However, despite such trends e-cigarette use remains very low (Jiang et al. 2017; Wang et al. 2019), there being an estimated one million regular vapers in China compared to 350 million smokers (McNeice 2018). The reasons for this are unclear but most likely reflect the fact that since China is by far the largest world producer of e-cigarettes, most are exported and, until recently, there has been no need to cultivate the Chinese market. Consequently with few exceptions, awareness of e-cigarettes remains low. However, this is likely to change as tobacco control initiatives gain momentum and the promotion of local e-cigarette consumption becomes a more important economic priority on the part of the CNTC. This is already occurring with a recent (2018) agreement between BAT and China Materialia being a sign of this (McNeice 2018). To date most e-cigarette marketing has targeted youth and women using well-tried marketing methods of promoting e-cigarettes as a fashion item and a more healthy form of smoking, although the health risks of e-cigarette use remain contested. Although China has no national regulations on e-cigarettes, these do exist in some cities such as Hangzhou and Shenzen.

4.3 Economic Reliance on Tobacco and Smoking Prevalence

Despite a large number of studies of cigarette consumption in China and differences between population groups, few studies have explored the link between the location of the tobacco industry and the effect of different contextual factors on consumption. This section builds on an earlier study of the impact of economic reliance on the tobacco industry on smoking prevalence (Yang et al. 2015). While other studies have examined both urban (e.g. Huang et al. 2014; Yang et al. 2011) and provincial variation (Pan and Hu 2008) in cigarette purchasing behaviour, awareness of tobacco advertising and smoking prevalence, their analyses have been confined to a limited number of cities or provinces. By contrast the study by Yang et al. (2015) was based on 20,601 participants from 27 cities located in all but five of China's municipalities/provinces. Economic reliance on cigarette manufacturing was measured by provincial cigarette production per 100 yuan GDP, and associations were examined with individual smoking prevalence after controlling for a variety of individual and regional confounders.

The main finding was that regional cigarette production was an important contextual variable affecting levels of current smoking. Controlling for confounding factors, including both individual and regional income, the study found that individuals who lived in the areas with the highest levels of production (measured by cigarettes produced per 100 yuan GDP) were almost 46% more likely to be current smokers than people who lived in areas of low production. This confirms earlier Chinese research which showed that areas cultivating tobacco had a higher prevalence of current smokers than those that did not (Cai et al. 2012, 2013). The result that more people are likely to smoke in regions where the industry is strongest, as has been found in the United States (Morley and Pratte 2013), is unsurprising but nevertheless important because it suggests that the regional geography of production, in itself, is an important factor influencing smoking prevalence and may have implications for the success of tobacco control initiatives. Given this, there are a number of possible pathways which can help to explain why individual smoking prevalence, after controlling for confounding factors, should be greater in regions with higher levels of cigarette production (Fig. 4.4).

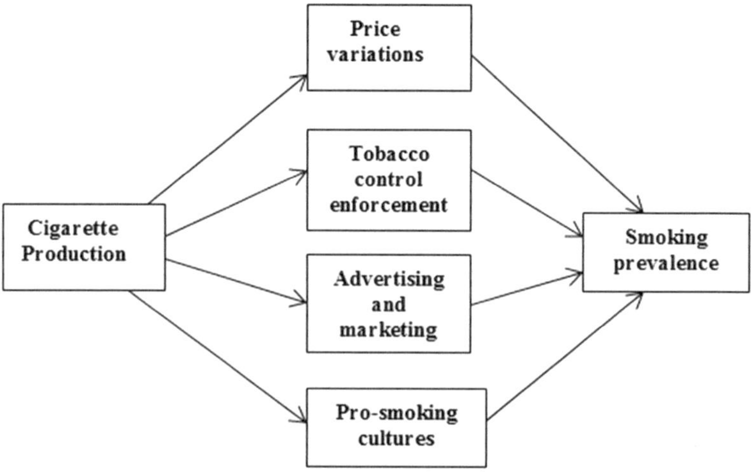

Fig. 4.4 Pathways between regional cigarette production and smoking prevalence. (Credit: Authors)

4.3.1 Price Variations

First, for consumers, higher production could mean lower prices which, in turn, results in higher consumption. This is reflected in the retail system, with the main cigarette-producing provinces also being those with greater numbers of (licensed) retailers per capita (Table 4.1). Elsewhere existing research has shown that tobacco outlet density is a good predictor of both youth and adult smoking (Riches et al. 2018). Following the introduction of stricter limitations on advertising, the Chinese tobacco industry has increased its investments in retail outlets and point-of-sale advertising, which have been shown to be effective in creating pro-smoking attitudes and encouraging youth to smoke (Cheung et al. 2018).

Despite centralised control of tobacco prices, it is possible that a greater industry presence and higher density of retailers results in more competition and lower prices. Although price setting of all tobacco products is a feature of the Chinese tobacco monopoly, the STMA nevertheless allows retailers to set their retail prices within a limited range (usually 10–15%). This also applies to cigarette factories with respect to producer and wholesale prices (Gao et al. 2012). Consequently some local variation in cigarette prices is likely to occur, but by how much and the extent to which such variations reflect tobacco retailer outlet densities is not known. Nevertheless there is some evidence of price discounting. For example, Yang et al. (2010), in their ITC Survey of six cities, found that in cities with lower levels of reported tobacco advertising, people had a higher awareness of retailer tobacco promotion activities such as free gifts or special discount offers. For instance, Yinchuan, a relatively poor city, had more direct price discounting, while in Guangzhou, a richer city, people reported more value-added gifts. In addition, as Zhou (2001) has noted, retail sales of cigarettes have never been nationalised and are mostly individual and family businesses. With such a large number of non-state-owned retailers, it is hard for state agencies to maintain strong controls on prices.

Cheng et al. (2015) report a wide range of cigarette prices nationally, from 1.5¥ per pack to up to 150¥ per pack for top end brands. Li et al. (2010), in the ITC survey of six cities, found that the median prices paid per pack varied from 7.50¥ in Shanghai to 3.70¥ in Shenyang. Access to lower-priced cigarettes is especially important for heavier smokers and for those on low incomes (Huang et al. 2014). Cheng et al. (2015), for example, found that the majority of blue collar workers in

Table 4.1 Correlations between measures of provincial cigarette production and tobacco retailers per 10,000 population, 2011[a]

Cigarette production (100 m pieces)	0.56**
Cigarette production per capita (100 m pieces per 10,000)	0.45*
Cigarette production per 100 yuan GDP	0.64**

** $p < 0.001$; * $p < 0.01$; [a]Licensed retailers only

Credit: Authors. Data derived from State Tobacco Monopoly Administration (various years) *China Tobacco Yearbook* (Science and Technology of China, Beijing)

Fig. 4.5 Retail display of lower-priced (¥16–25; US $2.32–$3.62) cigarettes, Hangzhou, 2019. (Credit: Authors)

Kunming bought low-priced cigarettes, usually less than 4¥ per pack, while Yao et al. (2014) also found that young and low-income smokers were more likely to purchase cigarettes from cheaper sources. However, because the profit margin on cheaper cigarettes is low, then tobacco companies are less likely to produce these than higher-cost brands. These different price margins help contribute to China's large price spread (White et al. 2013) and will encourage retailers, many of whom are part-time traders (Hu et al. 2010), to sell higher-priced brands. Because of this the CNTC requires local tobacco companies to produce a certain quota of cheaper cigarettes each year (Fig. 4.5) and subsidises them to compensate for lower profit margins. Thus in some respects the CNTC is no different from US tobacco companies which used discount cigarettes in the 1980s and 1990s to retain price-sensitive smokers (Cummings et al. 1997). The point is that retailers in localities where there is more competition from other retailers will be more likely to engage in price discounting and for consumers to purchase less expensive brands.

4.3.2 *Enforcement of Tobacco Control*

Although easier access to cigarettes in areas of cigarette production may stimulate higher rates of smoking, a second important pathway relates to the lack of enforcement of tobacco control policies in such areas. Given that cigarette manufacturing is a key generator of local government revenue, the conflict between economic and social objectives is likely to produce less enforcement of tobacco control initiatives (Warner 2000). In Yunnan, the 'tobacco kingdom' of China, tax revenues from tobacco production and cigarette manufacturing accounted for nearly 70% provincial government revenue, with almost half of this being used for rural development, education, social welfare and industry development (FAO 2003). Anti-smoking campaigns directly challenge the revenue earnings of local governments and industry profits, and thus these parochial local concerns traditionally have tended to outweigh any thoughts of tobacco control and China's international FCTC commitments (Li 2012).

In China the one study that examined provincial differences in the implementation of tobacco control (Astell-Burt et al. 2017) did not specifically examine its relationship to economic reliance on tobacco growing and manufacturing. However, it is likely that the geography of policy enforcement will be similar to the United States where the implementation of tobacco control measures tends to be less in states where tobacco revenue forms a higher percentage of state GNP (Morley and Pratte 2013). This is most evident in states with Republican control of state legislatures and where pro-business, anti-tax political interests are associated with lower excise taxes on cigarettes, weaker smoke-free policies and less spending on tobacco control. These trends are especially evident in states heavily dependent on tobacco manufacturing with tobacco-manufacturing states spending nearly 30 percentage points less than non-manufacturing states. Similarly Hoffer and Pellillo (2012) showed that tobacco control spending was negatively related to both the level of tobacco production and the level of campaign contributions from state tobacco lobbies. Not surprisingly, tobacco-producing states have much lower rates of smoking cessation (Dwyer-Lindgren et al. 2014). Similarly, in Japan, although a national health promotion law since 2003 mandated local prefectures to restrict exposure to second-hand smoke, significant regional disparities exist in terms of compliance with the legislation. In poorer countries, such as Tanzania (Kagaruki 2010) or Vietnam (Higashi et al. 2011), the implementation of tobacco control measures has been particularly poor in prefectures involved in growing tobacco (Yorifuji et al. 2011).

4.3.3 Advertising and Marketing

A third pathway linking higher rates of provincial cigarette production to increased smoking prevalence is the role of advertising. Yang et al. (2010) report that tobacco company lobbies have been aggressive in their marketing activities and that significant regional differences exist in awareness of tobacco advertising and promotion and in implementing a central set of laws designed to restrict tobacco promotion. In a study of tobacco advertising in six cities, they found that in the three cities with the greatest industry presence (Changsha, Shenyang and Shanghai), a higher proportion of people recalled noticing tobacco advertising on posters, billboards and newspapers/magazines and this was also true of promotional activities at the point of sale. At the time, awareness of tobacco advertising in China was reported to be high compared to other countries (Li et al. 2009). Yang et al. (2012), for example, found that 41% of adults reported 'sometimes' or 'very often' being exposed to tobacco advertising, with the highest exposures occurring through television (54%), billboards (28%), and visits to tobacconists (25%). Xiao et al. (2015) report similar results among Chinese middle school students, except that the internet emerged as a major additional factor.

Since 2015 direct advertising has been banned in China (Zheng et al. 2018), but it remains important particularly at the point of sale and through the sponsorships, community activities and engagement of the CNTC. Such findings point to the need for more research to be directed to both these advertising strategies, particularly since the extension of controls on one level almost inevitably leads to the tobacco industry utilising other advertising methods to circumvent such regulations. Even before the ban on direct advertising, cities who reported lower levels of advertising tended to report higher levels of point of sale activity (Yang et al. 2010).

4.3.4 Corporate Community Involvement and Pro-smoking Cultures

Like TTCs worldwide, the exercise of corporate philanthropy has been typical of the CNTC. Tobacco companies maintain a high community profile by directly making significant contributions to community development. For example, the Chinese Association on Tobacco Control (CATC) noted that 52 tobacco companies donated to, or sponsored, 79 welfare activities in 40 cities and counties during the last quarter of 2009 (Cheng 2014), including those, such as the Hope Project, designed to help poor children attend school (Jin 2014). The CNTC also regularly uses anti-poverty charities to promote its brands. This has been particularly important especially in poorer provinces such as Yunnan, where a recent report summarised its accomplishments over the last five years (Tobacco China Online 2020).

In addition, the industry has promoted competitions between cities in tobacco-growing areas. Recently in Yunnan 30 cities competed for the award of the 'most influential tobacco-growing area' in the province, an event which was reported in the national media. The event was held by the CNTC to honour the best tobacco products in the province. The last part of the competition included an art show which focused on the stories of tobacco-producing areas and brands. Events such as this glorify tobacco and celebrate the tobacco culture (Fig. 4.6).

These CNTC sponsored philanthropic projects are similar to those pursued by TTCs outside China and are important in promoting a positive public image of corporate social responsibility and maintaining support for the industry (Hu et al. 2013; Yang, Li et al. 2010). This was evident in a recent Yunnan study which showed that educators and parents did not perceive any negative impacts on children from exposure to pro-tobacco propaganda as a result of attending schools sponsored by the Honga Tobacco Group (Fang et al. 2019). They reported that local children's awareness of the Group and the general sense of gratitude reported by both children and

Fig. 4.6 Yunnan art show glorifying best tobacco products in Yunnan Province. (Credit: Wenshan News Network 17 January, 2020 Permission obtained. http://wsnews.com.cn/zsws_wstt/p/218509.html)

their parents showed that school sponsorship promoted the industry in a positive way and was likely to have produced a certain consumer loyalty to the company. Such 'tobacco schools' are not unique to China, having also been promoted by the CNTC in Zimbabwe (Fang, 2019).

As a result of such activities, provinces heavily dependent upon cigarette production are more likely to have a pro-smoking culture. A large body of work in western countries has documented the significance of cultural and social practices on smoking initiation and cessation (e.g. Bell et al. 2007; Christakis and Fowler 2008). This work has emphasised that smoking is a learned behaviour and that people's smoking habits will be influenced by local social norms with respect to its acceptability (e.g. Ross 1998). In China these norms have been heavily influenced by the tobacco industry, through its advertising and community activities, which have undermined the acceptance of anti-tobacco messages. Pro-smoking cultures have also been reinforced by the tobacco industry through its promotion of cigarette gifting (see Chap. 5). Ding and Hovell (2012) have argued that cigarette gifting should be conceptualised not as a Chinese custom but as a cultural product or social currency engineered by the tobacco industry to build brand identity and increase consumption. Although no studies have assessed the geography of cigarette gifting in China, the practice is likely to be especially important in tobacco-growing and cigarette-producing provinces where the industry is most powerful.

4.4 Tobacco Retailing and Consumption

4.4.1 Distribution Arrangements: Wholesalers and Retailers

Within the tobacco monopoly system, wholesalers and retailers make up an important part of the supply chain and have a significant influence on consumption (see Chap. 3, Fig. 3.4). Despite this, there have been very few studies of tobacco wholesaling and retailing in China, although this is likely to change now that stricter national bans on advertising have been put into place (Zheng et al. 2018) and POS advertising becomes more significant. It is important, therefore, to understand links between wholesalers and retailers, how each group functions within the rules of the tobacco monopoly system and the impact of different retailing advertising strategies on consumers themselves.

There are two sets of wholesalers: second-tier or provincial wholesalers, who purchase directly from cigarette factories, and third-tier wholesalers who buy from second-tier wholesalers and on-sell cigarettes to retailers within their respective localities (Zhou 2000). There are 33 provincial level commercial wholesaling enterprises designated by the CNTC who are responsible for cigarette marketing strategies within their respective provinces (Liao 2015b). Third-tier wholesalers, on the other hand, operate at the prefectural level and are the lowest level in the distribution hierarchy. Traditionally they have placed orders with second-tier wholesalers who have not always been able to meet the demands of the localities within which third-tier wholesalers operate. For a time, in the late 1980s and early 1990s, free wholesale markets operated, often being encouraged by local government. However, since these operated outside the tobacco monopoly and evaded taxes, they were soon discouraged by central government (Zhou 2000).

Zhou (2000, p. 124) has argued that under the tobacco monopoly system, buyers are at the mercy of sellers, whether these be second- or third-tier wholesalers or retailers. In the past, product shortages meant that third-tier wholesalers and retailers were the least powerful actors in the supply chain and could not always obtain the products they needed, particularly in rural areas. This is less true in more recent times of high cigarette production. Under China's tobacco monopoly system, tobacco producers offer cigarettes to wholesalers at a fixed price (the cigarette allocation price). Wholesale prices offered to retailers are also at a fixed price. In theory the allocation-wholesale profit margin and the wholesale-retail profit margin are both set by the STMA (Zheng et al. 2018). This should result, for given grades of cigarettes, in standard prices across China, but, as we argued previously, evidence suggests that this is not the case, given the increased pressures for profit maximisation among larger more corporate enterprises. What is not clear, however, is how these pressures manifest themselves in different areas of China and affect the availability and pricing of different classes of cigarettes as well as the type of tobacco advertising that occurs. Despite this, it is evident that central government still maintains a high degree of control over pricing and profit margins. The 2009 tax reform, for instance, did not affect retailers as prices remained unchanged, with the tax increases falling on tobacco producers and wholesalers (Xu et al. 2019a, b). Central government thus raised increased revenue but not at the expense of consumers who enjoyed continued low prices of cigarettes.

At the local level, retailers can only establish themselves if they obtain a valid monopoly licence and abide by certain rules regarding the location of their premises relative to other retailers and to schools. In theory they are also subject to rules regarding pricing and the display of tobacco products as set down by the CNTC, but administered by local Tobacco Monopoly Bureaus (Wang 2015, 2017). However, such rules have not always been enforced leading to outcomes that do not match central guidelines. Although tobacco products are sold through a wide range of outlets, the most common types of retailers tend to be small kiosks and convenience stores. In Hangzhou, for instance, kiosks (43.9%) and convenience stores (20.5%) accounted for almost two-thirds of tobacco outlets (Gong et al. 2013). Tobacco speciality stores, which often also sell alcohol, formed the other dominant type (27.5%). This group is most likely to sell higher-grade (classes I and II) cigarettes, which have been promoted by tobacco companies as part of their premiumisation strategy.

China has a high proportion of small retailers selling through kiosks, ranked second only to Argentina among 19 LMICs studied as part of the 2008–2012 GATS survey (Mbulo et al. 2019). The high proportion of kiosks has implications for tobacco control. Since these are usually one person or small family run businesses, with low overheads, they are less likely to conform to official guidelines especially regarding the location of their premises and the prices charged for cigarettes. By contrast, as in South Korea (Hwang et al. 2018), Chinese tobacco companies have close relationships with chain tobacco retail stores, owned by the companies and some second-level convenience stores which are certified by the companies. The stores are certified after store management is trained on how to display and advertise tobacco products in order to improve sales (Wang 2015). Since 2010 the STMA has placed great emphasis on developing customer-targeted marketing displays which are prominent enough to 'turn the stores into giant cigarette advertisements' (Zheng et al. 2018, p. 254). Stores which are certified can usually be identified by certification signs displayed at the entrance and counters of stores, the purpose being to inform customers that they are trustworthy retailers who sell high-quality company and not counterfeit products.

4.4.2 Tobacco Outlet Density and Smoking

The location of tobacco retail enterprises, along with variations in POS advertising, has important implications for smoking prevalence and levels of consumption. In western countries the effects of tobacco outlet density and spatial access to tobacco retailers on smoking (and alcohol) consumption has emerged as a significant research and policy issue. To a large extent this reflects the increased spatial concentration of smokers in more deprived neighbourhoods, and the targeting of such neighbourhoods by the tobacco industry. A large literature has emerged, particularly in Britain (Shortt et al. 2015; Siahpush et al. 2016; Riches et al. 2018; Caryl et al. 2020), North America (Lee et al. 2017) and New Zealand (Pearce et al. 2009), which shows that both tobacco and alcohol outlet densities are higher in more deprived neighbourhoods and around schools in such neighbourhoods. Cigarette prices have also been found to be lower at outlets near public compared to private schools (Cantrell et al. 2015). Measurements, using real-time data (Caryl et al. 2020), on youth activity spaces (Lipperman-Kreda et al. 2015a) show that exposure rates to tobacco retailers and other forms of unhealthy advertising (Isgor et al. 2016) are high in such areas.

Numerous studies have also shown that greater tobacco outlet density, or distance from a tobacco retailer, is associated with a variety of tobacco use outcomes, especially among youth and young adults. For example, increased density has been related to intention to smoke (Burton et al. 2015; Mennis et al. 2016), recent initiation of smoking (Cantrell et al. 2016), smoking prevalence (Ghenadenik et al. 2016; Lipperman-Kreda 2015b; Mistry et al. 2015; Novak et al. 2006; Pearce et al. 2009, 2016; Schleicher et al. 2016; Shortt et al. 2016) and decreased odds of making a quit attempt (Chaiton et al. 2018; Halonen et al. 2014; Kirchner et al. 2013; Pulakka et al. 2016; Reitzel et al. 2011). However, recent meta-analyses of adolescent smoking behaviour suggest that, while there is evidence for a positive relationship, the proximity effects are most evident for home rather than school environments (Finian et al. 2019). Nuyts et al. (2019) in a further meta-analysis argue that current literature does not provide consistent enough evidence to provide definitive conclusions. Nevertheless, the evidence is strongly suggestive that outlet density may be linked to increased smoking rates via a number of possible pathways: by directly increasing the availability of cigarettes and opportunities for purchase, by normalising smoking within the community and creating pro-smoking cultures and, as mentioned previously, by encouraging competition and reduced prices.

4.4.3 Relevance to China

In contrast to western countries, there has been little work on tobacco retailing in China. With some exceptions, most notably India (Mistry et al. 2015), Indonesia (Astuti et al. 2019) and South Korea (Hyun et al. 2017), this is also true of other LMICs in Asia. In China a limited number of studies have focused on tobacco advertising at the POS (Frick et al. 2013; Gong et al. 2013; Wang et al. 2017) with a particular focus on youth exposure near schools. To the best of our knowledge, only one study has specifically examined tobacco outlet density in relation to community socio-economic characteristics (Wang 2015), and only one has considered never smokers' exposure to POS tobacco displays and their effects (Cheung et al. 2018). Unfortunately, no Chinese studies have examined links between outlet density and consumption levels. Given the increased importance of POS marketing, this lack of attention is surprising. It could reflect the view that tobacco retailers are unimportant actors relative to other agents in the supply chain, or it may simply reflect the fact that retailers are so ubiquitous in Chinese cities and that local geographies of retailing are seen as unimportant influences on tobacco consumption.

In most western countries, although tobacco retailers may need to be licensed, their distribution is largely determined by market forces, such as the availability of affordable retail facilities and the distribution of demand. As demand from higher SES groups has shrunk, so too has the number of tobacco outlets, with their distribution becoming more socially and spatially concentrated. China presents a different picture. Given that the social gradient in smoking is much less evident than in western countries and that the location of tobacco retail outlets is far more ubiquitous, then a lack of a strong spatial concentration is to be expected. Nevertheless because the density of tobacco retailers is higher in China, the distances of retailers from neighbourhoods and their schools will be less and should still be an important factor in influencing social norms and smoking rates.

In China the density of tobacco retail outlets is supposed to conform to official local density regulations. Local Tobacco Monopoly Bureaus (TMBs) use two primary standards: (1) minimum distances allowed between the two closest retail outlets, which vary according to the type of street (e.g. busy business thoroughfares versus quieter suburban streets), and (2) a density measure relating to the number of households served by one store. In Changsha, Wang (2015, p. 17), who conducted an audit in the city in 2014, noted that this was one store per 200 households in residential areas and 1 to 2000 in non-residential areas. In Hangzhou, Gong et al. (2013) reported that the maximum allowed density was 3.5 outlets per 1000 population and that no store should be closer than 50 metres from its nearest neighbour. However, in 2007 although the average store density (of 2.4 stores per 1000 population) and average distance between stores (64.9 m) were well within official guidelines, nearly half (48.7%) of all outlets were located less than 50 metres from their nearest neighbour, indicating a certain degree of clustering.

Given the presence of TMB regulations, it is not surprising that tobacco outlet densities appear to be only weakly related to socio-economic status. In his study of 36 areas (12 neighbourhoods stratified by low, medium and high property prices) in Changsha, Wang found that neighbourhood differences in the density of tobacco outlets (per square km) were not significant. In contrast to the western context, outlet densities (per square km) were highest in neighbourhoods with medium (101.4 stores/km^2) and higher property prices (97.6 stores/km^2), compared to 84.4 stores/km^2 in low price neighbourhoods.

The lack of significant neighbourhood variation may reflect a number of factors. Because of the very high population densities in Chinese cities, neighbourhoods of different socio-economic status can be located quite close to each other and share the same stores particularly when tobacco retailers are situated on main thoroughfares separating different residential areas. Second, in western neighbourhood research, neighbourhoods are usually smaller more homogeneous areas such as census tracts or enumeration areas. By contrast in Wang's study 'neighbourhoods' were larger residential communities containing a group of similar residential buildings in gated or ungated areas. Third, in contrast to western cities, the lack of strong neighbourhood social differences in demand may also be a contributing factor. Fourth,

Changsha is the capital of a province heavily dependent upon the tobacco industry. It has no citywide smoke-free policy (see Chap. 9) and high rates of POS advertising (Yang et al. 2010). Thus, given the relatively strong relationship between the per capita provision of tobacco outlets and economic reliance on cigarette manufacturing at the provincial level, it is likely that tobacco retail facilities will be more ubiquitous in different types of neighbourhoods in such cities. By contrast, one might hypothesise that in cities with stronger tobacco control policies, neighbourhood differences in the location of tobacco outlets will be more marked.

Wang also considered store locations and outlet densities relative to schools in the city. Here local variations were more significant. Neighbourhoods close (within 100 m) to schools had significantly higher densities of tobacco retailers than those further away (100–200 m). Ninety-eight percent had at least one tobacco retailer within 100 metres. This was much higher than similar data collected in other Chinese cities such as Hangzhou (80.8%) (Gong 2013), Kunming (88%) (Frick et al. 2013) and even Denpasar (in Bali) (68.6%) in Indonesia (Astuti et al. 2019). These city differences are perhaps not surprising since Changsha is the capital of Hunan, the second most important cigarette manufacturing province after Yunnan (see Chap. 4).

With respect to schools, the high density of tobacco retailers found near school premises is likely to reflect a lack of enforcement. Although the TMB in Changsha in 2013 issued a regulation prohibiting tobacco sales within 100 m of the front entrance of schools, this was obviously ignored by the many tobacco retailers locating there (Wang (2017, p. 446). He also showed that most (60.4%) retail stores located within 100 m of schools did not have a tobacco licence, again showing a lack of compliance (Wang 2015, p. 51). This was also true with respect to compliance with the regulation requiring stores to display a ‘no sales to minors’ sign near POS counters. Most retailers (78%) did not, even if they were located close (within 100 m) to schools. As in Hangzhou (Gong 2013), many of these outlets were small kiosks run individually by their owners, who are unconcerned with the possibility of any threats of policy enforcement.

4.4.4 Point of Sale Advertising

As in western countries, POS advertising has become more significant in China as traditional advertising channels have become increasingly restricted. As a factor in consumption, POS advertising has traditionally been an important source of information about tobacco products (Li and Yong 2009). In the 2006 ITC Survey of regional differences in smokers and non-smokers awareness of tobacco advertising, POS advertising was recalled as the third most important source of tobacco advertising (by 29.2% respondents) after TV (34.5%) and billboards (33.4%). Advertising near street vendors was ranked fifth (20.3%) after posters (20.7%) (Yang, Li et al. 2010). However, despite its importance, POS advertising receives little attention in a recent book on tobacco control in China (Yang 2018). Most of the studies of POS advertising which have been undertaken were conducted before the amended *Advertising Law* came into effect in April 2015 (e.g. Zheng et al. 2012), so it is difficult to ascertain the likely effects of this change on advertising practices. Article 22 of the new law prohibits 'publishing tobacco advertising on mass media, in public places, on public means of transportation, or outdoors' (Zheng 2018, p. 269). It also bans any forms of advertisement to minors and prohibits publicising characteristics of tobacco products (e.g. name, packaging, etc.) through advertisements on any other goods, services or public service announcements. However, despite the law change, the Chinese Association on Tobacco Control, in a study of retailers in eight cities, has reported a rising number of cigarette advertisements and other forms of promotional behaviour since 2016 and that most (58%) retailers had failed to post notices against selling tobacco to minors (persons aged <18) (Fanxi 2019) (see Chap. 8).

Gong et al. (2013) reported that in Hangzhou most tobacco advertising occurred in specialist tobacco outlets which are more likely to be company owned or certified. This stands in contrast to North America, for example, where most POS advertising has occurred in convenience stores and especially to other countries, such as in Australia and New Zealand, where POS advertising has now ceased. In Changsha, Wang (2015) noted that outdoor advertising was more common than that carried out indoors. Approximately a quarter of all stores had at least one outdoor advertisement and that such advertisements were more prevalent in neighbourhoods than near schools. However, such signs were rated as being more obvious (as opposed to discreet or moderate) around schools. With respect to tobacco product displays, almost half (44%) of the stores displayed products in a way that they were visible from the outside. Marketing strategies targeting children were pervasive. For example, almost 83% stores displayed cigarettes within one metre of the floor and almost 60% located their displays adjacent to toys, candy and gum. These strategies were most common in stores close to schools compared to those in neighbourhoods. Although no analyses were conducted of advertising strategies by neighbourhood SES, Wang (2015) reported that no significant neighbourhood SES differences occurred in the display of tobacco licences and 'no sales to minors' signs. Most stores displayed neither; tobacco licences were visible in 43% stores and 'no sales to minors' signs were present in only 22% stores. Surprisingly, there were no significant differences, except for large cigars, in the types of tobacco products sold in different types of neighbourhoods.

4.4.5 Point of Sale Advertising and Consumption

Although many studies have examined the impact of advertising on consumption in China, as Cheung et al. (2018) have noted, few have examined never smokers' exposure to POS tobacco displays and their effects. In a study conducted in Hong Kong in 2015–2016, Cheung found that, among never smokers, younger groups, in particular those aged 15–29, were more likely to have noticed POS displays over the last 30 days. Those who often noticed displays reported that they encouraged them to smoke compared to those who sometimes noticed or did not notice. Respondents who often noticed displays showed less support for banning them than other non-smokers. Although they did not explore what attracted never smokers to the displays, as Zheng et al. (2012) found with Zhonghua tobacco advertisements in Shanghai, the cultural content of displays was important. In 2013 the Shanghai Tobacco Group marketed a new brand, *Zhonghua 5000*, to celebrate 5000 years of Chinese cultural heritage. In this way the company downplayed 'dangerous smoking behaviour' by associating the brand with patriotism and passion for China.

Although most research in China has not considered the effects of POS location and advertising on smoking, a considerable body of literature has nevertheless evolved which links advertising to smoking outcomes. Most of this literature uses crude recall measures which attempt to assess an individual's exposure to advertising over a given period, often the last 30 days. When measured in this way, higher exposure to advertising has been related to youth susceptibility to smoking (Xiao et al. 2015; Zhu et al. 2019a, b), smoking prevalence (Yang et al. 2012), difficulty of quitting (Yang et al. 2019) and exposure to second-hand smoke (Yang et al. 2015b). For example, in a study of urban male smokers in six cities, Yang et al. (2019) found that among adult males that exposure to tobacco advertising resulted in up to 60% fewer quit attempts and even minor levels of advertising resulted in much less success in efforts to quit. While exposure to anti-smoking information had the reverse effect, resulting in more quit attempts, these did not translate into smokers becoming smoke-free.

4.5 Social Media Marketing Strategies

As advertising restrictions are gradually imposed on traditional forms of communication, new media have emerged as hosts for tobacco advertising in China. This is not surprising as such trends have already occurred in western countries with stronger tobacco control policies, with TTCs increasingly resorting to social media to oppose policy, shape their public identity (Liang et al. 2015; Watts et al. 2019) and recruit a new generation of smokers (Kozinets 2019).

In China the potential for tobacco marketing is enormous. In 2006 only 3.8% of Chinese smokers were aware of internet advertising. However, this figure was much higher (16.2%) among younger smokers (those aged 18–29) but modest among other age groups (Li et al. 2009). As of 2019 there are over one billion users of the internet in China. By far the most important social media platform is *WeChat*, an all-in-one messaging app far more powerful than *Facebook*. *Weibo*, the Chinese equivalent of Twitter, is also significant with approximately 446 million monthly users (Dragon Social Net 2019).

Even before the advent of tighter restrictions on tobacco advertising, the Chinese tobacco industry quickly adapted its advertising strategies to embrace social media (Wang et al. 2020). For *WeChat* a number of companies developed accounts featuring their brand name and using their product packaging and slogans as account logos (Wang et al. 2014a). *WeChat* users have the option of joining an account, and once they have done so, they receive regular company briefings regarding promotions and new products. The tobacco industry also uses social media to further promote its products through targeting retailers and trade agencies who are able to submit purchase orders through the sites. Most companies also have active *Weibo* accounts. Wang et al. (2014b) note that these have proved useful in building relationships with users and for providing detailed information about products whether through text, images or video files.

To date there have been few studies of the impact of internet advertising in China and how new forms of social media advertising are affecting the smoking habits of particular groups. One of the few studies is by Wang et al. (2014b) who examined tobacco industry promotional activity on *Weibo*. Overall 33 Chinese tobacco companies owned *Weibo* accounts. These were ranked according to the number of 'fans'. From this total seven tobacco industry *Weibo* accounts, each having more than 5000 fans in 2012, were selected in order to examine the content of the *Weibos* established by the companies or their advertising agents. The postings were further classified under covert and overt marketing. The former included themes such as popular culture and linked tobacco use to content (e.g. clothes, ornaments, pop stars, cosmetics, entertainment) that was appealing to youth. Traditional culture was also considered and included statements about philosophies of life and used attractive words such as 'wonderful', 'life' and 'beautiful' to make the content of the *Weibos* more appealing. Posts under popular culture accounted for almost half (49%) of all posts, with traditional culture accounting for 11%. Overt marketing also featured and considered themes more directly related to tobacco consumption, including new product advertisements, public relations and tobacco culture and how tobacco is associated with history, literature and art. These three themes accounted for 35% total posts.

Tobacco company social media postings, such as *Weibo*, are significant for five main reasons. First, as a marketing strategy, they particularly aim to cultivate young smokers and this was evident in youth-friendly content and associations with youth culture. Second, they are an important venue for marketing electronic cigarettes which, by promoting a 'healthy' alternative, may encourage younger smokers to quit, although the evidence is far from clear (Yao et al. 2016; Wang et al. 2018; Zhu et al. 2019a, b). Third, social media postings often position smoking as a key element in Chinese culture. Tobacco brands, such as 'Mount Tai Club' named after important cultural sites, such as Mt. Tai in Shandong Province, homeland of the Chinese philosopher Confucius, featured strongly on the sites (Wang et al. 2014b). Fourth, the postings attempt to normalise tobacco use by promoting it along with the other necessities of everyday life. Finally, the use of such strategies exposes the weakness of existing tobacco control laws that do not consider advertising on social media sites. As Zheng et al. (2018) concluded, this is a grey area, one not considered in the most recent advertising bans.

However, since the amendments to the *Advertising Law* came into effect in 2015, changes have occurred to advertising on *WeChat*. Following the release of a report in April 2019 by the Beijing Center for Disease Control and Prevention, which detailed the extent of tobacco advertising on *WeChat*, officials of the internet provider shut down most public accounts involving tobacco, including those held by the State Tobacco Monopoly Administration, regardless of whether or not they violated relevant laws or regulations (Tobacco China Online 2019b). The outcome of this action is still unclear, however, as the article concluded:

> A large amount of embedded tobacco sales information marketing online still goes beyond the arm of the law under the cover of evaluation, tasting and sharing. Featured in a shopping story or appearing as a discussion of lifestyle, it is difficult to regulate and control this part of tobacco information.

The emergence of social media and its regulation is an important new research area. One aspect that has not been examined has been the links between social media use and smoking prevalence nor of the links between social media use, exposure to tobacco advertising through this source and smoking initiation. One of the few studies to address this issue is by Yang et al. (2020b) who examined the impact of the frequency of use of *WeChat* on different forms of health behaviour, including smoking. A cross-sectional multistage survey sampling process was conducted among 11,463 undergraduates from 31 universities in China. Over 90% students used the *WeCha*t app. Of these 19.3% were high-frequency users (used *WeChat* once an hour), and 31.2% used WeChat more than one hour each day. Both high frequency and time of use were related to higher rates of smoking, alcohol consumption and obesity-related eating behaviours. While such findings do not directly implicate online tobacco advertising, this is likely to have been a contributing factor in the heightened susceptibility to substance use and addictions among young people who engage in extensive social media use.

4.6 Conclusion

This chapter has explored recent changes in tobacco advertising and marketing in China. The Chinese cigarette industry has faced new challenges in the last few decades, including having to respond to an increasingly saturated domestic market and the emergence of stronger policies of tobacco control. As in the case of western TTCs, in addition to increased globalisation, the CNTC has responded to such trends by adopting new marketing strategies involving new products aimed at increasing consumption among younger people, women, as well as middle- and higher-income earners. In this sense China stands in marked contrast to western countries where TTCs have targeted low-income and minority communities (Yerger et al. 2007; Cruz et al. 2019). It is ironic that these same communities are now an important market for Chinese counterfeit cigarettes. In China the illicit market has responded to tightening controls and taken advantage of the demand for cheaper products by supplying more deprived communities in western countries where tobacco control tax policies have increased cigarette prices (von Lampe et al. 2014).

Internally, the odds of smoking in China are almost 50% greater in provinces more economically dependent on the tobacco industry, reflecting the persuasive power of this sector in maintaining high rates of consumption. Traditional studies of consumption in China have largely focused on economic determinants of demand and have not consistently explored the importance of contextual factors. Four issues were discussed; the role of price competition, variations in tobacco control, the impact of corporate social interventions which have helped shaped social norms and the role of local smoking cultures. Of these factors, the issue of corporate social interventions, in terms of the community activities of the CNTC, perhaps deserves the greatest consideration. It is evident that, despite the adverse health effects of smoking, the Chinese tobacco industry enjoys a high degree of legitimacy and many of its activities are a disguised form of advertising. Under the guise of corporate social responsibility, the industry has been remarkably effective in sponsoring a variety of community projects and, in so doing, of influencing attitudes towards the CNTC and smoking in general. In affecting society norms, the industry seeks to strengthen its market prospects by encouraging smoking initiation and reducing smoking cessation.

The issue of corporate social responsibility is just one of many marketing strategies that the industry has used to counter tobacco control. Other important changes are the increased focus of the industry on developing 'healthy' low-tar products, electronic cigarettes, and emphasising POS advertising and internet marketing as other avenues have been restricted. Unlike the situation in western smoking research, both POS and internet advertising have been relatively neglected in China. This, however, is to be expected. In western countries where restrictions on tobacco advertising occurred much earlier than in China, a considerable research effort developed to examine the impact of new advertising venues. However, in countries like Ireland, New Zealand or Australia increased regulation of POS advertising has constrained the activities of tobacco companies, and this will also be the case with social media.

In China it is not clear how the amended Advertising Law will eventually affect internet and point of sale displays. In Shanghai, for instance, despite the passing of the new law, POS sale continues unabated in various disguised promotional forms (Wei et al. 2017). The 2015 Advertising Law prohibits publishing tobacco advertisements in public places, including points of sale. However, because the law does not contain a definition of 'tobacco advertising and promotion', it is not clear whether product displays and cultural symbols and slogans on cigarette packaging, such as 'Loving my China' on *Zhonghua* brand cigarettes would be considered a form of advertising under the law. Therefore, the effect of the law on regulating POS advertising and its effects on consumption is uncertain. Similarly, the recent action of *WeChat* in China echoes campaigns elsewhere. For example, in the US Campaign for Tobacco-Free Kids is joining more than 125 public health and other organisations from 48 countries in calling on Facebook, Instagram, Twitter and Snapchat to immediately end the promotion of cigarettes, e-cigarettes and all other tobacco products on their platforms, including prohibiting the use of social media influencers (Campaign for Tobacco Free Kids 2019). At some point, perhaps sooner than expected, similar trends will happen in China.

References

Astell-Burt T, Zhang M, Feng X, Wang L, Li Y, Page A, Zhou M, Wang L (2017) Geographical inequality in tobacco control in China: multilevel evidence from 98,058 participants. Nicotine Tob Res 20(6):755–765

Astuti PAS, Mulyawan KH, Sebayang SK, Kurniasari NMD, Freeman B (2019) Cigarette retailer density around schools and neighbourhoods in Bali, Indonesia: a GIS mapping. Tob Induc Dis 17(July):55

Bell JF, Zimmerman FJ, Almgren GR, Huebner CE (2007) Associations between residential segregation and smoking during pregnancy among urban African American women. J Urban Health 84(3):372–388

Burton S, Hoek J, Nesbit P, Khan A (2015) "Smoking is bad, it's not cool … yet I'm still doing it": cues for tobacco consumption in a 'dark' market. J Bus Res 68(1):2067–2074

Cai L, Wu X, Goyal A, Han Y, Cui W, Xiao X, He J, Zhao K, Song Y, Jiao F (2012) Patterns and socioeconomic influences of tobacco exposure in tobacco cultivating rural areas of Yunnan Province, China. BMC Public Health 12(1):842

Cai L, Wu X, Goyal A, Han Y, Cui W, He J, Xiao X, Zhao K, Jiao F, Song Y (2013) Multilevel analysis of the determinants of smoking and second-hand smoke exposure in a tobacco-cultivating rural area of Southwest China. Tob Control 22(Suppl 2):ii16–ii20

Campaign for Tobacco-Free Kids (2012) The Chinese tobacco market and industry profile. Campaign for Tobacco-Free Kids, Washington DC. https://www.tobaccofreekids.org/assets/global/pdfs/en/TI_Profile_China.pdf

Campaign for Tobacco Free Kids (2019) Over 125 organizations call on social media companies to end all tobacco advertising, Including by Paid Influencers. Campaign for Tobacco-Free Kids, Washington DC. May 22. https://www.tobaccofreekids.org/press-releases/2019_05_21_socialmedia_advertising

Cantrell J, Ganz O, Anesetti-Rothermel A, Harrell P, Kreslake JM, Xiao H, Pearson JL, Vallone D, Kirchner TR (2015) Cigarette price variation around high schools: evidence from Washington DC. Health Place 31:193–198

Cantrell J, Pearson JL, Anesetti-Rothermel A, Xiao H, Kirchner TR, Vallone D (2016) Nicotine Tob Res 18(2):130–137

Caryl F, Shortt NK, Pearce J, Reid G, Mitchell R (2020) Socioeconomic inequalities in children's exposure to tobacco retailing based on individual-level GPS data in Scotland. Tob Control 29(4):367–373

Chaiton MO, Mecredy G, Cohen J (2018) Tobacco retail availability and risk of relapse among smokers who make a quit attempt: a population-based cohort study. Tob Control 27(2):163–169

Cheng K-W, Tsoh J-Y, Cui W, Li X, Kohrman M (2015) Smoking intensity among male factory workers in Kunming, China. Asia Pac J Public Health 27(2):606–615

Cheng TO (2014) Smoking in China: can or should China kick the habit? Int J Cardiol 175(2):219–221

Cheung YTD, Ho SY, Wang MP, Kwong A, Lai V, Lam TH (2018) Pro-smoking responses and attitudes due to pint of sale tobacco displays in never smokers: a cross-sectional study in Hong Kong. Tob Induc Dis 16(July):32

Christakis NA, Fowler JH (2008) The collective dynamics of smoking in a large social network. N Engl J Med 358:2249–2258

Chu A, Jiang N, Glantz SA (2011) Transnational tobacco industry promotion of the cigarette gifting custom in China. Tob Control 20:e3

Cruz TB, Rose SW, Lienemann BA, Byron MJ, Meissner HI, Baezconde-Garbanati L, Huang L-L, Carroll DM, Soto C, Unger JB (2019) Pro-tobacco marketing and anti-tobacco campaigns aimed at vulnerable populations: a review of the literature. Tob Induc Dis 17:68

Cummings KM, Hyland A, Lewit E (1997) Use of discount cigarettes by smokers in 20 communities in the United States, 1988–1993. Tob Control 6(Suppl 2):s25–s30

Ding D, Hovell MF (2012) Cigarettes, social reinforcement, and culture: a commentary on "Tobacco as a social currency: cigarette gifting and sharing in China". Nicotine Tob Res 14(3):255–257

Dragon Social Net (2019) 10 Most Popular Social Media Sites in China (2019 Updated). https://www.dragonsocial.net/blog/social-media-in-china/

Dwyer-Lindgren L, Mokdad AH, Srebotnjak T, Flaxman AD, Hansen GM, Murray CJL (2014) Cigarette smoking prevalence in US counties: 1996–2012. Popul Health Metr 12:Article No. 5

Euromonitor International (2012) Industry Overview. Retrieved from: https://www1.hkexnews.hk/listedco/listconews/sehk/2012/0629/01335_1381371/e113.pdf

Fang J (2019) China's tobacco industry is building schools and no one is watching. The Conversation, August 11 Retrieved from: https://theconversation.com/chinas-tobacco-industry-is-building-schools-and-no-one-is-watching-120961

Fang J, Lee K, Sejpal N (2017) The China National Tobacco Corporation: from domestic to global dragon. Glob Public Health 12(3):315–334

Fang J, Yang G, Wan X (2019) 'Pro-tobacco propaganda': a case study of tobacco industry-sponsored elementary schools in China. Tob Control 16(Suppl 1):A196

Fanxi T (2019) Despite Ban, Cigarette Ads on the Ride in China Sixth Tone, September 5. https://www.sixthtone.com/news/1004526/despite-ban%2C-cigarette-ads-on-the-rise-in-china

Finan LJ, Lipperman-Kreda S, Abadi M, Grube JW, Kaner E, Balassone A, Gaidus A (2019) Tobacco outlet density and adolescents' cigarette smoking: a meta analysis. Tob Control 28(1):27–33

FAO (Food and Agricultural Organization) (2003) Issues in the global tobacco economy: selected case studies. FAO, Rome

Frick M, Xiaodan W, Hongxia S, Hong C, Wenchun W, Weidong W, Xiaoliang L (2013) Assessing youth exposure to tobacco retailing near schools in Kunming, China using geographic information systems and social observation. Paper presented at the 141st Annual Meeting of the American Public Health Association, November 2–6, Boston, Ma. Downloaded from: https://apha.confex.com/apha/141am/webprogram/Paper277467.html

Gao S, Zheng R, Hu TW (2012) Can increases in the cigarette tax rate be linked to cigarette retail prices? Solving mysteries related to the cigarette pricing mechanism in China. Tob Control 21(6):560–562

Gao J (2013) What are the sales practices of internet cigar vendors in China? Tob Control 24(3):298–302

Ghenadenik AE, Frohlich KL, Gauvin L (2016) Beyond smoking prevalence: exploring the variability of associations between neighbourhood exposures across two nested spatial units and two-year smoking trajectory among young adults. Int J Environ Res Public Health 13(1):106

Gong T, Lv J, Liu Q, Ren Y, Li L, Kawachi I (2013) Audit of tobacco retail outlets in Hangzhou, China. Tob Control 22(4):245–249

Halonen JI, Kivimäki M, Kouvonen A, Pentti J, Kawachi I, Subramanian SV, Vahtera J (2014) Proximity to a tobacco store and smoking cessation: a cohort study. Tob Control 23(2):146–151

Higashi H, Khuong TA, Ngo AD, Hill PS (2011) The development of tobacco harm prevention law in Vietnam: stakeholder tensions over tobacco control legislation in a state owned industry. Subst Abuse Treat Prev Policy 6(1):24

Hoffer AJ, Pellilo A (2012) The political economy of tobacco control spending. Appl Econ Lett 19(18):1793–1797

Hoffman SJ, Mammone J, Rogers Van Katwyk S, Sritharan L, Tran M, Al-Khateeb S, Grjibovski A, Gunn E et al (2019) Cigarette consumption estimates for 71 countries from 1970 to 2015:

systematic collection of comparable data to facilitate quasi-experimental evaluations of national and global tobacco control interventions. Br Med J 365:I2231
Hu TW, Lee AH, Mao Z (2013) WHO framework convention on tobacco control in China: barriers, challenges and recommendations. Glob Health Promot 20(4):13–22
Hu TW, Mao Z, Shi J, Chen J (2010) The role of taxation in tobacco control and its potential economic impact in China. Tob Control 19(1):58–64
Huang J, Zheng R, Chaloupka FJ, Fong GT, Li Q, Jiang Y (2014) Chinese smokers' cigarette purchase behaviours, cigarette prices and consumption: findings from the ITC China survey. Tob Control 23(Suppl 1):ii67–ii72
Hwang J, Oh Y, Yang Y, Lee S, Cho S (2018) Tobacco company strategies for maintaining cigarette advertisements in retail chain stores: in-depth interviews with Korean convenience store owners. Tob Induc Dis 16(October):46
Hyun GS, Yan G, Kulbok PA (2017) South Korean adolescents' intention to smoke. Am J Health Behav 41(2):163–170
Isgor Z, Powell L, Rimkus L, Chaloupka F (2016) Associations between retail food store exterior advertisements and community demographic and socioeconomic composition. Health Place 39(May):43–50
Jin J (2014) Why FCTC policies haven't been transferred in China: domestic dynamics and tobacco governance. J Health Polit Policy Law 39(3):633–666
Jiang N, Ho SY, Lam TH (2017) Electronic cigarette marketing tactics in mainland China. Tob Control 26(2):230–232
Kagaruki LK (2010) Community-based advocacy opportunities for tobacco control: experience from Tanzania. Glob Health Promot 17(Suppl 2):41–44
Kirchner TR, Cantrell J, Anesetti-Rothermel A, Ganz O, Vallone DM, Abrams DB (2013) Geospatial exposure to point-of-sale tobacco. Am J Prev Med 45(4). https://doi.org/10.1016/j.amepre.2013.05.016
Kozinets (2019) How social media is helping Big Tobacco hook a new generation of smokers. The Conversation, March 27. https://theconversation.com/how-social-media-is-helping-big-tobacco-hook-a-new-generation-of-smokers-112911
Lee JGL, Sun DL, Schleicher NM, Ribisl KM, Luke DA, Henriksen L (2017) Inequalities in tobacco outlet density by race, ethnicity and socioeconomic status, 2012, USA: results from the ASPiRE study. J Epidemiol Community Health 71(5):487–492
Lee K, Eckhardt J (2017) The globalisation strategies of five Asian tobacco companies: a comparative analysis and implications for global health governance. Global Pubic Health 12(3):367–379
LeParkour (2015) Zhongnanhai Parkour Rally, June 30, 2015. http://www.leparkour.cn/content/?707.html
Li C (2012) The political mapping of China's tobacco industry and anti-smoking campaign. In: Thornton JL (ed) China Centre Monograph Series Number 5. Brookings Institution, Washington, DC
Li L, Yong HH, Borland R, Fong GT, Thompson ME, Jiang Y, Yang B, Sirirassamee B, Hastings G, Harris F (2009) Reported awareness of tobacco advertising and promotion in China compared to Thailand, Australia and the USA. Tob Control 18(3):222–227
Li L, Yong HH (2009) Tobacco advertising on the street in Kunming, China. Tob Control 18(1):63
Li Q, Hyland A, Fong GT, Jiang Y, Elton-Marshall T (2010) Use of less expensive cigarettes in six cities in China: findings from the International Tobacco Control (ITC) China survey. Tob Control 19(Suppl 2):i63–i68
Liang Y, Zheng X, Zeng DD, Zhou X, Leischow SJ, Chung W (2015) Exploring how the tobacco industry presents and promotes itself in social media. J Med Internet Res 17(1):e24
Liao A (2015a). China's slim cigarette market fattens up. Tobacco Asia, July 1. https://www.tobaccoasia.com/features/china%E2%80%99s-slim-cigarette-segment-fattens-up/
Liao A (2015b) China's STMA and the tobacco monopoly. Tobacco Asia, July 1. https://www.tobaccoasia.com/features/china%E2%80%99s-stma-and-the-tobacco-monopoly/

Lipperman-Kreda S, Morrison C, Grube JW, Gaidus A (2015a) Youth activity spaces and daily exposure to tobacco outlets. Health Place 34(July):30–33

Lipperman-Kreda S, Grube JW, Friend KB, Mair CB (2015b) Tobacco outlet density, retailer cigarette sales without ID checks and enforcement of underage tobacco laws: associations with youths' cigarette smoking and beliefs. Addiction 111(3):525–532

Mbulo L, Kruger J, Hsia J, Yin S, Salandy S, Orlan EN, Agaku I, Ribisl KM (2019) Cigarettes point of purchase patterns in 19 low-income and middle-income countries: Global Adult Tobacco Survey, 2008–2012. Tob Control 28(1):117–120

McNeice A (2018) British tobacco company looks to China for smoke-free future. China Daily 22 May. https://www.chinadaily.com.cn/a/201805/22/WS5b02f6bba3103f6866ee9b47.html

Mennis J, Mason M (2016) Tobacco outlet density and attitudes towards smoking among urban adolescent smokers. Subst Abus 37(4):521–525

Mistry R, Pednekar M, Pimple S, Gupta S, McCarthy WJ, Raute LJ, Patel M, Shastri SS (2015) Banning tobacco sales and advertisements near educational institutions may reduce students' tobacco use risk: evidence from Mumbai, India. Tob Control 24(e1):e100–e107

Morley CP, Pratte MA (2013) State-level tobacco control and adult smoking rate in the United States: an ecological analysis of structural factors. J Public Health Manag Pract 19(6):e20–e27

Novak SP, Reardon SF, Raudenbush SW, Buka SL (2006) Retail tobacco outlet density and youth cigarette smoking: a propensity modelling approach. Am J Public Health 96(4):670–676

Nuyts PAW, Davies LEM, Kunst AE, Kuipers MAG (2019). The association between tobacco outlet density and smoking among young people: a systematic methodological review. Nicotine & Tobacco Research (August). https://doi.org/10.1093/ntr/ntz153

Pan Z, Hu D (2008) Multilevel analysis of individual and community predictors of smoking prevalence and frequency in China: 1991–2004. J Public Health Policy 29(1):72–85

Pearce J, Hiscock R, Moon G, Barnett R (2009) The neighbourhood effects of geographical access to tobacco retailers on individual smoking behaviour. J Epidemiol Community Health 63(1):69–77

Pearce J, Rind E, Shortt N, Tisch C, Mitchell R (2016) Tobacco retail environments and social inequalities in individual-level smoking and cessation among Scottish adults. Nicotine Tob Res 18(2):138–146

Pulakka A, Halonen JI, Kawachi I, Pentti J, Stenholm S, Jokela M, Kaate I, Koskenvuo M, Vahtera J, Kivimäki M (2016) Association between distance from home to tobacco outlet and smoking cessation and relapse. JAMA Intern Med 176(10):1512–1519

Reitzel LR, Cromley EK, Li Y, Cao Y, Mater RD, Mazas CA, Cofta-Woerpel L, Cincirpini PM, Wetter DW (2011) The effect of tobacco outlet density and proximity on smoking cessation. Am J Public Health 101(2):315–320

Riches E, Whitehead R, Rennick L, Armour G, McAteer J, Laird Y, Reid G (2018) Rapid evidence review: what is the causal link between tobacco outlet density and smoking prevalence. NHS Health Scotland, Edinburgh, p 21pp

Ross NA, Taylor SM (1998) Geographical variation in attitudes towards smoking: findings from the COMMIT communities. Soc Sci Med 46(6):703–717

Schleicher NC, Johnson TO, Fortmann SP, Henriksen L (2016) Tobacco outlet density near home and school: associations with smoking and norms among US teens. Prev Med 91(October):287–293

Shortt NK, Tisch C, Pearce J, Mitchell R, Richardson EA, Hill S, Collin J (2015) A cross-sectional analysis of the relationship between tobacco and alcohol outlet density and neighbourhood deprivation. BMC Public Health 15:Article No. 1014

Shortt NK, Tisch C, Pearce J, Richardson EA, Mitchell R (2016) The density of tobacco retailers in home and school environments and relationship to adolescent smoking behaviours in Scotland. Tob Control 25(1):75–82

Siahpush M, Farazi PA, Kim J, Michaud TL, Yoder AM, Soliman G, Tibbits MK, Nguyen MN, Shaikh RA (2016) Social disparities in exposure to point of sale marketing. Int J Environ Res Public Health 13(12):1263

Tobacco China Online (2016) Profile: China tobacco Jiangxi industrial striving to be more competitive. Tobacco Asia, September 7. https://www.tobaccoasia.com/features/profile-china-tobacco-jiangxi-industrial-striving-to-be-more/
Tobacco China Online (2019a) The segmentation of China's tobacco market. Tobacco Asia, March 29. https://www.tobaccoasia.com/features/the-segmentation-of-china-tobacco-market/
Tobacco China Online (2019b) Control over online tobacco advertising in China gets stricter. Tobacco Asia July 22. https://www.tobaccoasia.com/features/control-over-online-tobacco-advertising-in-china/
Tobacco China Online (2020) Assisting poor people to create a happy new life. Tobacco China January 31. http://www.tobaccochina.cc/shangye/20201/202012394210_797157.shtml
von Lampe K, Kurti M, Bae J (2014) Land of opportunities. The illicit trade in cigarettes in the United States. In: van Duyne P, Harvey J, Antonopoulos GA, von Lampe K, Maljević A, Markovska A (eds) Corruption, Greed and Crime-money. Sleaze and Shady Economy in Europe and Beyond. Wolf Legal Publishers, Oisterwijk, Netherlands, pp 267–290
Wang F, Zheng P, Freeman B, Chapman S (2014a) Chinese tobacco companies' social media marketing strategies. Tob Control 24(4):408–409
Wang F, Zheng P, Yang D, Freeman B, Fu H, Chapman S (2014b) Chinese tobacco industry promotional activity on the microblog Weibo. PLoS One 9(6):e99336
Wang L (2015) Retail environment features that affect smoking behaviour in Changsha, China. PhD thesis, Public Health, Ohio State University, Columbus. https://etd.ohiolink.edu/!etd.send_file?accession=osu1437651533&disposition=inline
Wang L, Lu B, Wewers ME, Foraker RE, Xe M, Ferketich AK (2017) Are retailers compliant with zoning regulations than ban tobacco sales near schools in Changsha, China? Tob Control 26(4):446–451
Wang W, He Z, Feng N, Cai Y (2019) Electronic cigarette use in China: awareness, prevalence and regulation. Tob Induc Dis 17(April):30
Wang X, Xiong Y, CZhao W (2020) Tobacco control challenges in China: big data analysis of online tobacco marketing information. Int J Nurs Sci. https://doi.org/10.1016/j.ijnss.2020.07.002
Wang X, Zhang X, Xu X, Gao Y (2018) Electronic cigarette use and smoking cessation behavior among adolescents in China. Addict Behav 82(July):129–134
Warner KE (2000) The economics of tobacco; myths and realities. Tob Control 9(1):78–89
Watts C, Hefler M, Freeman B (2019) 'We have a rich heritage and, we believe, a bright future': how transnational tobacco companies are using Twitter to oppose policy and shape their public identity. Tob Control 28(2):227–232
Wei X, Ming L, Cai Y, Gao J (2017) Effects of strict control of tobacco advertisement after implementation of the advertising law. J Shanghai Jiaotong Univer (Med Sci) 37(2):137–140. (In Chinese)
Wenshan News Network (2020) The first most influential tobacco area in Yunnan. Wsnews, 17 January. http://wsnews.com.cn/zsws_wstt/p/218509.html
White JS, Li J, Fong GT, Jiang Y (2013) The effect of cigarette prices on brand-switching in China: a longitudinal analysis of data from the ITC China survey. Tob Control 23(Suppl 1):i54–i60
Xiao L, Yang J, Zhao L, Jiang Y, Chen X (2015) Exposure to tobacco advertisement and promotion programs among Chinese middle school students: a cross-sectional survey. Zhonghua Liuxingbingxue Zazhi 36(4):309–313
Xu SS, Gravely S, Meng G, Elton-Marshall T, O'Connor RJ, Quah ACK, Feng G, Uiang Y, Fong GT (2019a) Impact of China National Tobacco Company's 'Premiumization' strategy: longitudinal findings from the ITC China surveys (2006–2015). Tob Control 28(Suppl 1):s68–s76
Xu X, Zhang X, Hu TW, Miller LS, Xu M (2019b) Effects of global and domestic tobacco control policies on cigarette consumption per capita: an evaluation using monthly data in China. BMJ Open 9:e025092
Yang G (2014) Marketing 'less harmful, low tar' cigarettes is a key strategy of the industry to counter tobacco control in China. Tob Control 23(2):167–172
Yang G (ed) (2018) Tobacco control in China. Springer, Singapore, 323p

Yang T, Barnett R, Rockett IRH, Yang YY, Wu D, Zheng W, Li L (2015) The impact of regional economic reliance on the tobacco industry on current smoking in China. Health Place 33(May):159–171

Yang T, Barnett R, Peng S (2020b) Prevalence and behavioral associations of excessive *WeChat* use: a representative nationwide study in China. (Unpublished paper)

Yang T et al. (2020a) Anti-tobacco information exposure, environmental smoking restrictions and Price-induced smoking behaviour reduction (PSBR) among urban males in China (Unpublished paper)

Yang T, Jiang S, Oliffe JL, Feng X, Zheng J (2015a) Environmental smoking restrictions and light cigarette adoption among Chinese urban smokers. Prev Sci 16(6):801–810

Yang T, Jiang S, Barnett R, Peng S, Yu L (2015b) Individual and city-level determinants of second-hand smoke exposure in China. Int J Health Geogr 14:36

Yang T, Rockett IRH, Li M, Xu X, Gu Y (2012) Tobacco advertising, environmental smoking bans, and smoking in Chinese urban areas. Drug Alcohol Depend 124(1–2):121–127

Yang T, Xu X, Rockett IRH, Guo W, Zhou H (2011) Effects of household, workplace, and public place smoking restrictions on smoking cessation. Health Place 17(4):956–960

Yang T, Zhu Z, Barnett R, Zhang W, Peng S, Jiang S (2019) Tobacco advertising, anti-tobacco information exposure, environmental smoking restrictions, and unassisted smoking cessation among Chinese male smokers: a population-based study. Am J Mens Health 13(3):1557988319856152

Yang Y, Li L, Yong H-H, Borland R, Wu X, Li Q, Wu C, Foong K (2010) Regional differences in awareness of tobacco advertising and promotion in China: findings from the ITC China survey. Tob Control 19(2):117–124

Yao T, Huang J, Sung H-Y, Ong MK, Mao Z, Jiang Y, Fong GT, Max W (2014) Who purchases cigarettes from cheaper sources in China? Findings from the ITC China survey. Tob Control 23(Suppl 1):i97–i101

Yao T, Jiang N, Grana R, Ling PM, Glantz SA (2016) A content analysis of electronic cigarette manufacturer websites in China. Tob Control 25(2):188–194

Yerger VB, Przewoznik J, Malone RE (2007) Racialized geography, corporate activity, and health disparities: tobacco industry targeting of inner cities. J Health Care Poor Underserv 18(4 Suppl):10–38

Yorifuji T, Tanihara S, Takao S, Kawachi I (2011) Regional disparities in compliance with tobacco control policy in Japan: an ecological analysis. Tob Control 20(5):374–379

Zhang X, Xu X, Xu M, Hu T (2019) The impact of tobacco taxation policy on slim cigarette use and the growing popularity of slim cigarettes in China since 2014. Health 11(06):711–720

Zheng P, Ge X, Qian H, Wang F, Fu H, Berg CJ, Kegler MC (2012) 'Zhonghua' tobacco advertisement in Shanghai: a descriptive study. Tob Control 23(5):389–394

Zheng P, Xiao L, Wang F, Yang G (2018) Comprehensive ban on tobacco advertising, promotion and sponsorship. In: Yang G (ed) Tobacco control in China. Springer, Singapore, pp 245–284

Zheng R (2018) Tobacco tax and tobacco economy in China. In: Yang G (ed) Tobacco control in China. Springer, Singapore, pp 285–309

Zhou H (2000) Fiscal decentralization and the development of the tobacco industry in China. China Econ Rev 11(2):114–133

Zhou H (2001) Implications of interjurisdictional competition in transition: the case of the Chinese tobacco industry. J Comp Econ 29(1):158–182

Zhu J, Li J, He Y, Li N, Xu G, Yu J (2019a) The influence and interaction of exposure to pro-smoking and anti-smoking messaging on youth smoking behaviour and susceptibility. Tob Induc Dis 17:86

Zhu J, Shi F, Xu G, Li N, Li J, He Y, Yu J (2019b) Conventional cigarette and e-smoking among school personnel in Shanghai, China: prevalence and determinants. Int J Environ Res Public Health 16(17):3197

Chapter 5
Geographical Context and Cultural Practices Affecting Smoking

Xiazhao Y. Yang, Ross Barnett, and Tingzhong Yang

Abstract This chapter examines how cultural practices have influenced smoking in China with specific reference to regional and rural contexts. Particular attention is paid to cultural norms and practices and regional cultural differences. Consideration is also given as to why smoking is higher in the countryside and the importance of cultural and socio-economic factors in this context. Finally, given the rapid urbanisation of China, an assessment is made of the impacts of rural-urban migration and the discriminatory *hukou* system on the smoking behaviour of rural residents in new urban destinations.

Keywords Confucianism · Cigarette gifting and sharing · Cultural norms · Hukou · Patriarchy · Patriotism · Rural smoking · Rural-urban migration · Self-exempting beliefs · Structure-culture

5.1 Introduction

Beyond individual health hazards, the public health and economic toll from tobacco consumption have led scholars and policy makers to regard tobacco smoking as a multilevel health emergency. The course of development and impacts of this are rooted in structural factors beyond individual health behaviours and psychological

X. Y. Yang (✉)
School of Sociology and Anthropology, Sun Yat-sen University, Guangzhou, China
e-mail: yangxzh27@mail.sysu.edu.cn

R. Barnett
School of Earth and Environment, University of Canterbury, Christchurch, New Zealand
e-mail: ross.barnett@canterbury.ac.nz

T. Yang
School of Medicine, Zhejiang University, Hangzhou, China
e-mail: tingzhongyang@zju.edu.cn

R. Barnett et al. (eds.), *Smoking Environments in China*,
Global Perspectives on Health Geography,
https://doi.org/10.1007/978-3-030-76143-1_5

processes. Based on research into the socioecological determinants of health, the mechanisms of substance use behaviour lie at three distinct levels: the macro-level factors as manifested by GDP, urbanisation and population growth, which reflect the historical and social context of tobacco smoking; meso-level factors that operate in intermediate settings and connect individual persons to the broader macro-level structures of society, such as local public resources or the siting of tobacco outlets in the community; and, finally, individual-level social, demographic and psychological characteristics. Apart from this last level, the other two major aetiological platforms of substance use reflect structural forces that may operate without individuals' conscious realisation.

This chapter examines how cultural practices have influenced smoking in China with specific reference to regional and rural contexts. It can be considered a preliminary to Chap. 6 which follows on by examining the impact of urbanisation on smoking. With this in mind, the chapter is organised into three main sections. First, we focus on the cultural context of smoking in China paying particular attention to cultural norms and practices and regional cultural differences (Sect. 5.2). Second, we focus on rural areas and consider why smoking is higher in the countryside and the importance of cultural and socio-economic factors in this context (Sect. 5.3). Third, in Sect. 5.4, we examine the process of rural-urban migration and the impact of the discriminatory *hukou* system on the smoking behaviour of rural residents in new urban destinations. This section provides a link to Chap. 6 which specifically focuses on the impacts of urbanisation on smoking, the processes involved and how these are altering its social distribution.

5.2 Smoking and the Chinese Cultural Context

Cultural perspectives are important within contemporary health geography, and provide another lens on how people interpret the world and respond to social practices, including smoking and other forms of consumption. Despite debate over what constitutes culture, in a general sense, geographers are likely to understand the term as representing a whole litany of ideas and behavioural traits that include cultural beliefs, value systems, lifestyle practices and even material artefacts (Pow 2017). Culture can also be seen as a manifestation of the power relations between various groups, and the structure through which social change is constrained and enabled. This is now recognised in western contexts where tobacco control efforts are increasingly adapted away from 'one size fits all' to encompass the specific needs and desires of different cultural groups.

Engaging with cultural concerns is particularly important. Of course the more general question can be raised regarding the extent to which cultural factors influencing smoking are unique to China or are present in the wider Asian region or LMICs in general. Certainly differences in sex role norms and general expectations concerning gender-appropriate behaviour have resulted in marked gender differences in smoking in China as well as in other South and East Asian countries (see

Chap. 2). But other social practices, such as cigarette gifting, appear to be more an outcome of the Chinese cultural context and reflect specific political and institutional factors. The following section, therefore, discusses a variety of cultural practices in China that encourage the persistence of smoking and locates them within the national and wider global setting.

5.2.1 Smoking as an Extension of Confucian Culture

The culture of tobacco smoking can be traced from historical to contemporary Chinese society. The infusion of smoking as a gesture of hierarchical deference into Chinese culture is intimately related to the Confucian idea of 'knowing one's place' that is evident also in much of the rest of Asia (Kim et al. 2005). Confucian tradition still provides the standard mode of thinking and way of life in China. The writings of Confucius (558–479 BC), who lived during the 'Spring and Autumn Period', considered to be a glorious time in Chinese philosophy (Park and Chesla 2007), can be seen as a code of ethics that prescribes modes of behaviour in families and communities. As Confucius wrote at a time of conflict and social disintegration in China, Confucianism, therefore, is often seen as a set of values promoting greater peace and trust and higher levels of civilisation.

Within Confucianism a set of basic virtues were important: *ren* (altruism, humanity), *yi* (integrity and uprightness), *li* (etiquette and propriety), *zhi* (wisdom) and *xin* (trust) (Park and Chesla 2007). Embedded within these values are five basic relationships: between father and son, between husband and wife, between sovereign and subject, between the old and the young and between friends. Through these relationships Confucianism provided a simple guide for ordering family and society.

However, the values that guide Confucianism have been criticised on many counts, particularly for their impacts both on personal and state behaviours. The virtue of *li*, for example, has been seen as largely a hierarchical concept in which individuals should know their social position and act accordingly. This has had ramifications for the growth of authoritarian values in Chinese culture both within the family and the state. The patriarchal nature of gender roles and between husband and wife and within the extended family has had behavioural implications which persist today. Similarly Confucius' belief in the importance of ritual in orienting people, so that social institutions and everyday relationships could flourish, similarly institutionalised different forms of behaviour. Important here has been cigarette gifting and sharing. This is also reflected in the concept of *xin*, where respect and the development of proper social etiquette are important components of social relationships.

5.2.2 Adoption of Smoking and Cultural Resistance

While one can see Confucian values underpinning some of the cultural forces promoting contemporary smoking in China, they also shaped the early social distribution of cigarette smoking after its rapid adoption following the Treaty of Tianjin in 1858 (Hermalin and Lowry 2012). The forces of modernisation and colonial control weakened the traditional Confucian values that had guided China for so long, particularly the role of women in Chinese society and their freedom to smoke. However, it could be argued that, at the same time, they also encouraged the development of a modern social etiquette, so that by the time of the Qing dynasty in the eighteenth century, tobacco had become incorporated into the higher echelons of Chinese society in culturally specific ways (Du 2000).

Among the elite, tobacco was regarded by merchants, artists and scholars as a substance that could express hospitality, concentrate thoughts, enhance sexual pleasure and maintain or restore health. High-end tobacco, grown in particular regions, and marketed by specialised tobacco retailers, was distinguished by particular brands, such as *Yongding*, in the eighteenth century. This became popular among the elite after being declared as a 'tobacco fit for imperial consumption' (Benedict 2011, p. 58). The infusion of tobacco first among the Chinese elite thus reflected a common theme in the diffusion model (See Chap. 2) of promoting smoking as not only a desirable social attribute but also one that enhanced health and other personal qualities. Zhang (2005) has employed Bourdieu's notion of distinction to show how the status of opium, as an exemplary symbol of the gentry class, was vigorously defended by the social group that possessed it in late Imperial China. Dikötter et al. (2002) also argued that, besides its assumed medical benefit, the quick and widespread acceptance of tobacco in China was closely associated with the phenomenological meaning of smoke (air, or *qi*) found in Chinese folk religion's evil-dispelling ceremony. This mythical form of belief persists, and culminated when, in the 2003 SARS outbreak, people circulated the message that smoking cigarettes and burning incense could protect individuals from SARS (Tai and Sun 2011). This may have also have been true of reactions to COVID-19. Early reports from China on the clinical characteristics of COVID-19 patients admitted to hospital found that the proportion of smokers was less than expected based on the estimated national prevalence of smoking (Huang et al. 2020).

Although Confucian values of social harmony and etiquette may have enhanced tobacco consumption among the elite, this was not the case with gender distinctions in smoking. After 1800, although women smoked, they were encouraged to do so in the privacy of their own homes and to avoid strong tobacco smoke by using longer water pipes (Brook 2004). By the late 1920s and early 1930s, the desire to reassert Confucian values, coupled with a resentment against foreign domination and western values, resulted in the beginnings of countervailing forces especially with respect to cigarette use by women. As Hermalin and Lowry (2012) have indicated, the forces of modernisation were not unidirectional and, for women in particular, traditional Confucian values reasserted themselves in affecting the subsequent diffusion of smoking in Chinese society (see Chaps. 2 and 3).

5.2.3 Cultural Processes in Contemporary China

Despite rapid economic and political change in China since the 1930s, cultural factors, many associated with traditional Confucian values, remain important. Thus, an understanding of these, in addition to the socio-economic processes discussed in previous chapters, is important in helping explain the rise, and continued high prevalence, of smoking in China. It is to a more detailed analysis of some of the different cultural patterns that we now turn.

5.2.3.1 The Role of Gender and Patriarchy

Confucianism, which values family harmony and considers a patriarchal family order as the basis of harmonious gender and intergenerational relations, has traditionally dominated Chinese family life (Mao 2013). Under such arrangements women were designated as socially inferior and subject to their husband's wishes, but also pressured to prioritise their parents' needs above their own. Mao (2013) showed that young mothers generally had limited control over their household space and limited ability to influence the smoking practices of their husbands and extended family. The consequences were increased exposure to second-hand smoke both for mothers and their children (see Chap. 9). Only in some cases have women, as bearers and carers of children, been able to use this status to subvert the pre-eminence of men in domestic environments (Mao and Robinson 2015). Confucian values have not only limited women's power within traditional households but also affected their own propensity to smoke. As we showed in Chap. 2, during the early twentieth century, Confucian cultural traditions came into conflict with emerging women's identities, restricting women's ability to smoke. That a similar pattern of smoking is evident in Japan and Korea, two countries with strong cultural affinity to China, is used to buttress the argument (Hermalin and Lowry 2012).

The strength of patriarchal family relationships, however, varies internally within China according to different regional and urban-rural contexts (for a discussion of the latter, see Sect. 5.3.2.1). From a regional perspective, due to a long history of ethnic and racial amalgamation between local residents and various nomadic groups, the northern Han culture boasts a machismo ethos that highly values loyalty to friends and leaders, solidarity among unrelated 'brothers', bravery and, ultimately, risk-taking behaviours. As documented in the literature, this hyper-masculinity culture can facilitate a wide range of deviant behaviours, including substance use (Zheng 2006). Southern Chinese do smoke, but, lacking the machismo motifs found in the northern culture at the same level of smoking, southerners may face less pressure to smoke from peers who are less embedded in patriarchy.

Therefore, the intended sense behind smoking as a social action, the cultural motivation masked by the expressed discourse, may be entirely different between smokers from the two broad regional cultures. Certainly, to say that southern culture is qualitatively different from the north risks the danger of overgeneralisation and ethnocentrism, and there are socio-organisational causes beyond the cultural ones why southerners may smoke less when controlling for socio-economic influences. For example, the north tends to have smaller units of blood-related kin organisations compared to the south. Under the current collective ownership with pieces of the land contracted to individuals (Household Contract Responsibility System), the farms in the north scatter into small units due to the smaller family sizes. Comparatively, the southern villages are often single-named, inhabited by people with the same surname except for a few immigrant households. Thus, social relationships among northerners tend to rely on unrelated non-kin peers, while for southerners' familial hierarchical embeddedness tends to be stronger. As a result, smoking behaviour in the north is more likely to be a more important part of non-kin peer networks.

5.2.3.2 Sharing and Gifting Cigarettes

As a cultural practice, the gifting and sharing cigarettes is a well-accepted practice in China where it is rooted in Confucian expressions of courtesy and respect (Chu et al. 2011; Peng and Lingling 2017). Rarely seen in other cultures, and uncommon in western countries, sharing is a long-standing and well-accepted cultural practice which dates back to the Qing dynasty (1644–1912) when officials would regularly share pipe tobacco with each other (Benedict 2011). Collectivist slogans, from the early years of the People's Republic of China, such as 'men should smoke from the same pack', prompted the offering and sharing of individual cigarettes. This became popular following the 1949 Communist victory, when cigarettes were part of the rations issued to family units, leading to an expansion of male smoking and sharing as an expression of solidarity (Kohrman 2007).

Sharing, largely a male practice, is most widely practised in rural areas (see Sect. 5.3.2) (Wang et al. 2014b), and has been important in developing and maintaining social relationships, particularly during 'tough times' (Hu et al. 2012). During the Cultural Revolution (1966–1976), 'gifting' (i.e. giving packets of cartons of cigarettes, sometimes extravagantly presented) became popular, particularly during festivals and holiday periods, as expensive cigarettes were used as methods of facilitating access to scarce goods and services or avoiding persecution (Rich and Xiao 2012).

The sharing and gifting of cigarettes has also been important in urban China, but, in contrast to rural areas, the primary goal of cigarette sharing was the formation of business relationships (Wank 2000). Perhaps two main factors were important here. First, China's market reforms created a new competitive economic environment in which the gifting of premium brand cigarettes became more important in business and social interactions (Rich and Xiao 2012). This practice was stimulated by market processes such as premiumisation (see Chap. 4), with the production of higher priced and higher-quality cigarettes, meaning that coveted brands, some costing more than 1000 Yuan (China Daily 2017), served as important status symbols. Luxurious gifts for public officials or medical practitioners, gatekeepers to health services, became commonplace, leading to an anti-corruption drive in 2012. Prior to this, smoking was integral to Communist Party membership. As Pan (2004, p. 314) writes:

> Communist Party members, permanent employees and state-owned enterprise employees tend to be more likely to smoke because they are more heavily influenced by the traditional culture of building and using connections than others in the transitional economy. Connections are much more important for these people because most of the benefits they receive, including housing benefits, are obtained through the influence of their connections within their work units … Chinese live in a web of social relations that often determine the attainment and allocation of various resources

Second, during the 1980s and 1990s, the ritualised gifting of cigarettes was appropriated by domestic and foreign tobacco companies who promoted the gifting of culturally tailored packages especially during important holidays and festivals such as Chinese New Year (Chu et al. 2011; Ding and Hovell 2012; Rich et al. 2014). These trends resulted in both sharing and gifting becoming ubiquitous across China. The few studies that have examined correlates of gifting in urban settings suggest that, in accord with expectations, gift givers are younger, more affluent and more highly educated and work in private companies or in sales and management positions. Gifts are mainly given to friends and colleagues (61%), bosses and supervisors (14%), business partners (13%) and family members (10%) (Ark Marketing Research and Consulting 2002). A study conducted in six major cities in 2007–2008, of smokers *receiving* cigarette gifts, showed that they were more likely to be female and older, have higher educational attainment and live in Beijing and the people who smoked fewer cigarettes per day (Huang et al. 2012).

Overall the evidence suggests that the cultural practice of cigarette sharing and gifting has been modified by urbanisation and social change. It also appears to have become less common as such cultural practices have increasingly been modified by tobacco control messages, particularly in larger cities. Although the evidence is sparse, there is some indication that both sharing and gifting have declined in larger cities, such as Beijing and Guangzhou (Hu et al. 2012), as well as in some eastern provinces. Xu et al. (2016) found that both sharing and gifting declined in smoke-free households in Zhejiang Province between 2010 and 2012. In households which are not smoke-free, gifting (but not sharing) also became less common.

5.2.3.3 Self-Exempting Beliefs

That tobacco smoking causes a variety of morbidities is well known to a majority of citizens in a highly literate society like China (Xu et al. 2013b; Zheng et al. 2013; Cheng et al. 2015), although there are limitations in community understanding of the actual harms. First proposed by Festinger (1957), self-exempting beliefs can be used to explain why people knowingly perpetuate deeds harmful to themselves. Culture, broadly defined, provides a rich reservoir to sustain and reinforce self-exempting beliefs (SEBs) about smoking (Chapman et al. 1993; Oates et al. 2004; Heikkinen et al. 2010). Sociologists such as Swidler (1986) and Di Maggio (2002) have argued that culture consists of an elastic reservoir of cultural beliefs and practices that influence how people of different cultural and political backgrounds consume tobacco and other substances. These perspectives enable us to see how, in circumstances of conflicting behaviour and belief, culture steps in as a repertoire of justification to reinforce a smoker's behaviour.

In China several studies confirm this. Ma's exploratory study (Ma et al. 2008) informs us that the most common self-exempting beliefs about smoking among Chinese males include the importance of cigarettes in social and cultural etiquette. Ma's study suggests that a social norm tied to the high valuation of economic success and social promotion in a patriarchal society could function as direct support for men to maintain smoking behaviour. While concrete beliefs in the legitimacy of tobacco consumption may have varied, they all demonstrated that traditional culture in many cases could provide the fundamental basis of self-exempting beliefs among men in China. Such self-exempting beliefs have been tested through research. In their study of male smokers in Hangzhou, Yang et al. (2014) concluded that while involvement in family, friend and co-worker networks had a direct effect on the likelihood of smoking, SEB, as measured by a standard SEB scale, also had a significant indirect effect, but only in the case of involvement with co-workers. Male smokers, despite reporting positive responses to tobacco policy, also accepted tobacco consumption as a necessary part of social etiquette. Some writers go even further to claim that the importance of smoking for socialisation makes it a unique part of Chinese culture when compared to western countries (Huang et al. 2020, p. 481).

5.2.3.4 Shanghuo as a Self-Exempting Belief

In traditional Chinese medicine (TPM), the concept of *Shanghuo* is one form of self-exempting belief. Literally meaning 'raised fire', *Shanghuo* describes a state of Yin-Yang (hot versus cold) imbalance resulting in an increased susceptibility to disease. This traditional concept has no specific set of physiological symptoms and varies from person to person (He and Kurihara 2008; Pan et al. 2020). The status of *Shanghuo* is said to disrupt the homoeostasis of human body and to be an outcome of stress arising from emotional factors arising from the pace and competitive pressure of modern life.

When a person consumes too many hot items, he/she will suffer from *Shanghuo*, which has to be remedied gradually by consuming more cold things. Thus, even though smoking cigarettes constitutes an activity that elevates much hotness and disrupts the bodily balance of Yin and Yang, traditional beliefs emphasise that the consumption of certain food items, along with tobacco, can rebalance the equilibrium. Alternative Chinese medicine advocates often mention and recommend the use of *yumcha*, citrus tea fruit, for supposedly purging and discharging toxins from tobacco. Such lay belief, although without much established support from modern medicine, may provide a delusionally safe method of smoking and inadvertently serve as an excuse for continuing smoking.

Such cultural ideas are not unique to China (Huang et al. 2020). Jackson et al. (2004) report similar practices in Kelantan, a rural state in Malaysia. Respondents were asked whether they could do something to make smoking safe. Drinking water and eating sour fruit were frequently mentioned, the latter, for example, being described as 'cleaning' and able to 'scrape out the essence of cigarettes'. Such self-exempting false beliefs were widespread and, like *Shanghuo* in China, point to the need to address such ideas in antismoking campaigns. Self-exempting beliefs have also been reported in western contexts such as Australia. Oates et al. (2004, p. 779) found that they acted as shield particularly for older and less-well-educated smokers who were more likely to view smoking as a risk rather than a probable cause of illness. Almost a third of the respondents believed that eating healthy food and exercising regularly could overcome the harms of smoking.

5.2.3.5 Patriotism and Smoking

Since tobacco production and consumption became important to the Chinese economy, both the tobacco industry and governing bodies, including the Communist Party, have encouraged the identification of smoking with patriotism, a powerful cultural influence on the persistence of smoking. During the early era of cigarette production in the late Qing and Nationalist time, foreign cigarettes dominated the Chinese market (see Chap. 3). However, waves of democratic and patriotic movements, including the famous May 4 Movement (Chen 1970), increasingly demanded citizens to support domestic industries and shun imports. As a result, buying cigarettes became a part of the highly mobilised ambitious movement of 'industrial salvation' (*shiye jiuguo*). Thus, Chinese smokers were strongly encouraged to smoke locally produced cigarettes, and the government explicitly prioritised regime stability and security over health (Liu 2018). Even in the twenty-first century, those advocating tobacco control were accused of 'betraying China' and being under the influence of 'western ideological thought' and subject to 'the manipulation of foreign investors and transnational tobacco companies' (Yang 2018, p. 69) (see Chap. 7). Redmon et al. (2013) report that international tobacco control advocates have been criticised on the basis that outside investment will 'bring down' Chinese tobacco companies to the advantage of American industry. The cultural proclivity that domestic products should gain a stronger foothold over foreign imports may have a trenchant impact on how tobacco control is variously perceived and implemented in China. The continued strength of the CNTC reflects not only a quest for economic security but also the importance of national pride and a strong reaction to the shameful experience of past economic colonisation.

5.2.3.6 Regional Cultural Influences

At a national level, cultural differences have frequently been emphasised to explain differences in smoking behaviour between Asian and Western societies. Lee et al. (2020), for example, contrast the collective culture of Japan which encourages conformity, to the more individualistic culture of the USA. In China, as in Japan, there are intense social pressures on males to smoke as an expression of polite manners, compared to the USA where such social processes are much less important.

In a vast country like China, of heterogeneous linguistic and racial/ethnic composition, regional cultural practices influencing smoking and other health behaviours will not be uniformly distributed. Reminiscent of the north-south divide in Britain, regional differences in health in China have attracted increased attention. The geographical cutoff between the north and the south is not a trivial matter. In spite of variations between provinces, there are certain regional similarities in the substance use culture across the north and differences with the rest of China also evident for a variety of dietary and health indicators. Chen et al. (2019), for example, showed that, for female cancer deaths, smoking was the leading risk factor in six northern provinces (Heilongjiang, Tianjin, Inner Mongolia, Jilin, Liaoning and Anhui). Other health indicators related to smoking, including age standardised prevalence of stroke, also show distinct north-south variations (Xu et al. 2013b; Ru et al. 2019). Xu et al. (2013a) attributed such variations not to smoking but to the impacts of a higher northern prevalence of hypertension and excess body weight associated with the northern grain-based diet. However, Gao et al. (2017), using a more sensitive age-adjusted life course measure of smoking, showed that the latter was positively associated with the risks of hypertension, risks that increased with age.

While many factors contribute to regional differences in smoking and other forms of substance use, especially in low- and middle-income countries (Grittner et al. 2019), the effect of the cultural context cannot be overlooked. Most Chinese research, however, if it considers culture at all, largely views it as a residual factor in terms of the significance of regional variables in multilevel models. A number of studies have commented on the broad regional divisions in smoking; however, most do not use multilevel modelling strategies which control for other demographic, health and socio-economic factors and urban-rural status. Typical of the latter is Astell-Burt et al. (2018) who found that age-adjusted smoking rates were highest in a belt of north and northwestern provinces, but were also high in the main tobacco-growing areas of Southern China. Interestingly north-south differences were strongest for females, rather than males, with the prevalence of female smoking being highest in the north and in Tibet (Fig. 5.1).

One of the few multilevel studies is that of Lee et al. (2019). Using the 2011 China Health and Nutrition Survey, they found that smoking prevalence was higher in Northern China, even after accounting for urban-rural differences and a variety of other individual socio-economic and health status indicators. Compared to the Northeast, current smoking prevalence was significantly lower in the East Coastal (OR = 0.78) and Central Regions (OR = 0.72), but no significant difference occurred

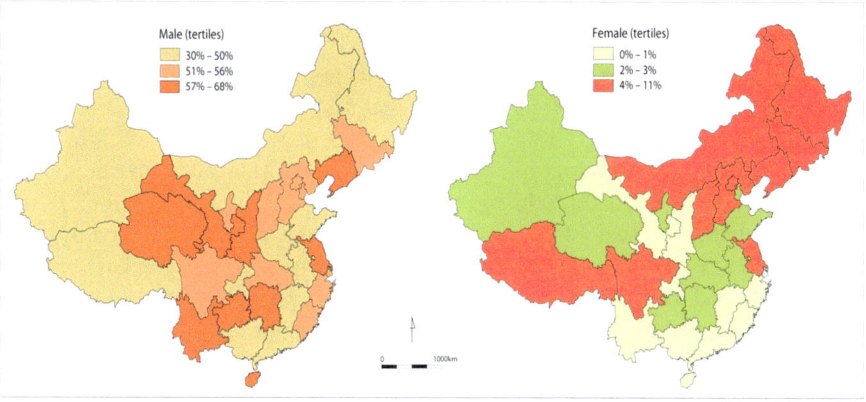

Fig. 5.1 Current smoking by province (2010). (Credit: Modified from Astell-Burt et al. (2018), p. 758. Reproduced courtesy of Oxford University Press (Rightslink Licence No. 4878561188341))

between the Northeast and the Southern Regions (which included Yunnan) or with the megacities. However, as Liao et al. (2017) found, regional distinctions between the Northeast and other regions were not present for levels of consumption (number of cigarettes smoked per day), except for the Southern Region where rates were slightly higher (OR = 1.09). In addition to the presence of regional effects, both current smoking and smoking intensity were also significantly higher in rural areas (urban ORs = 0.83 and 0.93, respectively). Current smokers were also more likely to be older age and of lower social status and were almost three times more likely to be heavier consumers of alcohol. However, other than commenting that the regional 'geographic disparity should be examined further' (Lee et al. 2019, p. 10), there was little discussion of the likely impact of cultural and other factors.

A greater attempt to unravel the influence of cultural factors was made in an earlier study by Li et al. (2015) on regional differences in smoking, drinking and physical activities. Like Lee et al. (2019), they found that smoking prevalence was highest in the Northeast (28.2%) and Northern regions (27.7%) followed closely by the tobacco-growing Southwest (26.3%), with the lowest rates in the East (24.3%), Central-South (24.2%) and Northwest (22.1%). The correlation between smoking and drinking (0.84) was also highest in the Northeast region (Liaoning, Jilin, Heilongjiang and Dalan provinces). The authors speculated that there is a northeastern subculture characterised by 'naturalism, which encourages adventure, eating and drinking freely, and collectivism which stresses hospitality, close social ties, frequent gathering, sharing, and no refusal on drinking/smoking proposals' (Li et al. 2015, p. 236). Anecdotal evidence also suggests that the Northern residents have an infamous reputation of drinking more heavily than people in other regions. The region's colder climate and nomadic tradition were also considered as a possible explanation as was the plains physical environment which was thought to have enabled greater social interaction and regional conformity in historical times.

As north-south cultural differences are a common theme in much of Chinese literature (Xi 2018), clearly more consideration is necessary of these differences and their implications for smoking and health and to guide public health prevention programmes. For instance, the more patriarchal machismo culture of northern and western China, which partly reflects its ethnic mix, can be hypothesised to be an important factor in explaining differences in smoking. Men may use smoking to display their highly valued machismo expression of carefreeness and brotherhood. A clue to the higher rates of smoking of women in the north is provided by Xiao Hong's novel, *Biography of Hulan River*. Xiao notes the harsh natural environment of the north and the importance of the 'fire kang' for rural women, 'victims of a patriarchal society' (p. 1595). The kang is a space for them to meet and where they can release their burdens to each other in their leisure time. Xi (2018, p. 1595) laments, 'they talk about everything here … tell about the difficulties in life and comfort each other" … and their suffering as women'.

North-south distinctions in China are also part of the popular discourse (Talhelm et al. 2014, 2018). There is a strong view among the Han majority that there is a division between the southern and northern populace (Ma et al. 2016). There are abundant genetic studies indicating that, despite some common elements, Han Chinese from the south and the north have different ancestral origins. Many southern Han may be closer to southern minorities, while many northern Han genetically cluster with Mongolians, Uyghurs and Manchurians (Du et al. 1997). For centuries after the establishment of the Han dynasty, the Chinese South (below the Yangtze River) had been a safe haven for northern royalty when they were driven from their lands by various waves of nomadic nations. From their southern refuge, they fought to reclaim their lands from steppe nomads. In the north there was repeated integration between Han and different nomadic groups of Tungusic, Mongolian and Turkic origin. During the Jurchen and Khitan periods from the ninth to the thirteenth centuries, many Han Chinese along the northern borders defected to, or seasonally migrated between, the nomadic territories and Chinese agrarian society (李月新 2007). A famous satirical poem, *Hehuang Yougan*, also depicted a picture of Han 'turncoats' in the late Tang dynasty (618–907) along the northwestern border. As Situ Kong wrote: 'war dust rising from Fort Xiaoguan on the Upper Yellow River interrupts the springtime in a foreign land. Han fellows all adopted the barbarian language, and turned to curse their Han folks'. These different cultural origins and practices, including the more recent promotion of regional identities (Oakes 2000), deserve further study for their impact on modern patterns of health behaviour (Kong 2010).

5.3 Cultures of Rural Smoking

Many of the cultural factors discussed previously have their roots in ancient China and in a society that was, until recently, largely rural (Unger 2016). When the PRC was formed in 1949, the urban proportion of the nation's population was only 10.6%

(Quan 1991). Even though this share had doubled to 19.6% by 1961, it soon fell again to 13.4% in 1966 given the restrictions on urban employment imposed by Central Government in the years following the economic disaster of the 'Great Leap Forward'. This demographic concentration changed little in the next 20 years until urban growth increased following the start of Deng Xiaoping's economic modernisation in 1978. Nevertheless, as late as 1990, over 80% of the population were still classified as rural with the urban population not exceeding that of rural areas until 2015 (Fig. 5.2).

Despite the country's massive urbanisation, even today many Chinese people comment that China is still dominated by agrarian mores which include Confucian values of tradition and respect. The presence of rural-urban migration on a massive scale has meant that the rural fabric is still ingrained in Chinese society, symbolised by family reunifications at Chinese New Year and other holidays when many urban dwellers return to their rural roots. These are also important in the history of the Communist Party. The Maoist military strategy of using villages to encircle cities had in buttressing industrialisation in major cities such as Jinan, Zhengzhou and Chongqing, often at the expense of the survival of the peasants. For example,

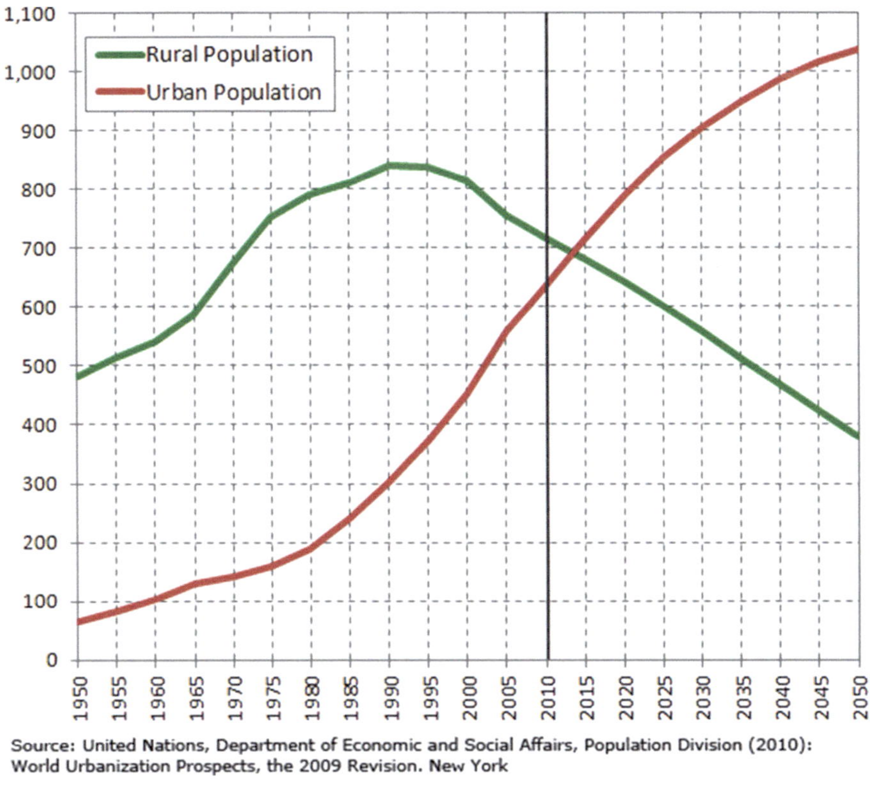

Fig. 5.2 Urbanisation of China's population (m). (Credit: United Nations (2010) Public domain)

Garnaut (2014) showed that during the natural disasters from 1959 to 1961, starvation- and malnutrition-related mortality was highest in rural areas with good infrastructure links to cities, suggesting that, despite official secrecy surrounding relevant data, the peasants from these rural places died because a substantial portion of their crop was likely have been directed to support nearby cities. The historical plight of the countryside, although much improved in recent years, still reflects a variety of health disadvantages, including higher rates of smoking, the causes of which are addressed in Sect. 5.3.2.

5.3.1 Rural Smoking Prevalence

High rural smoking rates reflect the persistence not only of cultural practices discussed previously but also of a rural-urban division that is deeply ingrained in modern Chinese social history. The massive urbanisation of China's population has been accompanied by a widening urban-rural gap in smoking prevalence that appears to have increased in recent years (Zhi et al. 2019). Many studies have shown that the high rates of smoking among rural Chinese men exceed those of urban men (Liu et al. 2017), especially at older ages (Mao and Wu 2007). People from rural areas tend to smoke (and drink) at an earlier age (Liao et al. 2017), are less likely to quit (Liao et al. 2017; Wang et al. 2014a) and have a poorer knowledge of tobacco harms (Xu et al. 2013b; Mao et al. 2014). Both prevalence rates (Yang et al. 2008, 2009) and consumption levels (Cai et al. 2019) are highest among those with the greatest income and household assets, but this social variation is likely to reflect other provincial factors such as health education and the extent to which tobacco control policies have been implemented in rural areas.

5.3.2 Why Smoking Is Higher in the Countryside

The continued high rural smoking rate can be attributed to a number of factors, which together have accentuated structural inequalities and increased the difficulties of quitting among rural smokers.

5.3.2.1 Gender Norms

In China's countryside, the gender norm is particularly traditional and retains a greater number of conventional familial arrangements than in the cities. The very issue of at-risk behaviours associated with the mythical ideas of masculine superiority and feminine submission again floats out of the pretension of patriarchy culture to assist speculation on the causes of smoking. In the countryside, due to slower development and the stable organisation around the division of arable lands into the hands of family-based productive units, the changes and shuffle of values in social life tend to lag behind the cities. Gender-based norms still dominate all aspects of rural society. Despite lower disposable income, the bride price is infamously high in the countryside and often makes headlines when bridegrooms commit suicide when they cannot 'afford' their beloveds. This is because traditional roles are strongly embedded in these matrimonial unions based on the idea that engaged girls become 'owned property', in stark contrast to the romance-based marriages in urban settings. Men are also expected to show valiant conduct for the proof of their masculine qualities, which includes prioritising nonfamilial relationships over the domestic sphere and treating male friends as more important than their own families. This external-oriented value is observed among men from many cultures of both the West and the East, with the cultural disposition of genders reflecting two structurally distinguished value systems: the domestic and the external (Bourdieu 2001). In such a cultural system, at-risk behaviours are exclusively reserved as the rights and duties of men.

5.3.2.2 Cigarette Sharing and Gifting

As outlined previously, in rural areas an important cultural norm has been the role of cigarette sharing which forms a significant part of community identity (Hu et al. 2012; Wang et al. 2014b). Today both sharing and gifting permeate every aspect of rural family life from economic activities to leisure pastimes both in family and wider social interactions (Fig. 5.3). The extent to which gifting is so ingrained in rural society is evident in Mao et al.'s (2014) ethnographic study of households in a rural area of Central Jiangsu. Mao found that cigarette gifting was particularly important for family juniors who 'were expected to show filial piety to senior generations of family elders' (p. 7). Usually such cigarettes were almost always more expensive brands that were believed to be less harmful to the health of the recipient. Cigarettes also played a role in off-farm economic activities either as a type of payment to business partners or gifts to business customers. Thus, both gifting and sharing were important rituals for establishing and strengthening economic ties. They were also often used as bonus payments by employers to employees, or if local households hired craftsmen, then cigarette gifting would be expected by the latter.

With rural-urban migration, gifting has provided a substantial number of cigarettes to rural smokers, and can be seen as a form of 'remittance' from children in the city to their families back home (Rich et al. 2014). Such gifts are often a major source of cigarettes especially for poorer rural households. Not only do they underwrite the cost of smoking, but also allow limited funds to be devoted to other priorities, or cartons to be regifted or sold (Hu et al. 2012).

The cultural practices of sharing and gifting are so ingrained that it makes quitting very difficult, thus contributing to the higher smoking prevalence in rural areas. As a consequence, the smoking gap between rural and urban areas will probably

Fig. 5.3 A farmer lights up a cigarette for his fellow before himself in Laofen Village, Pingshan, Hebei, 2012. (Credit: Alamy Image ID: D05KG3 Stock photo)

further diverge, with rural smoking rates remaining relatively unchanged compared to the decline that is happening in cities.

5.3.2.3 Poverty and Income Gaps

Poverty and inequality can also affect rural smoking. Traditionally rural China has been characterised by high levels of poverty and disadvantage. In such situations smoking was an important form of relaxation providing relief from the burdens of daily life. However, the economic transition, initiated at the end of the 1970s, led to a dramatic increase in income for Chinese households. Rural income growth led to a considerable reduction in poverty levels: the official rate, based on the new poverty line issued in 2010, declining from 98% in 1978 to 10% in 2013 (Luo et al. 2020). Nevertheless income growth in rural areas lagged behind that of the cities, resulting in a widening gap, at least until 2007 (see Chapter 6). The drop in absolute rural poverty has been driven by economic growth, in part aided by government subsidies for agriculture, cash transfers aimed at lower-income groups and the out-migration of lower-income workers. As of 2014, Luo et al. (2020, p 4) estimated that approximately 70 million rural people still had incomes below the official poverty line. Within rural areas, income inequality also increased, but not at the rate seen in urban settings. Nevertheless, income gaps, which also influence smoking, were more apparent in the period after 1990 than before (Ravillion and Chen 2004).

5.3.2.4 Stress

Stress is a major physiological culprit for smoking (Slopen et al. 2013; Stubbs et al. 2017) and the use of other psychoactive substances including alcohol and betel nut. The bleak fact that China's rural residents suffer a disproportionate number of stressors in their daily life explains partially why the smoking rate is much higher there. As noted above, the standard of living in the countryside and the socio-economic status of rural residents are substantially lower than in the cities. In 2018, the disposable annual average income of rural residents (¥14,617) was only 37.2% of their urban counterparts (¥39,251). After deducting the cost of living, there is a difference of 5.4 times in the savings gap between urban and rural residents (Xinhua News 2019). This means that rural residents will have little savings in the event of major illness or accidents, for social events such as childbirth, marriage and funeral ceremonies, and for children's educational expenses. Any of these individual events, in conjunction with community factors, such as poor healthcare, are likely to increase stress levels among rural residents, making them vulnerable to health problems (Fu et al. 2018; Yang et al. 2019) and at increased risk of suicide (Zhang and Wang 2012). Using tobacco to elevate euphoria and block the stress response is prevalent among smokers (Twyman et al. 2014) and is one of the key reasons why smoking rates are higher and quit rates lower among rural dwellers.

5.3.2.5 Differential Access to Health Services

The recent restructuring of the Chinese healthcare system has provided another form of stress for rural residents. Until 1978, China was characterised by a health system modelled after the Soviet Union (*Semashko* system). Healthcare was the responsibility of the State, with funding for health services largely from government sources. Before 1978, Chairman Mao Zedong encouraged the masses to learn about medicine and use traditional Chinese medical knowledge to treat themselves. During this time, professional healthcare workers were sent to the countryside in large numbers, partly as a form of political re-education and partly to equip rural healers (barefoot doctors) with some modern medical knowledge (Fang 2012).

However, the 1978 neoliberal economic reforms resulted in an increased commercialisation of the health sector, with tightened hospital budgets, reduced healthcare funding and increased patient co-payments (Mullins-Owens 2015). The effects of the reforms were acutely felt in rural China. In the late 1970s, a variety of healthcare plans, under the Co-operative Medical System, covered approximately 90% rural citizens (Liu et al. 1999), but by the early 1990s, this had been reduced to approximately 10% (Tang and Bloom 2000). In addition to increased patient co-payments, the decentralisation of fiscal responsibility to provinces and lower levels of government funds accentuated the already uneven distribution of services between urban and rural areas, leading to a withdrawal of many preventive and patient services (Shi 1993). For poorer rural smokers, this resulted in difficulties of financing care, but also in geographically accessing services, including primary care and smoking cessation support (Lin et al. 2019, p. 7) (see Chap. 8).

The market-oriented reform from 1978 led to a near collapse of healthcare in rural areas because rural co-operative healthcare, based on the collective economy, lost its main source of funding. Since 2003, the New Cooperative Medical System (NCMS), offering three insurance schemes, has been developed, but has been criticised for not increasing access to care (Babiarz et al. 2012), especially among poorer patients who have chosen the least expensive insurance option (Ma et al. 2012). The effect of the reforms leading to problems of social insurance coverage is that urban-rural disparities in health have become increasingly larger with age (Jiang and Wang 2018).

5.3.2.6 Rural Ageing

Urban-rural migration has led to an increasing aged population in rural areas, with significance for smoking, given higher rates among older generations of Chinese (see Chap. 2). Like rural areas everywhere, the urban migration of younger family members has imposed additional burdens upon older Chinese especially since the former one-child policy has meant that the presence of just one offspring has made care of the elderly more difficult (Liu 2014; WHO 2015). Being 'left behind' has been shown to elevate the suicide risk among the rural elderly, through increasing life stresses, depressive symptoms, mental disorder and decreased social support (Zhou et al. 2019). Such trends have been compounded by the restructuring of healthcare services which has restricted both the amount and quality of healthcare support available.

A number of studies have demonstrated significant urban-rural differences in access to and utilisation of healthcare services (Liu et al. 2007). Zhang et al. (2017), for instance, showed that during the period 2005–2014, older adults in rural areas were nearly twice as likely as those in urban areas to report inadequate access to healthcare which, in turn, was associated with higher rates of disability and overall mortality among rural compared with urban dwellers. Controlling for health behaviours, such as smoking, attenuated but did not eliminate the significance of this relationship, suggesting the presence of other rural pathways affecting mortality outcomes.

5.3.2.7 Institutional Causes

In addition to the effects of health services restructuring, higher smoking rates in rural areas can also be attributed to a number of institutional causes. Perhaps chief among these are the effects of the discriminatory *hukou* system in affecting the welfare of rural migrants and their home communities. We address issues associated with rural-urban migration in the next section.

Urban-rural differences in smoking also reflect the differential impact of tobacco control policies on urban and rural areas (see Chaps. 8 and 9). The growth of tobacco control in China has largely been an urban process, with little policy transfer occurring between urban and rural locales. Progress at the provincial level has been slow and uneven with no specific policies aimed at rural areas with higher prevalence rates. The lack of positive intervention by state actors is also true of the health system where, as Wang et al. (2014a) found in Shandong Province, tobacco control activities at villages were rare and infrequent. A lack of knowledge of effective smoking cessation tools was characteristic of both smokers and village doctors, who themselves were likely to be smokers.

5.4 Culture as Institutionalised Discrimination: Rural-Urban Migration and the *Hukou* System

An important part of the Chinese cultural context has been massive rural-to-urban migration and the role of the *hukou* system. Indeed, as Hamnett (2020) has argued, the post-reform pattern of urbanisation in China, and the impact of the locality-based citizenship *hukou* restrictions on urban migrants, makes it unique and sets China apart from other countries. In this final section, we first deal with the scale of rural-urban migration, characteristics of the *hukou* system and its relationship to patterns of smoking among rural-to-urban migrants.

5.4.1 Rural-Urban Migration and Smoking

An important feature of the urban condition has been the enormous growth of internal migration. Rapid urbanisation and industrialisation, especially in cities in eastern China, created a huge demand for additional labour. From 1990 to 2006, it was estimated that the migrant population tripled from 50 million to an estimated 127 million, or approximately 10% of China's total population. In 2019, there were approximately 291 million migrant workers in China; 117 million were local migrant workers, working in the vicinity of their home, 99 million worked at more distant places in their home province, and 75 million had left their province and worked in regions far away (Textor 2020).

Rural-urban migration has had a powerful influence on the health behaviour and health outcomes of urban migrants. A growing number of studies have documented the negative effects of marginalisation and social exclusion on migrant stress (Fu et al. 2007; Luo et al. 2016; Zhong et al. 2016) and on the mental and physical health of urban migrants (e.g. for recent reviews, see Mou et al. 2015; Li and Rose 2017; Bengoa and Rick 2020; Wang 2020). Over the last few decades, a growing literature has also examined the health behaviour implications of rural-urban migration; only recently have the explicit effects of *hukou* status on smoking and other health outcomes been considered. That seems an afterthought and is strange given the pervasiveness of *hukou* in the Chinese cultural context. With respect to smoking, the results of this work are summarised in Table 5.1. A number of key themes are evident.

First, despite the many disadvantages faced by rural migrants, smoking prevalence among rural-to-urban migrants is commonly reported to be lower than the national prevalence rate. For instance Liu et al. (2016) reported the smoking prevalence for male migrants to be 46.7% in 2010 (lower than the national rate of 52.9%), but for higher female migrants (5.3%), higher than the general female population (2.4%). Similarly Chen et al. (2009) showed that rural migrants were not more likely than others, such as rural dwellers left behind, to engage in health-risk behaviours. Such conclusions, however, need to be tempered by the presence of the healthy migrant effect, the need to control for *hukou* status, the existence of very high stress levels and smoking rates among certain groups of migrants (Cui et al. 2012) and regional environmental factors which have been shown to affect the propensity of migrants to smoke (Xie 2017).

Second, in terms of the predictors of migrant smoking, studies have shown that cigarette smoking among migrants was related to a variety of environmental conditions such as years of migration, number of cities worked in, number of jobs held, type of job, job dissatisfaction and hours worked per day (Chen et al. 2004; Yang et al. 2009; Mou et al. 2013; Liu et al. 2015; Zheng et al. 2018). Cui et al. (2012) reported that migrants with high perceived stress had an excess likelihood of being current smokers; those with high perceived work stress a 75% excess likelihood; and those with a high perceived life stress a 45% excess likelihood.

Third, the effect of environmental factors often varies by gender. For example, employment in temporary construction jobs, with high risks and low benefits, was associated with higher smoking among females than males (Chen et al. 2008); in Shenzhen the absence of a job contract and physically demanding jobs were more important risk factors for smoking among women, whereas longer working hours and less rest were more important for men. Perceived discrimination was particularly important for female restaurant/hotel and commercial sex workers (Shin et al. 2013). Among women, higher rates of depression affected substance use, but the effects varied by income; for higher-income migrants, depression increased the likelihood of smoking, but the reverse was true among female migrants with lower incomes (Chen et al. 2008). This may reflect the social pressures associated with earning higher incomes in different work environments, such as nightclubs, where job-related smoking and drinking were common (Chen et al. 2008).

Fourth, migrant housing and neighbourhood conditions also affected smoking. Chen et al. (2004) found that living in rental properties was associated with an increased likelihood of smoking. Absence from home, family and familiar surroundings and living in overcrowded collective dormitories were also key risk factors (Yang et al. 2009). In Shenzhen, however, Mou et al. (2013) found that workers living in non-dormitory accommodation were more likely to smoke. This partly reflected dormitory smoking bans but also the likelihood of family accompaniment

Table 5.1 Recent studies of cigarette smoking among migrant populations in China

Study	Data	Location	Current/daily smoking rate (%)	Key findings
Chen et al. (2004)	MHBS 2002	Beijing	Total M 51.7 F 10.9 High income M 62.4 F 19.8	Smoking levels associated with age, income, migration years, number of cities worked in, number of jobs held and living in rental properties, job dissatisfaction & hours worked per day. Female migrants in particular were at an elevated risk of smoking the greater the number of cities they had worked in; male smoking prevalence increased from 48.1 to 67.2% (1–5 cities) vs 9.5%–31.65% for females
Hesketh et al. (2007)	Work unit survey 2004	Hangzhou	Total M 54.0 F 1.8	Compared to urban workers (M 58%, F 2%), migrant workers smoked less. Migrants took up smoking later, were less likely to become heavy smokers and were more likely to want to quit
Chen et al. (2008)	MHBS 2002	Beijing, Nanjing	Total M 56.5 F 10.5	Workplace, income and depression are associated with substance use (smoking, drinking and drugs), but effects were not constant by gender, e.g. construction jobs were associated with more substance use among females Depression affected substance use, but the effects varied by income; for higher-income female migrants, depression increased the likelihood of smoking, but the reverse was true among low-income female migrants
Chen et al. (2009)	Three samples: Urban residents, rural migrants, rural residents 2004–2005	Beijing, 8 provinces	Urban residents M 70.2 F 20.8 Urban migrants M 66.9 F 11.6	Compared to urban and rural residents, both migrant men smoked less. Migrant women smoked less than urban women but more than rural women. Similar patterns were evident for alcohol consumption

(continued)

Table 5.1 (continued)

Study	Data	Location	Current/daily smoking rate (%)	Key findings
Yang et al. (2009)	Migrant workplace surveys	Chengdu, Shanghai, Beijing	Pre-migration M 29.2 F 1.0 Post-migration M 44.4 F 1.3	Smoking prevalence increased following migration, with the number of cities of migration, higher personal income and housing type (highest in collective dormitories). Smoking initiation was related to the number of cities of migration and higher personal income
Finch et al. (2010)	Migrant women working in restaurants & commercial sex venues	Beijing	Restaurant workers 6.5 Sex workers 33.3	Current smoking rates were high compared to rates in the general population. Compared to non-smokers current smoking rates were related to education (protective) and job satisfaction. Exposure to female-branded cigarettes increased the risk of ever smoking
Wan et al. (2011)	Migrant women working in restaurants/ hotels & sex venues 2010	10 provincial capitals	Restaurant/ hotel workers 3.2 Sex workers 41.9	Those who initiated smoking after moving to the city were more likely to be current smokers than those who first tried before migration Exposure to female cigarette brands increased susceptibility to smoking Older age and higher education were negatively associated with having first tried smoking after migration
Cui et al. (2012)	Migrant workplace surveys	Hangzhou, Guangzhou	Overall 64.9 Age 40+ 74.3 Low educ 75.3 High stress Work 71.7 Life 71.0	Migrants with high perceived work stress had a 75% excess likelihood to be current smokers, while those with a high perceived life stress had a likelihood of 45%

(continued)

Table 5.1 (continued)

Study	Data	Location	Current/daily smoking rate (%)	Key findings
Mou et al. (2013)	Migrant factory workers 2008	Shenzhen	Daily smoking M 27.3 F 0.7	Prevalence rates significantly lower than national smoking rates Higher rates of male smoking associated with longer working hours, less rest & frequent alcohol consumption For women the key risk factors were absence of a job contract, physically demanding jobs, frequent internet use & alcohol consumption Accumulated working time in Shenzhen as did accommodation type also had an impact
Shin et al. (2013)	Migrant women working in restaurants/ hotels & sex venues 2010	10 Chinese cities	Restaurant/ hotel workers 3.2 Sex workers 41.9	Perceived discrimination was independently associated with current an ever smoking for both groups City smoking level increased the probability of smoking for both groups of workers as did life satisfaction Higher monthly income increased the probability for sex workers
Wang et al. (2014b)	Qualitative study of 39 male migrants 2013	Shanghai	54.8	Smokers were aware of the health risks of smoking but considered it to be an important part of their identity and social participation as did consuming more expensive cigarettes. Smoking was also seen as a way of relieving psychological stress caused by discrimination
Liu et al. (2015)	Migrant survey 2012	Shanghai	M 45.0 F 2.0	Among males current smoking was higher among those who worked in construction & entertainment, who were divorced/ widowed and who had poorer mental health. A longer duration of migration and number of migratory cities lived in were also important Among females current smoking was higher for construction, hotel/ restaurant & entertainment workers, higher monthly income, those in poorer health & number of migratory cities lived in

(continued)

Table 5.1 (continued)

Study	Data	Location	Current/daily smoking rate (%)	Key findings
Xie (2017)	RUMiC Survey 2008–2009	15 cities	M 53.2 F 3.2	The effect of mental stress on the likelihood of smoking depended on regional context, being most evident in eastern China
Zheng et al. (2018)	MDMS 2013	12 provinces	M 54.4% F 3.7%	Among males years of education and duration of stay affected smoking; among females household registration status and marital status had the most impact
Hou et al. (2018)	CHARLS 2011–2012 Persons aged 45 & over	China samples of migrants & nonmigrants	Ever smoking: Nonmigrants R 40, U 34 Migrants R to U 33 U to U 28 Rural return 63 Urban return 58	Strong association between smoking behaviours and migration status in later life: Compared with nonmigrants, migrants who have returned to their place of origin have the greatest % people who have ever smoked and the lowest % of smokers who have quit Migrants who have not returned had the lowest % ever smokers and the highest % quit
Song and Smith (2019)	CHARLS 2015	China urban & rural samples by *hukou* status	Range of health outcomes	Rural *hukou*ers in rural areas experienced the greatest level of hardships In urban spaces migrants who were *hukou* converters have better health behaviours than urban migrants with rural *hukou*
Bengoa and Rick (2020)	RUMiC survey 2008–2009	China urban & rural samples by *hukou* status	Migrants with rural *hukou* 27.6; urban *hukou* 20.3	Migrants with urban *hukou* maintain lower levels of blood pressure & are less likely to develop hypertension or obesity than rural *hukou* migrants. These differences persist after controlling for SES factors such as smoking
Zhang and Churchill (2020)	China Family Panel Studies 2012–2016	China 635 communities	Social Well-being	The effects of income inequality on social Well-being are stronger for rural *hukou* residents compared to urban *hukou*

CHARLS China Health and Retirement Longitudinal Study, *MDMS* Migrant Dynamics Monitoring Survey, *MHBS* Migrant Health Behaviour Survey, *RUMiC* Longitudinal Survey on Rural Urban Migration, *R* rural, *U* urban which decreased the risk of depression and provided a network where males could smoke socially.

Fifth, there were few qualitative studies which explored migrant knowledge of the risks of smoking and the ways in which smoking was a reaction to perceived discrimination, stigma and social marginalisation. Wang et al. (2014b) were one of the few studies that examined smoking as a component of the identity of the floating population and reported how cigarette sharing was important in establishing social relationships in new group situations. Despite the economic costs, buying expensive brands was seen as an important part of maintaining a desirable personal image and personal contacts.

Finally, while many studies have examined relationships between mental stress and health outcomes, Xie (2017) has shown that in the case of tobacco use, this relationship is moderated by regional differences. Migrants from the eastern area of China when under high mental stress were more likely to smoke, but this was not true of migrants from central or western China. In central China length of migration appeared to be the key factor, but for western China neither variable was significant. Xie (2017) hypothesised that different factors associated with the host society, such as smoking prevalence, attitudes towards and social support for migrants and the presence of local tobacco control policies, affected the extent to which mental stress was a significant determinant of tobacco use.

5.4.2 Migration and the Hukou *System*

Migrants have faced a significant institutional barrier, the *hukou* system, which has prevented their political and social integration into the cities to which they moved. The *hukou*, as a strategic element of Chinese urban development, creates risks for worsening health disparities, including smoking. The *hukou* has become entrenched in the institutional and social structure of Chinese society; it affects occupational and employment rights, healthcare eligibility, birthright and the place of burial. It is a strategic cause of health inequalities and reflects deliberate agency external to the affected individuals.

The *hukou* system originated in 1958, during the Maoist era. Modelled on the Soviet *propiska* internal passport system, it was designed to limit the growth of large cities and the demands placed on them, but nevertheless serve state needs for massive industrialisation. Most Asian societies, once under the influence of Confucianism, have practised some sort of *hukou*. However, as Chan and Buckingham (2008) have noted, the Chinese *hukou* system differs substantially from other Asian systems of household registration. By contrast, the main goal of the Chinese system has always been to regulate population mobility while at the same time producing enough cheap labour for industrial enterprises and minimising the burden on state welfare agencies. Within Asia only Vietnam's *hokhau* system that manages mobility is similar to China's model, as is Korea's former system of *hojuje* (戶主制), which ceased in South Korea in 2008, but still survives in North Korea.

Subsequent reforms in the late 1990s enabled the conversion of rural-to-urban *hukou* under certain conditions. However, from its inception rural-to-urban status conversion has been very selective with education, party membership and military experience being the three major factors facilitating *hukou* mobility (Wu and Treiman 2007). While the *hukou* system has been relaxed since the late 1970s, since the 1990s it has also been decentralised to local governments giving them greater autonomy to experiment with *hukou* reforms (Song and Smith 2019). Since one of the initial goals of *hukou* was to control the growth of urban populations, this has meant that larger cities most attractive to urban migrants also often have the strictest *hukou* policies (Colas and Ge 2019).

Even today, lacking a local *hukou* means a person from the countryside or another city may not buy houses or cars or have access to health insurance and unemployment and retirement benefits. People need employers or first-degree relatives (including spouses) to sponsor *hukou* conversion. Although purchasing urban properties to acquire *hukou* is possible for rural migrants, in most cities the purchase itself requires a local *hukou*. In megacities, such as Beijing and Shanghai, getting *hukou* for non-locals is only viable through employment in public institutes with sufficient *hukou* quota, marriage or investments exceeding one million Yuan per year in tax contribution. Consequently the *hukou* system has contributed to feelings of deprivation especially among the unskilled who are more likely to suffer economically as a result of being unable, compared to skilled workers, to obtain a local urban *hukou* (Qiu and Zhao 2019).

5.4.3 Hukou Status and Smoking

Results of research to date point to the negative effects of hukou on health behaviours such as smoking and drinking. By restricting the movement of migrants and eligibility for housing and healthcare, the ease of management for legislators has helped create an environment that facilitates risk-taking behaviours. Bengoa and Rick (2020), for example, showed that *hukou* status affected migrant smoking; urban migrants with rural *hukou* recorded higher rates (27.6%) of smoking than permanent urban dwellers with urban *hukou* (20.3%). In terms of ever smoking, the highest prevalence (63%) and lowest quit rates for older people (aged 45 years and over) were recorded for rural return migrants, perhaps reflecting the burdens of living and working in urban areas without the benefits of an urban *hukou*. Few studies, however, have examined the effect of residence and *hukou* status on smoking and other health behaviours. The main exception is Song and Smith (2019) who found that in urban spaces, migrants who were able to convert their *hukou* from a rural to urban status had better health behaviours, including smoking, than urban migrants who continued to have a rural *hukou*. Bengoa and Rick (2020) found similar patterns with respect to hypertension and obesity.

More work needs to be undertaken on the effects of *hukou* status on smoking and wider health outcomes. Not all migrants wish to obtain urban *hukou* and opt to straddle and between city and countryside. Despite evidence which shows that receiving an urban *hukou* substantially enhances subjective well-being (Tani 2017), many migrants see increased benefits from retaining their rural hukou including farming and housing land and more relaxed birth control. Indeed Wang and Schwart (2018) present evidence of increased *hukou* intermarriage, particularly between urban men and rural women, suggesting the potential childbearing benefits of retaining a rural *hukou* on the part of one partner. Other factors such as those relating to rural sending areas and migrant integration into the city (Tyner and Ren 2016) and the links to stress and smoking also deserve increased attention. Currently, a considerable portion of Chinese rural-urban migrants have already formed meaningful subcultures within their social world, often through strong native-place networks and hometown identities. The concentration of migrants in occupations and living spaces based on such networks has been found in previous studies to be associated with at-risk health behaviours (Yang et al. 2016; Yang and Yang 2019). In the social world of rural-urban men with the same place origins, they share cigarettes to enforce solidarity, as well as smoking to boost a sense of masculinity in the face of discrimination and belittlement from urban people. Even for Chinese females who rarely smoke, except for the very old generation born before 1949, migration experience increases the female smoking rate, which rises from 9.5% to 15% after 5 years of stay in the city (Chen et al. 2004).

5.5 Conclusions

In any society cultural factors are important for understanding health behaviour and health outcomes, and this is certainly true of China, where a cluster of cultural practices have influenced smoking. Unravelling these, however, presents many challenges; culture is often hard to define, and it is difficult to separate its influence from that of other factors. Although cultural practices are not easily quantified, there is sufficient evidence in China that they continue to be important in influencing smoking and health.

Culture is expressed in many ways and is affected by a variety of structural factors. Within the Chinese context, there are a number of important connections between economic, social and political factors and different forms of cultural expression that have affected smoking (Table 5.2). While not exhaustive, such linkages point to the need to clearly link cultural elements to wider forms of structural change and to distinguish both their temporal and spatial impacts. In terms of the three elements, economic, social and political, the effect of economic changes has perhaps been the most recent. As pointed out in Chap. 3, the CNTC has increasingly commodified cultural emblems as part of its marketing strategies. Similarly the

Table 5.2 Relationship between structural and cultural factors influencing smoking

Structural factors	Cultural expression
Economic	Commodification, *hukou*
Social	Patriarchy, sharing & gifting, self-exempting beliefs
Political	Authoritarianism, patriotism, *hukou*

implementation of *hukou* policy in 1958, while not exclusively Chinese, was largely introduced for economic reasons and become an important cause of higher smoking rates among certain migrant populations. Because of its authoritarian bias, somewhat relaxed in recent years, *hukou* can also be seen as a political expression of elements of Confucian culture.

Social changes have also influenced culture. Although the effects of patriarchy remain strong, these have been weakened as a result of urbanisation and greater freedom among women (see Chap. 6). Cigarette sharing and gifting too, while being more important in rural areas, has been transformed by urbanisation with gifting becoming more important in urban social interactions. Given the importance of social interactions as a route to urban social mobility, self-exempting beliefs have also been perpetuated as an important part of traditional Confucian culture, emphasising politeness and respect. Finally political factors have had a role to play, not just in the imposition of *hukou* policies but in the early emphasis on patriotism in encouraging smoking.

As implied above, cultural factors also have a temporal and spatial expression. Confucian culture itself has varied in its importance over time. It arose during a period of social disorder in China and was later subsumed as a result of modernisation processes in the late nineteenth and early twentieth centuries during which time cultural restraints on female smoking were partly relaxed. The rise of the PRC and greater authoritarianism is intimately related to cultural processes such as patriotism and *hukou* both of which helped increase consumption. Likewise in post-reform China, commodification has become a more important element. In a similar way, culture also has a spatial expression, whether this be in the form of rural-urban or regional differences in smoking prevalence between North and South China, or the effect of ethnic factors relating to non-Han minorities in some regions (see Chap. 2).

Taken together the above comments suggest that cultural influences can be seen as outcomes of different forms of structural change. This has been a familiar theme in western research on 'cultures of poverty', factors influencing their perpetuation and impacts on health (Wilson 1997, 2009; Small et al. 2010). As was evident then, and now in China, the challenge is to gain a greater understanding on why different cultures arise and the extent to which they continue to influence smoking and other health and social behaviours.

References

Ark Marketing Research & Consulting (2002) Characteristics of Chinese smokers and their smoking-related behaviors. Ark Marketing Research & Consulting, Guangzhou (In Chinese) Cited in: Huang et al. (2012)

Astell-Burt T, Zhang M, Feng X, Wang L, Li Y, Page A, Zhou M, Wang L (2018) Geographical inequality in tobacco control in China: multilevel evidence from 98,058 participants. Nicotine Tob Res 20(6):755–765

Babiarz KS, Miller G, Yi H, Zhang L, Rozelle S (2012) China's new cooperative medical scheme improved finances of township health centers but not the number of patients served. Health Aff 31(5):1065–1074

Benedict C (2011) Golden-silk smoke: a history of tobacco in China, 1550–2010. University of California Press, Los Angeles

Bengoa M, Rick C (2020) Chinese *hukou* policy and rural-to-urban migrants' health: evidence from matching methods. East Econ J 46(4):224–259

Bourdieu P (2001) Masculine Domination. Stanford University Press, Stanford

Brook T (2004) Smoking in Imperial China. In: Gilman SL, Xun Z (eds) Smoke: a history of global smoking. Reaktion Books, London

Cai L, Wang X-M, Fang L-M, Cui W-L, Golden AR (2019) Socioeconomic disparities in prevalence and behaviors of smoking in rural Southwest China. BMC Public Health 19(1):Article 1117

Chan, K. W., & Buckingham, W. (2008). Is China abolishing the hukou system? China Q 195:582–606. https://doi.org/10.1017/S0305741008000787

Chapman S, Wong WL, Smith W (1993) Self-exempting beliefs about smoking and health: differences between smokers and ex-smokers. Am J Public Health 83(2):215–219

Chen CW, Xia C, Zheng R, Zhou M, Lin C, Zeng H, Zhang S, Wang L, Yang Z, Sun K, Li H, Brown MD, Islami F, Bray F, Jemal A, He J (2019) Disparities by province, age, and sex in site-specific cancer burden attributable to 23 potentially modifiable risk factors in China: a comparative risk assessment. Lancet Glob Health 7:e257–e269

Chen JT (1970) The may fourth movement redefined. Mod Asian Stud 4(1):63–81

Chen X, Stanton B, Li X, Fang X, Lin D (2008) Substance use among rural-to-urban migrants in China: a moderation effect model analysis. Subst Use Misuse 43:105–124

Chen X, Stanton B, Li X, Fang X, Lin D, Xiong Q (2009) A comparison of health risk behaviors of rural migrants with rural residents and urban residents in China. Am J Health Behav 33(1):15–25

Chen X, Li X, Stanton B, Fang X, Lin D, Cole M, Liu H, Yang H (2004) Cigarette smoking among rural-to-urban migrants in Beijing, China. Prev Med 39(4):666–673

Cheng HG, McBride O, Phillips MR (2015) Relationship between knowledge about the harms of smoking and smoking status in the 2010 global adult tobacco China survey. Tob Control 24(1):54–61

China Daily (2017) High-price cigarettes sign that gift-giving corruption is back. China Daily, 17 February. http://www.chinadaily.com.cn/opinion/2017-02/17/content_28234071.htm

Chu A, Jiang N, Glantz SA (2011) Transnational tobacco industry promotion of the cigarette gifting custom in China. Tob Control 20(4):e3

Colas M, Ge S (2019) Transformations in China's internal labor migration and the *hukou* system. J Lab Res 40(3):296–331

Cui X, Rockett TH, Yang T, Cao R (2012) Work stress, life stress, and smoking among rural-urban migrant workers in China. BMC Public Health 12(1):Article 979

Dikötter F, Laamann L, Xun Z (2002) Narcotic culture. A social history of drug consumption in China. Br J Criminol 42(2):317–336

Di Maggio P (2002) Why cognitive (and cultural) sociology needs cognitive psychology. In: Cerulo KA (ed) Culture in mind: toward a sociology of culture and cognition. Psychology Press, New York, pp 274–281

Ding D, Hovell MF (2012) Cigarettes, social reinforcement, and culture: a commentary on "tobacco as a social currency: cigarette gifting and sharing in China". Nicotine Tob Res 14(3):255–257
Du R, Xiao C, Cavalli-Sforza L (1997) Genetic distances between Chinese populations calculated on gene frequencies of 38 loci. Sci China Ser C Life Sci 40(6):613–621
Du Y (2000) Chinese recognition on relationship between smoking and health in Ming-Qing dynasties. Chin J Mediev Hist 30(3):148–150
Fang X (2012) Barefoot doctors and Western medicine in China. University of Rochester Press, Rochester. 294 pp
Festinger L (1957) A theory of cognitive dissonance. Stanford University Press, Stanford
Finch K, Novotny TE, Ma S, Qin D, Xia W, Xin G (2010) Smoking knowledge, attitudes and behaviors among rural-to-urban migrant women in Beijing, China. Asia Pac J Public Health 22(3):342–353
Fu D, Wong K, Li CY, Song HX (2007) Rural migrant workers in urban China: living a marginalized life. Int J Soc Welf 16(1):32–40
Fu Q, Wu C, Liu H, Shi Z, Gu J (2018) Live like mosquitoes: *Hukou*, rural-urban disparity and depression. Chin J Sociol 4(1):56–78
Gao K, Shi X, Wang W (2017) The life course impact of smoking on hypertension, myocardial infarction and respiratory diseases. Sci Rep 7(1):43330
Garnaut A (2014) The geography of the great leap famine. Modern China 40(3):315–348
Grittner U, Willsnack S, Kuntsche S, Greenfield TK, Wilsnack R, Kristjanson A, Bloomfield K (2019) A multilevel analysis of regional and gender differences in the drinking behavior of 23 countries. Subst Use Misuse 55(5):1–15
Hamnett C (2020) Is Chinese urbanisation unique? Urban Stud 57(3):690–700
He R, Kurihara H (2008) Shanghuo syndrome in traditional Chinese medicine. World Sci Technol 10(5):37–41
Heikkinen H, Patja K, Jallinoja P (2010) Smokers' accounts on the health risks of smoking: why is smoking not dangerous for me? Soc Sci Med 71(5):877–883
Hermalin AI, Lowry DS (2012) The decline of smoking among female birth cohorts in China in the 20th century: a case of arrested diffusion? Popul Res Policy Rev 31(4):545–570
Hesketh T, Lu L, Jun YX, Mei WH (2007) Smoking, cessation and expenditure in low income Chinese: cross-sectional survey. BMC Public Health 7(1):Article No. 29
Hou B, Nazroo J, Banks J, Marshall A (2018) Migration status and smoking behaviors in later-life in China – evidence from the China health and retirement longitudinal study (CHARLS). Front Public Health 6:Article 346
Hu M, Rich ZC, Luo D, Xiao S (2012) Cigarette sharing and gifting in rural China: a focus group study. Nicotine Tob Res 14(3):361–367
Huang C, Wang Y, Li X, Ren L, Zhao J, Hu Y et al (2020) Clinical features of patients infected with novel 2019 novel coronavirus in Wuhan, China. Lancet 395(10223):497–506
Huang L, Thrasher JF, Jiang Y, Li Q, Fong GT, Quah AC (2012) Incidence and correlates of receiving cigarettes as gifts and selecting preferred brand because it was gifted: findings from the ITC China survey. BMC Public Health 12(1):Article 996
Jackson AA, Manan WA, Gani AS, Carter YH (2004) Lay beliefs about smoking in Kelantan, Malaysia. Southeast Asian J Trop Med Public Health 35(3):756–763
Jiang J, Wang P (2018) Health status in a transitional society: urban-rural disparities from a dynamic perspective in China. Popul Health Metrics (1):16, Article 22
Kim SS, Son H, Nam KA (2005) The sociocultural context of Korean American men's smoking behavior. West J Nurs Res 27(5):604–623
Kohrman M (2007) Depoliticizing tobacco's exceptionality: male sociality, death and memory-making among Chinese cigarette smokers. China J 58(July):85–109
Kong L (2010) China and geography in the 21st century: a cultural (geographical) revolution? Eurasian Geogr Econ 51(5):600–618
Lee C, Gao M, Ryff CD (2020) Conscientiousness and smoking: do cultural context and gender matter? Front Psychol 11:1593

Lee YH, Fang T, Ang A, Lin HC, Chang YC (2019) Rural-urban disparities in smoking patterns among Chinese adults: a socialcological approach. J Ethn Subst Abus. https://doi.org/10.1080/15332640.2019.1633980

Li J, Rose N (2017) Urban social exclusion and mental health of China's rural-urban migrants: a review and call for research. Health Place 48(November):20–30

Li R, Wang D, Chen J, Chai J, Tang M (2015) Regional differences in smoking, drinking and physical activities of Chinese residents. Asia Pac J Public Health 27(2):NP230–NP239

Liao Y, Chen X, Tang J (2017) Differences in cigarette smoking and alcohol consumption by sex and region in China: a population-based, multi-stage, cluster sampling survey. Lancet 390(Special Issue S54):54

Lin HL (2018) Lyricism, the veneration of feeling and narrative techniques in the poetry talks of the southern society. Fron Lit Stud China 12(2):324–350

Liu J (2014) Ageing, migration and familial support in rural China. Geoforum 51(January):305–312

Liu M, Zhang Q, Lu M, Kwon CS, Quan H (2007) Rural and urban disparities in health services utilization in China. Med Care 45(8):767–774

Liu S, Zhang M, Yang L, Li Y, Wang L, Huang Z, Wang L, Chen Z, Zhou M (2017) Prevalence and patterns of tobacco smoking among Chinese adult men and women: findings of the 2010 national smoking survey. J Epidemiol Community Health 71(2):154–161

Liu W (2018) Experimentation: cigarettes in the communist base areas during world war II. In: Kohrman M, Gan Q, Liu W, Proctor R (eds) Poisonous pandas: Chinese cigarette manufacturing in critical historical perspectives. Stanford University Press, Stanford, California, pp 35–55

Liu Y, Hsiao WC, Eggleston K (1999) Equity in health and health care: the Chinese experience. Soc Sci Med 49(10):1354–1356

Liu Y, Gao J, Shou J, Xia H, Shen Y, Zhu S, Pan Z (2016) The prevalence of cigarette smoking among rural-to-urban migrants in China: a systematic review and meta-analysis. Subst Use Misuse 51(2):206–215

Liu Y, Song H, Wang T, Yang H, Gong J, Shen Y, Dai W, Zhou J, Zhu S, Pan Z (2015) Determinants of tobacco smoking among rural-to-urban migrant workers: a cross-sectional survey in Shanghai. BMC Public Health 15(1):Article 131

Luo C, Li S, Sicular T (2020) The long-term evolution of national income inequality and rural poverty in China. China Econ Rev 62(April):101465

Luo H, Yang H, Xu X, Yun L, Chen R, Chen Y, Xu L, Liu J, Liang H, Zhuang Y, Hong L, Chen L, Yang J, Tang H (2016) Relationship between occupational stress and job burnout among rural-to-urban migrant workers in Dongguan, China: a cross-sectional study. BMJ Open 6(8):e012597

Ma S, Hoang MA, Samet JM, Wang J, Mei C, Xu X, Stillman FA (2008) Myths and attitudes that sustain smoking in China. J Health Commun 13(7):654–666

Ma X, Ren X, Xu J (2016) The difference of collectivism between north and South China and its cultural dynamics. Adv Psychol Sci 24(10):1551–1555

Ma Y, Zhang L, Chen Q (2012) China's new cooperative medical scheme for rural residents: popularity of broad coverage poses challenges for costs. Health Aff 31(5):1058–1064

Mao A (2013) Space and power: young mother's management of smoking in extended families in China. Health Place 21(May):102–109

Mao A, Robinson J (2015) Home smoking restrictions before, during and after pregnancy – a qualitative study in rural China. Health Promot Int 31(3):606–613

Mao A, Yang T, Bottorff JL, Sarbit G (2014) Personal and social determinants sustaining smoking practices in rural China: a qualitative study. Int J Equity Health 13:Article 12

Mao ZF, Wu B (2007) Urban-rural, age and gender differences in health behaviours in the Chinese population: findings from survey in Hubei, China. Public Health 121(10):761–764

Mou J, Fellmeth G, Griffiths S, Dawes M, Cheng J (2013) Tobacco smoking among migrant factory workers in Shenzhen, China. Nicotine Tob Res 15(1):69–76

Mou J, Griffiths SM, Fong HF, Daes MG (2015) Defining migration and its health impact in China. Public Health 129(10):1326–1334

Mullins-Owens H (2015) Inequities in Chinese health services: an overview of the recent history of Chinese health care and recommendations for reform. SAGE Open 5(1):1–7
Oakes, T. (2000). China's provincial identities: reviving regionalism and reinventing "Chineseness". J Asian Stud 59(3):667–692.
Oates W, Chapman S, Borland R, Balmford J, Trotter L (2004) 'Bulletproof sceptics in life's jungle': which self-exempting beliefs about smoking most predict lack of progression toward quitting? Prev Med 39(4):776–782
Pan M, Zhu S, Duan W, Ma X, Luo X, Liu B, Kurihara H, Li Y, Chen J, He R (2020) "Shanghuo" increases disease susceptibility: modern significance of an old TCM theory. J Ethnopharmacol 250:112491
Pan Z (2004) Socioeconomic predictors of smoking and smoking frequency in urban China: evidence of smoking as a social function. Health Promot Int 19(3):309–315
Park M, Chesla C (2007) Revisiting Confucianism as a conceptual framework for Asian family study. J Fam Nurs 13(3):293–311
Peng L, Lingling L (2017) A cultural reflection on China's tobacco technology. Case Stud J 6(2):10–14
Pow CP (2017) Cultural geography. In: Richardson D, Castree N, Goodchild MF, Kobayashi A, Liu W, Marston RA (eds) The international Encyclopedia of geography. Wiley, New York, pp 1–17
Qiu L, Zhao D (2019) Urban inclusiveness and income inequality in China. Reg Sci Urban Econ 74:57–64
Quan ZX (1991) Urbanisation in China. Urban Stud 28(1):41–51
Ravillion M, Chen S (2004) China's uneven progress against poverty. World Bank Policy Research Working Paper 3408. World Bank, Washington, DC
Redmon P, Chen L, Wood J, Li S, Koplan J (2013) Challenges for philanthropy and tobacco control in China. Tob Control 22(Suppl 2):ii4–ii8
Rich ZC, Hu M, Xiao S (2014) Gifting and sharing cigarettes in a rural Chinese village: a cross-sectional study. Tob Control 23(6):496–500
Rich ZC, Xiao S (2012) Tobacco as a social currency: cigarette gifting and sharing in China. Nicotine Tob Res 14(3):258–263
Ru X, Wang W, Sun H, Sun D, Fu J, Ge S, Wang L, Wang L, Jiang B (2019) Geographical difference, rural-urban transition and trend in stroke prevalence in China: findings from a National Epidemiological Survey of stroke in China. Sci Rep 9(1):Article No. 17330
Shi L (1993) Health care in China: a rural-urban comparison after the socioeconomic reforms. Bull World Health Organ 71(6):723–736
Shin S, Wan X, Wang Q, Raymond HF, Liu H, Ding D, Yang G, Novotny TE (2013) Perceived discrimination and smoking among rural-to-urban migrant women in China. J Immigr Minor Health 15(1):132–140
Slopen N, Kontos EZ, Ryff CD, Ayanian JZ, Albert MA, Williams DR (2013) Psychosocial stress and cigarette smoking persistence, cessation and relapse over 9-10 years: a prospective study of middle-aged adults in the United States. Cancer Causes Control 24(10):1849–1863
Small ML, Harding DJ, Lamont M (2010) Reconsidering culture and poverty. Ann Am Acad Pol Soc Sci 629(1):6–27
Song Q, Smith JP (2019) *Hukou* system, mechanisms, and health stratification across the life course in rural and urban China. Health Place 58(July):102150
Stubbs B, Veronese N, Vancampfort D, Prina AM, Lin PY, Tseng PT, Evangelou E, Solmi M, Kohler C, Carvalho AF, Koyanagi A (2017) Perceived stress and smoking across 41 countries: a global perspective across Europe, Africa, Asia and the Americas. Sci Rep 7(1):7597
Swidler A (1986) Culture in action: symbols and strategies. Am Sociol Rev 51(2):273–286
Tai Z, Sun T (2011) The rumouring of SARS during the 2003 epidemic in China. Sociol Health Illn 33(5):677–693
Talhelm T, Zhang X, Oishi S, Shimin C, Duan D, Lan X, Kityama S (2014) Large-scale psychological differences within China explained by rice versus wheat agriculture. Science 344(6184):603–608

Talhelm T, Zhang X, Oishi S et al (2018) Moving chairs in Starbucks: observational studies find rice-wheat cultural differences in daily life in China. Sci Adv 4(4):eaap8469
Tang S, Bloom G (2000) Decentralizing rural health services: a case study in China. Int J Health Plann Manage 15(3):189–200
Tani M (2017) *Hukou* changes and subjective Well-being in China. Soc Indic Res 132(1):47–61
Textor C (2020) Number of migrant workers in China 2019. Statista, June 24. https://www.statista.com/statistics/234578/share-of-migrant-workers-in-china-by-age/
Twyman, L., Bonevski, B., Paul, C., & Bryant, J. (2014). Perceived barriers to smoking cessation in selected vulnerable groups: a systematic review of the qualitative and quantitative literature. BMJ open 4(12):e006414.
Tyner A, Ren Y (2016) The *hukou* system, rural institutions, and migrant integration in China. J East Asian Stud 16(3):331–348
Unger J (2016) The transformation of rural China, 2nd edn. Routledge, New York. 188 pp
United Nations, Department of Economic and Social Affairs, Population Division (2010) World urbanisation prospects, the 2009 revision. York, New
Wan X, Shin SS, Wang Q, Raymond HF, Liu H, Ding D, Yang G, Novotny TE (2011) Smoking among young rural to urban migrant women in China: a cross-sectional survey. PLoS One 6(8):e23028
Wang J, Li C, Jia C, Liu Y, Liu J, Yan X, Fang Y (2014a) Smoking, smoking cessation and tobacco control in rural China: a qualitative study in Shandong Province. BMC Public Health 14(1):916
Wang JW, Cui JT, Ding N, Zhang CG, Usagawa T, Berry HL, Yu JM, Li SS (2014b) A qualitative study of smoking behavior among the floating population of Shanghai, China. BMC Public Health 14(1):1138
Wang P (ed) (2020) The health status of internal migrants in China. Springer, Singapore. 177 pp
Wang Y, Schwart CR (2018) *Hukou* intermarriage and social exclusion in China. Res Soc Stratif Mob 56:28–39
Wank DL (2000) Cigarettes and domination in Chinese business networks: institutional change during the market transition. In: Davis DS (ed) The consumer revolution in urban China. University of California Press, Berkley, California, pp 268–286
Wilson WJ (1997) When work disappears: the world of the new urban poor. Vintage Books, New York
Wilson WJ (2009) More than just race: being black and poor in the inner city. Norton, New York, 156 pp
WHO (World Health Organization) (2015) China country assessment report on ageing and health. World Health Organization, Geneva
Wu X, Treiman DJ (2007) Inequality and equality under Chinese socialism: the *hukou* system and intergenerational occupational mobility. Am J Sociol 113(2):415–445
Xi Z (2018) On the northeast regional culture in Xiao Hong's novels. J Lit Art Stud 8(11):1593–1596
Xie S (2017) The migration, mental stress, and tobacco use of internal migrants in China: the moderating effects of the social context of the host society. Subst Use Misuse 52(13):1733–1743
Xinhua News (2019) National residents' disposable per capita income 28228 Yuan. Xinhua News, 21 January, 2019. http://www.xinhuanet.com/fortune/2019-01/21/c_1210043114.htm
Xu G, Ma M, Liu X, Hankey GJ (2013a) Is there a stroke belt in China and why? Stroke 44(7):1775–1783
Xu Y, Xu S, Wu Q, Guo Y (2013b) Tobacco knowledge among adults in Zhejiang Province, China. PLoS One 8(3):e59172
Xu Y, Xu S, Wu Q, Guo Y (2016) Association between secondhand smoke exposure at home and cigarette gifting and sharing in Zhejiang, China: a repeat cross-sectional study. BMJ Open 6(3):e010058
Yang G (2018) Action subject and surroundings of tobacco control and anti-tobacco control in China. In: Yang G (ed) Tobacco control in China. Springer, Singapore, pp 47–71

Yang T, Li F, Yang X, Wu Z, Feng X, Wang Y, Wang X, Abdullah ASM (2008) Smoking patterns and socio-demographic factors associated with tobacco use among Chinese rural male residents: a descriptive analysis. BMC Public Health 8(1):248–255
Yang T, Wu J, Rockett IRH, Abdullah ASM, Beard J, Ye J (2009) Smoking patterns among Chinese rural–urban migrant workers. Public Health 123(11):743–749
Yang XY, Yang T (2019) Nativity homogeneity in social networks and prostitution patronage among male migrant laborers. AIDS Behav 23(5):1339–1351
Yang XY, Kelly BC, Yang T (2014) The influence of self-exempting beliefs and social networks on daily smoking: a mediation relationship explored. Psychol Addict Behav 28(3):921–927
Yang XY, Kelly BC, Yang T (2016) Together we have fun: native-place networks and sexual risk behaviours among Chinese male rural-urban migrants. Sociol Health Illn 38(4):559–575
Yang X, Jing W, Gao C, Attané I (2019) Smoking behavior of "marriage squeezed" men and its impact on their quality of life: a survey study of China. Am J Mens Health 13(3):1557988319859733
Zhang J, Wang C (2012) Factors in the neighborhood as risks of suicide in rural China: a multilevel analysis. Community Ment Health J 48(5):627–633
Zhang Q, Churchill SA (2020) Income inequality and subjective Well-being: panel data evidence from China. China Econ Rev 60(April):101392
Zhang X, Dupre ME, Qiu L, Zhou W, Zhao Y, Gu D (2017) Urban-rural differences in the association between access to healthcare and health outcomes among older adults in China. BMC Geriatr 17(1):151
Zhang Y (2005) The social life of opium in China. Cambridge University Press, Cambridge
Zheng PP, Qian H, Wang F, Sun S, Nehl EJ, Wong FY (2013) Awareness of tobacco advertising, perceived harms of smoking, and beliefs about tobacco control among a sample of Shanghainese in China. Health Educ Res 28(5):879–887
Zheng T (2006) Cool masculinity: male clients' sex consumption and business alliance in urban China's sex industry. J Contemp China 15(46):161–182
Zheng Y, Ji Y, Dong H, Chang C (2018) The prevalence of smoking, second-hand smoke exposure, and knowledge of the health hazards of smoking among internal migrants in 12 provinces in China: a cross-sectional analysis. BMC Public Health 18(1):655
Zhi K, Wang L, Han Y, Vaughn MG, Qian Z, Chen Y, Xie L, Huang J (2019) Trends in cigarette smoking among older male adults in China: an urban-rural comparison. J Appl Gerontol 38(6):884–901
Zhong BL, Liu TB, Huang JX, Fung HH, Chan SSM, Conwell Y, Chiu HFK (2016) Acculturative stress of Chinese rural-to-urban migrant workers: a qualitative study. PLoS One 11(6):e0157530
Zhou L, Wang G, Jia C, Ma Z (2019) Being left-behind, mental disorder, and elderly suicide in rural China: a case-control psychological autopsy study. Psychol Med 49(3):458–464
李月新.契丹统治下的汉人来源与分布[J].辽宁师范大学学报*(社会科学版)*, 2007(3):111–115.

Chapter 6
Income Inequality, Urban Development and Smoking

Xiaozhao Y. Yang and Ross Barnett

Abstract Neoliberal reforms have economically transformed China, increased income inequality and changed the nature of the built environment. This chapter first develops a conceptual framework for examining patterns of intra-urban change and their effects on health behaviour and health outcomes. Using this framework two major themes are addressed: the growth of income inequality and its association with health outcomes and smoking and ways in which the pattern of urban development has affected the social distribution of smoking and increased health inequalities. While the evidence is fragmentary, there is some suggestion that social differences in urban smoking are beginning to conform to those expected from the western diffusion model.

Keywords Income inequality · Migrant enclaves · Neoliberalism · New urban poverty · Pathways · Post-socialist housing transition · Smoking · Stigmatisation · Stress · Tunnel effect · Urban development

6.1 Introduction

Since the 1990s China has experienced unprecedented economic growth which raised its World Bank classification to an upper middle-income economy. This growth is most manifest in the remarkable change in the built environment of Chinese cities. In Shanghai, for example, the new global financial centre in Pudong stands in marked contrast to some remaining traditional 'hutong' neighbourhoods

---

X. Y. Yang (✉)
School of Sociology and Anthropology, Sun Yat-sen University, Guangzhou, China
e-mail: yangxzh27@mail.sysu.edu.cn

R. Barnett
School of Earth and Environment, University of Canterbury, Christchurch, New Zealand
e-mail: ross.barnett@canterbury.ac.nz

R. Barnett et al. (eds.), *Smoking Environments in China*,
Global Perspectives on Health Geography,
https://doi.org/10.1007/978-3-030-76143-1_6

which, as in Beijing, have largely been eradicated unless they have been deemed necessary for heritage and tourism purposes. Underlying this physical and economic transformation have been profound socio-economic changes which, despite notable reductions in regional levels and differences in poverty, have been accompanied by marked increases in income inequality. This increase has been so persistent in the last two decades that China is now among the most unequal countries in the world (Jain-Chandra et al. 2018).

Smoking, along with other health behaviours, needs to be considered in the context of the neoliberal reforms that have altered patterns of income distribution and the urban fabric of China. With this in mind, this chapter first develops a conceptual framework for examining patterns of intra-urban change and their effects on health behaviour and health outcomes (Sect. 6.2). Using this framework two major themes are addressed. First we examine the growth of income inequality in China, its major dimensions and contributing factors (Sect. 6.3). We then consider the links between income inequality, health outcomes and smoking (Sect. 6.4). Given the neoliberal emphasis of the Chinese economy in recent decades, widening divisions between rich and poor come as no surprise, but, as Wan and Zhang (2006) first noted over a decade ago, there has been a lack of academic attention to the urban and health implications of such trends.

The second theme explores some of the different ways in which the pattern of Chinese urban development has contributed to smoking and increased health inequalities. Two particular questions are addressed: (1) within cities what particular urban socio-economic processes are likely to lead to an increased social concentration of smoking? (Sect. 6.5); (2) to what extent has increased urbanisation affected the social distribution of smoking? (Sect. 6.6) Is there any evidence that urban smoking patterns are beginning to conform to those expected from the western diffusion model? Because changes in health behaviour have not been central to research into either of these concerns, many of our comments are speculative, but highlight the need for more research into the links between changing urban environments, smoking and health. In order to understand the importance of these two issues, we first briefly provide a general framework for understanding the growth of urban inequality and its links to changes in the built environment and behavioural and health outcomes.

6.2 A General Framework on Urban Income Inequality, Urban Change and Health Outcomes

Income inequalities have increased in many parts of the world, leading to highly politicised debate concerning the link between these inequalities and poor health outcomes (Townsend and Davidson 1992; Wilkinson and Pickett 2011). Since the 1980s a voluminous literature has evolved on the links between the growth of state neoliberal policies and their social impacts, especially by theorists such as David

Harvey (1989) and Bob Jessop (1994) and more recently by researchers such as Brenner and Theodore (2002) and Storper (2016). Much of this work has focused on the impact of neoliberalism in deregulating markets, on the privatisation or marketisation of public services and on curbing union power in a variety of institutional settings. But it has also been highly ethnocentric, largely focusing on urban changes in more mature western economies with long histories of state intervention. Pinson and Journel (2016) are critical of the 'shallow attention' given to (other) 'contexts' that are equally deserving of analysis. There have also been a variety of theoretical and methodological debates, not only over the pathways linking increased inequality to health but also over the relative influence of the effects of absolute and relative inequality on health outcomes (Truesdale and Jencks 2016; Kragten and Rözer 2017).

Such debates are relevant to China which underwent its own economic reforms beginning in 1978. Typical of many other transition economies, the impacts of neoliberalism were manifest in the dismantling of former state enterprises, the deregulation of markets, an increased focus on marketisation and improved efficiency in productive sectors. The push for increased economic growth saw cutbacks in state support for public services, such as health care, with increased self-reliance increasingly replacing state support (Mullins-Owens 2015). These policies were not uniformly applied but affected rural regions to a greater extent than their urban counterparts. As shown in Fig. 6.1, such reforms led to adverse social consequences, particularly increased poverty and income inequality. While both had health and other social impacts (Yang et al. 2019a), it is only recently that increased attention has focused on the effects of income inequality on health behaviour.

It would, however, be a simplification to attribute adverse social outcomes purely to the impact of neoliberal policies. As Le Galès (2016) and others have noted, neoliberalism will not explain everything about the current transformation of urbanisation processes as cities may change for a variety of reasons, of which neoliberalism is only one. For example, urban change can also occur as a result of strategic policy actions which predate neoliberalism. In China the classic case is the *hukou* system, discussed in the previous chapter, which was a fundamental force in limiting state responsibility for migrant welfare in the face of rapid urbanisation (Lu and Wan 2014).

Nevertheless there is a close relationship between neoliberal reform and the nature of urban development. The partial dismantling of state monopolies, housing privatisation and the encouragement of an individualistic consumerist culture, for example, have all had impacts on health. In recent years there has been a renewed interest in the relationship between urbanisation-associated changes in patterns of human activity and lifestyles and the rise of chronic diseases such as cancer and diabetes (Gong et al. 2012; Miao and Wu 2016; Li and Wu 2016). Many studies have commented on the health penalties of urbanisation (Fang et al. 2009; Li et al. 2016; Hou et al. 2019), but apart from a focus on dietary change (Zhou et al. 2017) have seldom considered the impacts of urbanisation on smoking (Chen et al. 2018b), in particular how patterns of neighbourhood change may have an impact on the smoking prevalence of different social groups. Because neoliberalism is a social polarizing process, it might be expected that the negative effects of urban change

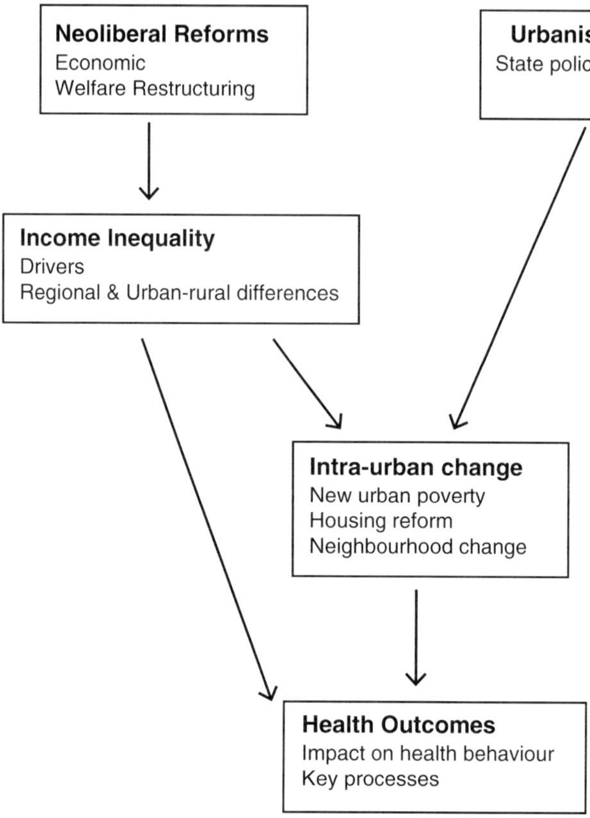

Fig. 6.1 Linkages between inequality, urban change and health outcomes. (Credit: Authors)

will have a greater impact on smoking prevalence of more disadvantaged groups, a pattern so common in the western world (Barnett et al. 2017). With these broader processes in mind, we now turn to an analysis of the growth of income inequality, internationally and in China and its health and smoking outcomes.

6.3 The Growth of Income Inequality

6.3.1 Global Trends

Research on income inequality and health is not new. Since the Black Report in the United Kingdom in the 1980s (Townsend and Davidson 1992), internationally a wealth of studies have documented the growth of health inequalities and noted the deleterious effects of globalisation, neoliberalism and increased income inequality upon health and other social outcomes. As a universal global issue, income inequality's toll on health was initially highlighted particularly by the work of Richard Wilkinson and colleagues (Wilkinson and Pickett 2006, 2011) who showed that countries with more unequal income distributions had more adverse health and social outcomes. However, despite a considerable volume of research over the last few decades, knowledge of the mechanisms linking income inequality and health is still incomplete especially since, until recently (Anwar et al. 2017; Vincens et al. 2018; Massa et al. 2018; Odusanya and Agboola 2017; Gugushvili et al. 2020), most work has occurred in western countries.

There is some consensus that there are three pathways linking income inequality and health outcomes: social comparisons, disruption in social capital and lack of investment in public resources. Negative perceptions of social position may be reflected in anti-social behaviour, reduced civic participation, reduced trust and less social cohesion. The loss of social capital can be both detrimental in the sense that more socially isolated individuals may smoke more (Graham 1993) and protective in the sense that it plays an important part in social interactions and the formation of community identity. The latter may encourage increased smoking particularly in those communities trying to cope with higher levels of stress (Stead et al. 2001; Stubbs et al. 2017). Similarly lack of social resources may also have an impact. Neoliberal economic regimes prioritise economic growth and financial performance. One direct consequence of neoliberal positions in governments is austerity and the lack of public institutional resources (Abramovitz et al. 2014; Viens 2019).

6.3.2 The Growth of Income Inequality in China

The growth of urban inequality can be seen as a structural cause linked to the broader processes of globalisation and neoliberal economic and social policies that have transformed social structures and reshaped much of the global economic order (Wan 2008). In China, following the introduction of the 'Open-Door Policy' in 1978, the initial impacts were both social and spatial. Social polarisation rapidly increased, but so did geographic differences in development with coastal provinces experiencing much more rapid growth than inland areas (Kanbur and Zhang 2005).

In recent years, sharp reductions in poverty and improved health outcomes, including life expectancy, have led to greater interest among Chinese economists in income disparities and their effects. In 2018 the International Monetary Fund (IMF) published a working paper on *Inequality in China – Trends, Drivers and Policy Remedies* (Jain-Chandra et al. 2018) which focused on the implications of income inequality for public services investment and economic growth. Consistent with other reports at the time, there were four main conclusions.

First, China has moved from being a moderately unequal country to being one of the most unequal (Piketty et al. 2019). Between 1980 and 2012, the Gini coefficient, an index documenting the level of the unequal income distribution, almost doubled, increasing from 0.30 in 1980 to 0.55 in 2012 (Xie and Zhou 2014). Although levels of inequality have plateaued in recent years (Zhou and Song 2016), the growth in income inequality has far exceeded the average for other countries at similar levels of development, and even those of some wealthier neoliberal countries, such as Germany, the United Kingdom and the United States (Xie and Zhou 2014; Chan et al. 2019). The reported recent decline in income inequality may be more rhetorical than real, with researchers being suspicious of the accuracy of official data (Chen et al. 2016, p.9).

Second, when viewed in terms of population subgroups, differences in education and skills were significant drivers of the increase and subsequent modest decline in income inequality. Better educated and more highly skilled persons disproportionately benefitted especially during the early years of China's economic boom, but such differences slowly became less apparent after 2007. This was also true of differences in income based on the sector of employment. In the 1990s income inequality rose most for those without work, suggesting an uneven provision of social protection. Recent policy moves, however, such as the *Dibao* system, providing income protection for rural residents, have resulted in a reduction in such differences (Jain-Chandra et al. 2018). More fundamentally, the price equilibrium in the global market has driven up the labour cost of China's shrinking pool of industrial workers. Combined with the oversupply in white-collar labours and an intensive cultural demand placed on education, in recent years China watchers and scholars have started to notice in some areas there is a greater financial return to skilled blue-collar training than professional credentials.

Third, while urban-rural income differences have widened since the mid-1980s (Wang et al. 2014), the contribution of regional disparities to this inequity has declined. In 1995 provincial differences accounted for 35% total income inequality but only 11% by 2013 (Jain-Chandra et al. 2018, p. 9). By contrast the share of total inequality accounted for by the rural-urban gap was 44% in 1995 and, following a further increase up to 2007, declined slightly to 34% by 2013. Compared to the United States, income inequality in China is still far more explainable by urban-rural and regional disparities, whereas such influences are negligible in the United States where family structure and race are more important factors (Xie and Zhou 2014).

Fourth, in marked contrast to most developing countries, relative income inequality was initially higher in rural compared with urban areas. Since the mid-1990s there has been some convergence, with the gap between rich and poor rising much faster in cities than in the countryside. Ravillion and Chen (2004, p. 53), for example, showed that by the early 2000s, the absolute Gini index, relative to the 1990 mean, was approximately three times higher in urban compared to rural areas. Chen et al. (2016) have also shown that while urbanisation increased income inequality, its effects were lagged, suggesting that its causal effects on the growth of inequality were not artefactual. The greater growth of income inequality in urban areas is likely to reflect a number of factors including the greater growth of productivity, higher returns to education and increased skills in urban areas and the ballooning of urban-rural migration resulting in a surplus and 'captive' low-wage population. In addition, the effects of institutional biases imposed by the *hukou* system have trapped migrant groups in segregated places and long-term poverty.

6.4 Income Inequality and Health in China

Of course the important question is to what extent has China's embracing of neoliberal reforms produced social consequences similar to those experienced in western countries (Li et al. 2019; Luo and Xie 2020). We approach this question by first examining general effects of income inequality on different measures of health before turning to a more detailed analysis of smoking.

In recent years an increasing number of studies have revealed the adverse impacts of income inequality on health outcomes (Table 6.1). Apart from three exceptions (Wong et al. 2009; Bakkeli 2016; Ding et al. 2017), two of which relied on medical biomarkers, the evidence suggest that the Chinese experience is similar to that of western countries. Income inequality is generally negatively related to different types of health outcomes, whether these be self-rated health status (Pei and Rodriguez 2006; Feng et al. 2012; Lin et al. 2017) or mental health, well-being and happiness (Sun et al. 2012; Wang et al. 2015a; Du et al. 2019; Fan et al. 2020). Studies using biometrics, such as blood pressure and cholesterol level (Bakkeli 2016; Ding et al. 2017), however, suggest that the relationship between inequality and health is more nuanced and subject to the specific measures of health outcomes.

Second, the relative effects of inequality on health are group dependent. As in western countries, there is increasing evidence that inequality has greater impacts on the health of the poor compared to the nonpoor (e.g. Du et al. 2019; Feng et al. 2012; Tan et al. 2018; Yang et al. 2019c; Zhang and Churchill 2020). The effects are not consistent, however, with some studies indicating that high-income earners are

Table 6.1 Recent studies of income inequality and health in China

Health indicator	Studies 2006–2020
Self-rated health status	Li and Zhu (2006), Pei and Rodriguez (2006), Jin and Yangyang (2007), Wong et al. (2009), Feng et al. (2012), Sun et al. (2012), Fang and Rizzo (2012), Lin et al. (2017), Ding et al. (2017), Yang et al. (2019c) and Inoue et al. (2019)
Subjective Well-being, quality of life/subjective social status	Hu and Lu (2012), Lin et al. (2017), Wu and Li (2017), Rarick et al. (2018); Tan et al. (2018)[a], Du et al. (2019) and Yu et al. (2019)
Happiness	Smyth and Qian (2008), Oshio et al. (2011), Wang et al. (2015b), Huang (2019) and Cheung (2016)
Depression/mental health	Sun et al. (2012), Lin et al. (2017), Zeng and Jian (2018), Fan et al. (2020) and Zhang and Churchill (2020)
Medical biomarkers	Bakkeli (2016) and Ding et al. (2017)
Health behaviours	Li and Zhu (2006), Ling (2009), Sun et al. (2012) and Yao and Asiseh (2019)

[a]Based on the EQ-5D index of health status, including anxiety and depression

also adversely by income inequality (Yu et al. 2019; Fan et al. 2020). Du et al. (2019) suggest that in growing economies, the richest people often suffer from a heavy workload and are more likely to be highly stressed in order to maintain their high socio-economic ranking. Thus, such pressures may exacerbate unhappiness and distress, with higher rates of smoking prevalence and intensity being likely outcomes.

Third, there is also evidence that the relationship between income inequality and health outcomes is curvilinear (Li and Zhu 2006; Jin and Yangyang 2007; Wang et al. 2015a). Known as the 'tunnel effect', this suggests that income inequality, as a marker of economic progress, may initially be viewed positively by certain groups who view it as an economic opportunity. This opportunity, however, may fail to occur, as has been the case in some transition economies (Grosfeld and Senik (2010) as well as in less economically developed regions of China with fewer chances of social mobility (Yao and Asiseh 2019). In such circumstances, the effects of income inequality on social well-being are likely to be reversed and become negative (Zhang and Churchill 2020).

6.4.1 Impacts of Income Inequality on Smoking

Despite an increased interest in links between inequality and health, this has not transferred to studies of health behaviour. Internationally, countries, such as China, that have higher levels of income inequality tend to have higher rates of youth smoking (Li and Guindon 2012). Kuo and Chiang (2013) report similar results for Taiwan, where relative deprivation was significantly associated with a higher prevalence of poor self-rated health, depressive symptoms and current smoking in both genders.

Within Mainland China Li and Zhu (2006) found that both smoking prevalence and smoking intensity were higher in more unequal communities; a one standard deviation increase in the community Gini increased the probability of smoking by 2.1% points. Levels of alcohol consumption were also considered but were not significant. Communities were defined as either neighbourhoods within cities or villages within counties. The authors acknowledged that the scale at which inequality is measured is likely to be important, implying that there is a need to consider other geographic scales of income inequality given the likelihood of different pathways linking the latter to smoking outcomes.

In contrast to the standard approach using Gini indexes as a measure of income inequality, Sun et al. (2012) examined perceptions of inequality and their relation to smoking. This is a common approach in western research (e.g. Siahpush et al. 2006). The authors examined the effect of two measures of self-perceived relative income (SPRI) inequality on smoking among Chinese youth: household income relative to peers and relative to their own past (Sun et al. 2012). The relationship between SPRI and smoking was not consistent and differed between girls and boys. Higher income relative to others was a risk factor for smoking among girls but not for boys. By contrast, both higher and lower incomes relative to the past were risk factors for smoking among boys but not girls. These differences perhaps suggest that rapid changes in family income in either direction caused greater stress among boys and that social comparisons with peers were more salient for girls. This will be particularly evident for men who have traditionally been the main income earners. However, for younger women, smoking may represent a sign of conspicuous consumerism or success.

The effects of relative inequality have also been shown to be important among older adults. In a study of adults over the age of 50, Ling (2009) showed that while changes in relative deprivation increased the probability of ever, but not current, smoking, the former was most typical of the highest-income quartile of older adults compared to other income quartiles. This suggests that higher income inequality was a risk factor for persons with higher incomes, reflecting either peer pressures to smoke or the stresses associated with maintaining a high social position among this group. There was also a much larger effect of relative deprivation on smoking in the urban compared to the rural samples suggesting that the pressures of maintaining one's social position were more acute in urban settings.

The lack of research on the effects of income inequality on smoking is partly compensated by the strong links found between inequality and other health status variables such as depression, which is often higher among smokers (Ni et al. 2017; Luo et al. 2018; Stubbs et al. 2018). Until more research is undertaken on changes in smoking on the part of different social groups in contexts where income distributions have changed significantly, the conclusions drawn will largely have to depend on secondary health indicators which show strong associations to smoking. Given the structural changes affecting Chinese society and the stresses they have generated, there is considerable scope for further research between increased social inequality and smoking and other health behaviour outcomes.

6.4.2 How Relevant Are Western Explanations of the Health Effects of Income Inequality?

Most studies of the effects of income inequality on health have been conducted in western countries especially at a time when socio-economic conditions were markedly different from China. Since lower-income groups and certain ethnic minorities in western countries were disproportionately affected by state economic and welfare restructuring, it is not surprising that the health and well-being of such groups were more affected by high rates of inequality. China presents a different picture. Decades of economic growth and increased social mobility, which helped reduce absolute levels of poverty, have meant that the 'rising tide' did indeed lift all boats, but that some boats lifted faster than others. Under such circumstances the negative effects of inequality on health and smoking may not only exhibit threshold effects but also be context dependent. With this in mind, we briefly evaluate the importance to China of the three pathways identified previously. Given the paucity of research on smoking, our comments pertain more to general health outcomes, but bearing in mind that many of these such as self-rated health or depression are themselves related to smoking (Stubbs et al. 2018).

6.4.2.1 Social Comparisons

According to the social comparison viewpoint, people compare themselves with others in terms of their place in the income distribution (Festinger 1954). Such comparisons may occur with respect to a variety of reference groups, be these be structured by locality, by age, by occupation/labour force position or by some combination of these factors (Kondo et al. 2008). The extent to which social comparison occurs is also likely to vary according to national context and the history of labour and ethnic relations which, in many western countries, have involved fractious battles for hard-won gains. In China higher egalitarian preferences may also be important, influenced by Chinese traditional culture and past socialist egalitarianism (Yu et al. 2019). Under the influence of Confucian culture, income inequality may be disliked and homogeneity valued to a much greater extent. Yang et al. (2019c) have argued that while being an individual of lower status is inherently hurtful and has detrimental consequences for health and well-being, the context within which the status is produced and maintained matters even more. In particular, lower status individuals in a *homogenous context*, that is, where everyone is more or less similar in status, suffer fewer of the undesirable health effects exerted by low status. Some indication of the impact of social comparisons in provoking stress and increased smoking is evident in Kuo et al.'s (2013) study of working-age Taiwanese men. While higher income inequality produced increased depressive symptoms, it also produced higher rates of smoking, thus suggesting that psychosocial stress was an important factor in these relationships.

Social comparisons are also affected by other contextual variables such as the degree of economic development of a region or level of rurality. For example, a number of studies have shown that the effects of income inequality on health are most damaging for lower-income groups and that the effects are worse in rural areas (e.g. Wang et al. 2015a; Tan et al. 2018). In such situations it could be argued that the social context is less fluid and chances of social mobility are lower, thus leading to greater negative perceptions on the part of lower-income groups of their life chances. The harmful effect of social comparison may be also amplified for low-income individuals in rural areas because of the paucity of supportative resources and agencies that are otherwise available in larger financially healthier cities to selectively assist with poor people's living.

In more fluid urban situations, with greater social mobility, it is likely that income inequality will have less impact on perceptions of well-being, future life chances and health behaviour outcomes. This is the basic tenet of the 'tunnel hypothesis' which posits that the poor may not build up a sense of dissatisfaction when they compare their incomes to the rich. This hypothesis suggests that high inequality associated with rapid economic development may initially be tolerated in a society where economic and social changes are accompanied by positive expectations about the future (Cheung 2016), greater life chances and increased social well-being (Zhang and Churchill 2020). This situation perhaps will be most typical of urban migrants and those in provincial situations undergoing rapid change. However, if income disparities persist and little change occurs, then the impact of the 'tunnel effect' on health can soon change from positive to negative. The tunnel effect also appears to apply to rural areas. Knight et al. (2009) found a positive correlation between both past and expected future relative incomes of rural residents and social well-being, suggesting that people saw higher levels of income inequality as an opportunity for social advancement.

To date there have been insufficient studies of the long-term impact of the 'tunnel effect' on health outcomes, but two pieces of evidence suggest that its effect is transitory. First, compared to other national contexts, income earning potential is rather unstable in China, and incomes earned this year are a poor predictor of incomes earned in subsequent years (Chan et al. 2019). Second, while some cross-sectional studies show support for a 'tunnel effect' (among migrants) (Yu et al. 2019), as a general explanation, this is not sufficient given that the positive effects of income inequality on health can soon turn negative. This suggests that migrant expectations have not been met and that other contextual influences may be more important in influencing smoking and health outcomes. The 'tunnel effect' has been shown to be less important in Western China indicating that the rate of economic development also has an effect on perceptions of income inequality. As actual social mobility has also been shown to be lower in more economically developed cities (Chen et al. 2018a), where competition is fierce, this suggests that the relationship between income inequality and health outcomes is curvilinear. Such findings thus point to the need for more detailed work to be undertaken on the significance of psychosocial processes and smoking, not just among different types of migrant populations but also among lower-income Chinese in a variety of contextual settings.

6.4.2.2 Social Capital

The loss of social capital has also been advanced to explain the adverse effects of income inequality on health. In transitional societies, such as China, with high rates of population mobility, changes in community social structure and social connectedness, such factors assume even more importance. High rates of social mobility may cause an erosion of mutual trust, once so common in rural communities, but may also affect positive expectations about the future especially in urban situations where new social networks need to be established. While an increasing literature has devoted itself to the evolution of different forms of social capital in China and their links to general, physical and mental health, there has been little attention to how social capital has been a mediating factor in limiting the adverse impacts of income inequality. Generally it has been shown that while social capital is protective of general and mental health (Meng and Chen 2014; Wang et al. 2018; Zhang and Jiang 2019), this depends very much upon social position. The transient lifestyle and difficulty to establish stable networks at a place led migrants to be frequently characterised by lower levels of trust both within their communities and with local residents (Wang et al. 2009). Levels of trust also reduce levels of happiness especially in places where income inequality is higher (Chen et al. 2012).

Western social scientists have frequently proposed that inequality damages social relations, but few Chinese studies have explored the extent to which social capital mediates the relationship between inequality and well-being. Du et al. (2019), for example, found that income inequality, when measured at the provincial level, was not associated with the quality of social relationships and levels of trust. However, when measured at the urban scale, Dai et al. (2020) report that residents in cities with greater inequalities (in housing assets) had lower levels of trust. Nevertheless, there is increasing evidence that, at the individual level, those with higher rates of trust will have healthier lifestyles, and are more likely to be non-smokers (Xue and Cheng 2017; Sun and Yuan 2019). Overall it is surprising that, given recent concerns with high rates of income inequality and its effects on health in China, more attention has not been directed to the mediating role of social capital.

6.4.2.3 Lack of Institutional Resources

Institutional resources that serve the poor tend to be lacking in unequal economies. The importance of the institutional arrangements, however, will depend upon the relative power of central and subnational governments in addressing inequality concerns. In China the State Council of the PRC in the Twelfth Five-Year Plan in 2011 emphasised the need to reverse the widening income gap as soon as possible (State Council 2011). Its subsequent actions, relating to income tax reform, labour market minimum wage policies, rural poverty relief and expansion of medical care coverage, have all contributed to the reduction of income inequality observed in recent years (Jain-Chandra et al. 2018). Despite such moves on the part of central government, subnational governments also affect the degree of inequality. China, as part of its economic reforms, devolved fiscal authority from central to provincial governments which are now responsible for more than half of total spending (Cevik and Correa-Caro 2019). However, Hao et al. (2020) using data for 23 Chinese provinces showed that fiscal decentralisation increased income inequality and was detrimental to public health. This reflects the nature of public spending, especially in less economically developed provinces, and in rural areas, where the focus was largely on infrastructure investment and public administration rather than social protection or health care. These results confirm the findings of Shi and Starfield (1993) in the United States where they found that higher state investment in primary care mediated the negative impacts of income inequality on health. Such findings are particularly important for smoking in China especially given lower health spending in rural areas.

Thus, certain types of subnational government spending appear to worsen inequality more than is compensated for by central government transfers. What is not clear, however, is the extent to which the presence of high social inequality at the provincial level affects local political decisions regarding the nature and location of public spending. The extent to which this occurs is likely to reflect the nature of the local political leadership and the extent to which income inequality and its social consequences are major policy concerns. Zhang et al. (2020) found that cities that had 'local' leaders were more (almost 6%) likely to have redistributional policies that targeted social transfers to low-income households, compared to cities where leaders had been transferred from other places. They suggest that while mass cadre transfers of non-local personnel between localities may have limited the formation of local interest groups and enhanced government efficiency, they also appear to have prioritised economic over social concerns, compared to situations where 'local' leaders were more attuned to local needs. The effect of income inequality on patterns of local spending is also likely to reflect the ideological orientations of local leaders and the extent to which they attribute poverty and disadvantage to individual 'failure' or wider structural patterns of disadvantage (He et al. 2020).

The effect of institutional explanations on the link between income inequality and health also extends to the urban-rural divide where differences in smoking prevalence are obvious. Despite increased public intervention since 2003, public financial support to rural residents for health care is meagre compared to that for their urban counterparts, resulting in higher smoking rates and other adverse health outcomes. To illustrate this point, Fan et al. (2020) examined how the effects of income inequality on depression among older adults of differing social status from both urban and rural areas were mitigated by public health investment. They found that while income inequality resulted in increased depression, public health investment tended to reduce it, but the effects were different between urban and rural areas and by social group. Public health investment had its greatest impacts in reducing depression in urban areas and among the nonpoor. By contrast in rural areas public health investment had no significant impact, suggesting that lack of rural resources may have played a part in this outcome.

6.5 Chinese Urban Development and Its Implications for Smoking

The previous section has shown that the rise of income inequality is an important process affecting smoking in China. Following our model in Fig. 6.1, however, we also need to consider the interrelationships between neoliberal changes and patterns of urban growth. In western countries the advent of neoliberal governments in the 1980s and 1990s had profound implications for the health of cities, especially their lower-income populations. Heightened urban inequality, coupled with increased disadvantage, raised levels of urban stress and helped entrench smoking in many poor and minority communities. These issues have been discussed in detail elsewhere (Barnett et al. 2017), but have received little attention in China.

Before examining the social distribution of smoking in Chinese cities (Sect. 6.6), we step back to take a broader perspective on the urbanisation processes in China that may have implications for smoking. Using four brief case studies, we then demonstrate how the growth of urban inequality, and other associated political and social processes, is likely to invert traditional patterns of smoking resulting in higher consumption on the part of lower-income groups.

6.5.1 *Patterns of Urban Development*

As was emphasised in Chap. 4, studies of the impact of neighbourhood conditions on health and smoking are sorely lacking in China. Yet the pattern of neighbourhood development has created many of the preconditions for poorer health outcomes along with high rates of smoking and other forms of substance abuse. Particularly important here have been the growth of urban poverty and the social polarisation of urban space. While attention has been increasingly directed to the health effects of these different trends, smoking has received little attention.

In western market economies, increased poverty and income inequality were intimately related to neoliberal policies of economic restructuring and the contraction of the welfare state. These trends were instrumental in leading to the growth of spatial concentrations of poverty and extreme disadvantage (Wilson 1997). The effects of such policies were evident in numerous multi-level analyses which showed that area-level deprivation had emerged as a significant predictor of smoking prevalence above and beyond the significance of individual-level factors. For example, in a case study of Scotland, Barnett et al. (2017, pp. 93–94) showed that area-level differences in deprivation were most evident for current smokers compared to ever-smokers and particularly for females.

In transitional economies, such as China, the question remains regarding the extent to which patterns of urban development are leading to a spatial concentration of disadvantage and associated patterns of smoking. In the following sections, we briefly review the extent to which spatial concentrations of poverty may be occurring with respect to different local urban processes: the growth of urban poverty, the extent to which there is an increased social polarisation of urban space and the role of housing booms in limiting the residential choices of Chinese urban residents.

6.5.1.1 A New Urban Poverty

The adoption of neoliberal policies in China's new economic order has been evident with the dismantling of state-owned enterprises which, like many of their counterparts in the western world, were overstaffed and lacked technological innovation and market efficiency. A sharp rise in urban poverty and income inequality was the inevitable result of these changes.

The new urban poor include two main groups, affected by different institutional arrangements: the permanent urban poor and the migrant poor (Liu et al. 2008c). The permanent poor largely consist of a group of workers made redundant by the collapse of former state-owned enterprises, but who may have been partially absorbed into the informal sector (Liu and He 2017). Many of these workers were once part of state work units (*danwei*), which prevailed from 1949 until 1998 and were, at the time, the basis of urban community formation and governance (Wang and Murie 2000; He 2013). Many *danwei* built their own residential compounds creating live-work communities that encompassed employment, housing and welfare (Kan et al. 2017). Each *danwei* compound was managed by a state-owned enterprise. Thus, as in the former Soviet Union, housing segregation reflected different economic sectors rather than social segregation. The marketisation of the Chinese economy, however, resulted in the gradual dismantling of the work-based welfare system (Liu et al. 2008a, 2008b). Today many of these former *danwei* areas are comprised of poorer-quality public housing with many units rented on a short-term basis (Kan et al. 2017).

As was pointed out in Chap. 5, the migrant poor is differentiated from permanent urban poor because of the effects of the discriminatory *hukou* system which excludes them from social services and from formal urban employment. Many are, therefore, forced into the informal sector, often competing with laid off workers (Liu et al. 2008a). In common with the situation in many LMICs, they live in urban villages, commonly on the urban periphery and known for their substandard housing and low-quality built environments. In a study of Nanjing, Liu et al. (2008a) found that, compared to the urban poor, the migrant poor tended to be younger, have less education, have higher job turnover and residential mobility and were more likely to live in the private rental housing sector. Despite these material differences, the authors report that they had a higher level of satisfaction with their living conditions, often comparing their current situation to that of their previous living environment in rural areas. However, such conclusions need to be treated with caution given extensive evidence of the marginalisation and poorer health of rural migrants especially those with rural *hukou* status (Fu et al. 2007, 2018; Yang et al. 2018). In addition, younger migrants are also those more likely to suffer a decline in health following migration, a trend partly reflecting changes in their emotional state and residence in high turnover neighbourhoods lacking social capital and trust (Lu et al. 2020). It is not surprising that such groups experience higher levels of life and work stress and consequently higher levels of smoking (Cui et al. 2012).

6.5.1.2 The Social Polarisation of Urban Space

The question immediately arises regarding the extent to which the emergence of poor neighbourhoods in Chinese cities represents the formation of a new polarised urban structure. This is an important question, since new patterns of segregation might affect perceptions of inequality, social capital formation, smoking and other health outcomes. As Liu and He (2017, p. 296) have noted, ‘to date there have been few efforts to form a spatial understanding of China's new urban poverty and the significance of their neighbourhoods to different poverty groups remains unclear'. For example, in western cities disadvantaged neighbourhoods are frequently stigmatised resulting in increased feelings of social alienation and an oppositional relationship to tobacco control initiatives (Factor et al. 2011). Smoking thus often becomes part of a local neighbourhood identity with the result that, as was emphasised in the previous section, both individual- and community-level socio-economic status have independent effects on smoking and its continuance. However, area stigmatisation, as yet, does not appear to be significant in Chinese cities. Liu and He (2017), in a study of 25 poor neighbourhoods in 6 cities, found that most residents (94%) did not perceive their neighbourhood to be stigmatised and that there was little variation in this perception between the 4 different types of neighbourhoods studied (old city neighbourhood, workers' village and inner and peri-urban migrant neighbourhoods). The lack of stigmatisation of poorer neighbourhoods was attributed to their recency of development compared to more deprived neighbourhoods in western cities. Of course, differences in stigmatisation also reflect the spatial concentration of poverty which is likely to be much greater in western cities.

High levels of spatial social segregation, comparable to western cities, are not yet evident in China. Instead, as Li and Wu (2008) demonstrated in the case of Shanghai, residential segregation is mainly tenure based, especially in terms of public housing rentals and commodity housing, rather than socio-economically based. Although spatial segregation (as measured by the index of dissimilarity) existed between groups with the highest and lowest levels of education, the difference was modest (0.325) with substantial mixing occurring between high and middle levels of education (0.276) and especially between middle and low education groups (0.078). By contrast much higher levels of segregation occurred between different housing tenures, particularly between public rental and purchased housing (0.682) and rented commodity housing (0.506).

Traditionally housing and neighbourhood differences have reflected the importance of *danwei* work units, but these no longer play a significant role in post-reform China with cities transformed as a result of economic restructuring and the privatisation of many former state-owned enterprises. Consequently new forms of commodity housing, built by real estate developers, have emerged as the dominant tenure (He and Wu 2007). Between 1985 and 2010, the proportion of public rental housing dropped from 70.7 to 3.4%, while that of privatised public housing and newly built commodity housing increased from zero to 42.6% and 50.6%, respectively (He 2013, p. 253). Of the houses built from 2000 to 2007, 80% were in gated commodity housing estates. This new structure has sustained former patterns of privilege with leaders, managers and professionals benefitting most, industrial workers less and migrants the least. The reconstitution of housing markets is likely to accentuate urban social-spatial polarisation as workers remaining in poorer public housing are differentiated from those doing better. For migrants not eligible for public housing, the result is likely to be the same. Currently the post-socialist housing transition in China is closely related to pre-existing institutional patterns which have yet to reconstitute themselves. However, as Li and Wu (2008), and others (Hao

2015; He et al. 2015; Wu 2016), have speculated, in the long term tenure-based segregation is likely to find a greater spatial expression as developers focus on more preferred urban locations and different social groups begin to congregate in different residential areas.

6.5.1.3 Housing Booms and Increased Inequality

The recent restructuring of urban space also reflects the influence of housing booms as a contributing factor to urban stress and the growth of urban inequality. Rampant expansion of real estate ventures since the 1990s has fuelled the Chinese economy and GDP growth. Monetary policy after the 2009 global recession injected four trillion Yuan into the Chinese market to help business with leverage and consumers with expenditure. However, instead of reviving manufacturing, much of this money flowed into real estate, including both commercial land use and residential housing. Consequently residential property prices tripled or quadrupled across much of China (Wang et al. 2012). Soaring prices were experienced mainly by properties in 'good areas' with access to schools, whereas the development of houses in less desirable areas merely kept up with market performance.

The health implications of the housing boom are receiving increased attention. Compared to income inequality, a perhaps more serious but often neglected aspect of inequality is property inequality (measured by the Gini coefficient in housing values) which rose dramatically following the economic stimulus package applied by central government in the wake of the 2008 global financial crisis (Dai et al. 2020). The authors found that higher levels of property inequality lessened trust, with this most evident among individuals whose fathers were Communist Party members, who saw inequality as unfair and a signal of exploitation. While the authors did not specifically address personal responses, such as smoking or drinking, it is likely that high housing prices are implicated as a result of the increased stress of living in an increasingly unequal urban environment.

High housing prices are also implicated in other changes in the built environment, particularly the loss of green space. Public green spaces have not been considered an integral part of the public infrastructure of Chinese cities, with commercial and residential developments often seen as more attractive by boosting local economic growth (Chen and Hu 2015). Chen and Hu also found that local government dependence on land finance exerted a negative influence on the provision of urban green space, especially in cities in the eastern coastal and central regions where land demand was highest. While evidence for the impact of urban green space on smoking in China is largely non-existent, important relationships have been shown between green space and stress (Yang et al. 2019b), itself related to smoking and a range of other negative health outcomes (Chen et al. 2020; Leng et al. 2020; Wang et al. 2017).

6.5.1.4 Landscapes of Change

To illustrate the new social differentiation which appears to be emerging in urban China, participant observation fieldwork was undertaken in neighbourhoods in four Chinese cities. While not exhaustive, the four case studies nevertheless give some indications of recent neighbourhood trends and their likely smoking consequences. The case studies are Sanhe, an urban migrant neighbourhood (in Shenzhen), a previous migrant neighbourhood which has undergone some gentrification (in Changsha); Gudang, a socially mixed established neighbourhood (in Hangzhou); and Songshan Lake, a more affluent neighbourhood (in Dongguan).

The first image (Fig. 6.2) shows a lane in the Sanhe migrant enclave on the outskirts of Shenzhen City, part of the special economic zone. The area, now undergoing gentrification, still has many contractors who hire day labourers en masse. At the turn of the century, the city's migrant worker population was located downtown, but as manufacturers, such as Foxconn Technology Group, relocated closer to the urban periphery, recruiting stations were established at Sanhe in Longhua District. A new congregation point for migrant workers emerged, making it a magnet and transfer station for unskilled labourers across Shenzhen and all over the country (China Daily 2017).

Sanhe is known for harbouring transients who cannot, or will not, find stable jobs elsewhere, and people who live in exile because of criminal records or bad credit ratings. It is the stereotypical migrant enclave where the neighbourhood is crowded and unsanitary. Most buildings are multiple-family households rented as packed dimly lit dormitories. When you enter these dormitories, you are likely to see many strangers, along with mice and cockroaches and in some places up to eight people share the same dark smelly room. Amid the smell of human odour, pungent shoes, rancid food and mould emanating from the crevices of aged buildings, a catch of tobacco smoke is actually quite refreshing and aromatic. The environment itself invites the use of tobacco and alcohol to numb the senses after a day's hard work. In addition, packing into a room with strangers is itself a stressful routine experience. Under such circumstances people often resort to substances like tobacco to relieve their stress, and consequently smoking prevalence rates are very high among certain groups of migrants. Living in a neighbourhood like the one pictured also fully exposes you to a smoking environment. Most of your friends will be male smokers, people spend their leisure time smoking, and friendships are built and strengthened by both smoking and drinking.

The scene depicted in Fig. 6.3 was taken during fieldwork on urban development in Changsha. Changsha is a developed provincial capital but does not enjoy the same level of foreign investment in manufacturing and salaries comparable to mega-cities such as Beijing and Shanghai. The photo shows an old community uplifted and refurbished from a migrant urban village. Compared to the Sanhe migrant enclave, the Changsha neighbourhood caters more to lower- to middle-income family groups. The infrastructure in this neighbourhood is still very poor with power wires haphazardly laid in the air. The road pavement is uneven with lots of potholes and untidy piles of garbage on the roadside, but building quality is solid. Apartments

Fig. 6.2 Sanhe migrant enclave, Shenzhen. (Day labourers resting and smoking in an urban alley. Empty tobacco packages were littered on the ground. Credit: Authors)

are spaced at a standard distance from each other. Some apartment owners even decorated their doors and windows and beautified them with ferns. Shops and stalls on the roadside also operate in compliance with local hygiene and housing ordinance. In this type of community, it is possible that the environmental stress resulting directly from the unpleasant living space and building structure is much alleviated. Due to a greater physical distance between strangers and an appropriate social space reserved for residents, the transmission of smoking by sight and smell is weaker. Tobacco smoking is still endemic in this neighbourhood, but the causes of smoking likely differ from that in urban slums. Social capital will be higher in this type of neighbourhood with higher levels of trust and public participation (Zhu and Fu 2017). In such situations social contact between neighbours may encourage smoking and the sharing of cigarettes.

The Hangzhou neighbourhood of Gudang (Fig. 6.4) is more affluent than that of Changsha, but is still relatively socially mixed. The area's development is likely to

Fig. 6.3 A smoker passing an old neighbourhood in Changsha. (Credit: Authors)

have benefitted from its close proximity to Zhejiang University. Smaller apartments are frequently seen above shops along the main commercial throughfares which also contain many tobacco outlets (see Chap. 1). As it is an older established neighbourhood, with good access to green space and characterised by some pleasant tree-lined streets, property prices will be higher than average as will social stability and neighbourhood social capital. The neighbourhood contains important role models of tobacco control, and in Hangzhou such networks have been shown to be particularly important in enlisting support for smoking bans (Yang et al. 2014). It is noteworthy that in 2019 Hangzhou was one of the first cities in China to ban the use of e-cigarettes in public places (CGTN 2019). Because of its neighbourhood educational profile, smoking rates are likely to be much lower than the average for the city as a whole (26.2% in 2011) (Yang et al. 2015, p. 165). Lv et al. (2011) showed for Hangzhou that adults with the highest levels of education were those most likely to have healthy lifestyles, including lower levels of smoking and being more physically active. Because Hangzhou has paid particular attention to improving access to greenspace areas (Wolch et al. 2014), neighbourhoods such as Gudang will become

Fig. 6.4 Gudang neighbourhood, Hangzhou. (Credit: Authors)

more desirable to Chinese adults seeking healthy lifestyles. Such trends thus are likely to accentuate urban social differences in smoking prevalence.

Not all more affluent neighbourhoods, however, have lower levels of smoking. As a stark contrast, to traditional views of much of urban China, the fourth picture (Fig. 6.5) was taken from the spectacular balcony in an upscale villa in Dongguan City north of Shenzhen. The city's growth owes much to the growth of low value-added manufacturing and being the largest importer and exporter of goods in China. In 2012 Guangdong Province had more billionaires than any other province in China. It indicates how disparately urban development has benefitted people of privilege compared to those without. In 2000 current smoking rates among males (65.2%) were recorded as being among the highest in China with many smokers indicating that their main reasons for starting were their responses to social pressure, to get rid of feelings of fatigue and increase the efficiency of their work (Zeng and Lin 2000). The picture in Dongguan emphasises the different impacts of income and education in influencing smoking (see Chap. 2). In Hangzhou the urban residents with the most healthy lifestyles were from higher educational groups. By contrast adults with the highest asset index had only moderately healthy lifestyles (Lv et al. 2011).

In summary, the four case studies illustrate some of the divergent patterns of neighbourhood change in Chinese cities. It is evident that spatial differences in neighbourhood social status exist, but it is still not clear how much urban spatial segregation is occurring in different socio-economic groups. In western cities this was often a post-World War II phenomenon reflecting state production of social housing followed by its subsequent privatisation which helped entrench social disadvantage. In China urban housing changes are more recent; given the importance

Fig. 6.5 Patio view from a luxury old villa in Dongguan. (Credit: Authors)

of private agents affecting urban development, it is likely that social differences between more and less preferred neighbourhoods will become increasingly apparent.

6.6 Urbanisation and the Social Distribution of Smoking

Historically, urbanisation has been a process that has powerfully reshaped ways of living. It has been intimately associated with the radical economic restructuring that transformed China over the last few decades and which has produced its own set of inequalities. However, it would be a mistake to simply attribute changes in the social distribution of smoking to urban economic changes. Rather it has occurred as a result of multiple socio-economic and political factors, including the effects of urban poverty, stresses associated with urban migration and discriminatory institutional policies, as well as the effects of rising incomes and tobacco advertising that encourage consumption. Such trends, however, have been countered by increased education and awareness of tobacco harm, better access to health advice for smoking cessation and increased levels of tobacco control, particularly in larger cities. Thus, it is necessary to examine economic and social aspects of urbanisation that have affected the social distribution of smoking.

In terms of economic processes, scholars have suggested that urbanisation may directly or indirectly increase population-level substance use as a result of the economic expansion and the laissez-faire marketisation experienced during economic reforms in transition economies such as China. Some studies have observed a positive correlation between economic growth and substance use (Ruhm 2005; Stevens and Caan 2008). Increased demand may occur because of rising incomes and the increased availability and affordability of goods. Commercial outlets quickly multiply during economic booms as the production and transaction of goods seeks to meet demands in an enlarged market, especially if the area under consideration has undergone a period of commodity shortage. In China nightclubs, restaurants, dive bars, vendor stalls and supermarkets appeared in multitudes in the post-reform period, and acquiring alcohol and tobacco had been never easier, particularly as more citizens became able to afford such goods. In addition, as we saw in Chap. 3, many provincial and local governments, in pursuit of faster economic growth, manipulated local demand by encouraging consumers to buy local cigarettes. Other transition economies, such as countries in Eastern Europe and the former Soviet Union, also witnessed large increases in smoking, but unlike China, this was largely due to the activities of transnational tobacco companies and weakened regulatory government involvement than the effects of economic growth itself (Gilmore and McKee 2004; Perlman et al. 2007).

In China, the literature has consistently shown that individual socio-economic status is associated with smoking (see Chap. 2). However, such an association is likely to depend on community-level social and economic development. According to rational choice theory, compared to the poor, people with more disposable income will be less likely to spend money on recreational substance use when the economy thrives, since the opportunity to allocate their resource has been diversified (Buchmueller and Zuvekas 1998). Therefore, the social distribution of smoking will be different in more urbanised communities compared to rural areas. In more traditional societies, we would expect to see substance use to be higher among people of more advantageous social status, for whom smoking, drinking or dietary patterns are markers of conspicuous consumption.

Despite the arguments presented above, other aspects of urbanisation, such as the development of social services, health facilities and sanitary improvements, may counter the positive effects of economic urbanisation on smoking. The modern welfare system and investment in public goods have not only produced visible material outcomes but also increased awareness of tobacco harm (Anderson 1990; Bambra 2007; Bayer and Stuber 2005). Tobacco control efforts, which have had a largely urban focus, have banned smoking from the public spaces of many Chinese cities (see Chap. 9). Despite reservations at local administrative levels, public health practitioners and social workers at universities and centres for disease control have long advocated for smoke-free legislation (Ngok and Li 2010). The goal of both education and regulatory actions is to promote denormalisation of smoking and increase its stigmatisation. However, because of their lower levels of education and income, the poor are less likely to be aware of smoking-related harms or to be able to afford, and access, smoking cessation services (see Chap. 8). They are also more likely to suffer from greater psychological stress and have lower life satisfaction.

The social dimension of urbanisation also includes lifestyle that depends on positive relationships, instead of a cohesion consolidated by participation in risk-taking rituals. It could be argued that urbanised and more individualistic lifestyles are less supportive of the types of ritual drinking and smoking, typical of many rural communities.

In summary, as a community develops, economically and socially, we expect to see changes in social patterns of smoking. While urban economic developments may provide increased opportunities for substance use, the impact of social and regulatory changes is likely to do the reverse. However, despite the importance of these processes, Yang (2017) has observed that they have received little attention in the global literature. As was highlighted in Chap. 2, models of the smoking transition, for example, do not explicitly examine subnational variations and largely ignore processes of urbanisation and socio-economic change, the effects of which will vary across different communities. Thus, it is important to consider how community socio-economic change affects smoking, independent of individual characteristics. We turn to this issue in the following section.

To test the theoretical expectations on urbanisation's impact on individual smoking behaviour, Yang (2017) used longitudinal data from the China Health and Nutrition Survey (CHNS) that tracked 12,432 respondents from 9 Chinese provinces over the course of 20 years from 1991 to 2011. The study used multi-level growth curve modelling to test how economic and social service urbanisation affected individuals' frequency of tobacco and alcohol use. For economic urbanisation, the study adopted several indicators related to the modernisation of the economy, including the typical wage of a manual worker and the number of people working in nonagricultural sectors. At the same time, urbanisation in social services was measured by indicators reflecting the adoption of modernised modes of public resource distributions, such as health-care coverage, child-care provision and welfare provision, among other items that a traditional agrarian community tends to lack. Besides household and individual surveys, separate data was collected at the community level to enable a multi-level analysis.

When community-level social and economic measures of urbanisation were introduced into the model, Yang (2017) found that, of the two measures net of the effects of other control variables, economic urbanisation had no independent effect on individual smoking. However, improvements in social services in urban areas significantly dampened the association between income and smoking. This was also true for drinking. This means that in communities of low social development (the dotted line in Fig. 6.6), rich people are more likely to smoke (and drink) than less affluent people. By contrast, higher-income people living in an economically modernised community with better social services (the solid line in Fig. 6.6) are much less likely to smoke than their lower-income counterparts in those communities.

The gender pattern of smoking was also reversed by social development. Yang (2017) showed that, in contrast to males, females tend to smoke (and drink) more in communities of high levels of social development. Smoking and drinking in transition economies can help construct an independent and liberalised female identity,

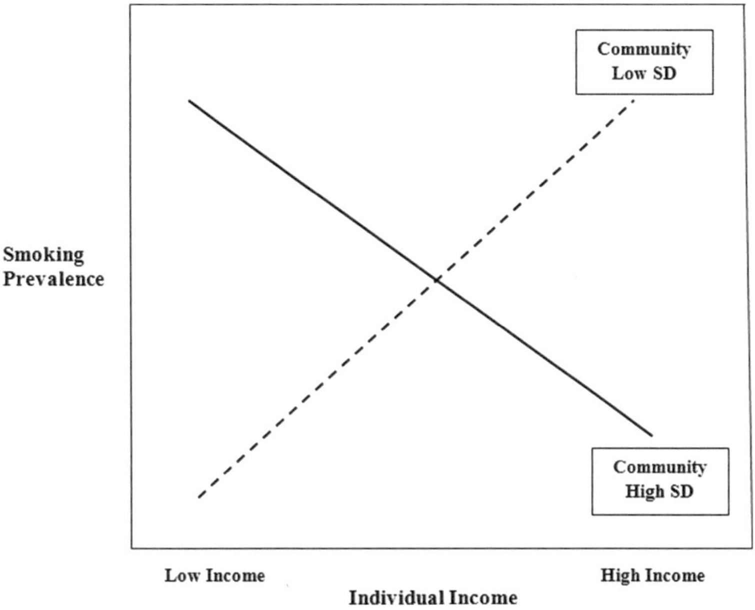

Fig. 6.6 Cross-level interactions between individual income and smoking prevalence by level of community social development (SD = social development. Credit: Authors. Modified from Yang 2017, Fig. 2)

while on the other hand, the need to demonstrate masculinity through risk-taking behaviour and substance use becomes less pressing.

Taken together, these findings stress the importance of urbanisation, as a process of community-level social development, in reversing traditional patterns of smoking. Community-level development has contributed to less smoking overall, but its effects are divergent, depending on gender and income. This is also true of other health behaviours such as drinking, and complements other work on obesity in China which has challenged uniform descriptions of the social patterning of the growing obesity epidemic (Tang et al. 2019; Zhou et al. 2017). As in the case of smoking (and drinking), within less educated areas of China, higher BMI is characteristic of those of higher social status with the opposite being true in more educated (urbanised) areas. Such findings are indicative of the diversity of health transitions at different stages across China's vast landscape, just as they are internationally. By virtue of changing the cultural norms, improving health education and promoting an individual-oriented market-based rational behavioural mode, in which other forms of consumption become more important than substance use, social urbanisation has tended to reverse the traditional relationship between socio-economic status and smoking. Tian and Wang (2013) have similarly noted that the deterrent effects of education on smoking were far more pronounced in urban compared to rural areas, an effect which became more apparent over time (1993–2009). This suggests that traditional western patterns of increased smoking among more deprived and socially disadvantaged groups are likely to become more common in China, in a similar fashion to other Asian UMIC countries, such as Korea and Malaysia (Chang et al. 2019; Mariapun et al. 2019), especially in urban contexts.

6.7 Conclusions

This chapter has explored the growth of urban inequality and urbanisation on smoking and the likely impacts of current trends in urban development. With respect to all three processes, research on their health behaviour impacts is in its infancy. This is surprising given the significance of urban change in post-reform China, not only in terms of health outcomes but also because of the way state policy is implicated. The results of this chapter suggest three main findings.

First, the growth of income inequality is a structural process intimately related to urbanisation and macroeconomic policies typical of transition economies. The evidence suggests that higher income inequality in China is related to smoking along with a series of adverse health outcomes, including depression, poorer self-rated health status and lower levels of subjective well-being. In this respect the findings for China are very similar to the past experience of western countries. However, there are also suggestions of differences between China and the western experience. The rise of income inequality in China has occurred at a time of rapid rural-urban migration and economic growth, the reverse of the experience in many western cities. The presence of a 'tunnel effect', indicating that migrants and other urban residents perceived high levels of income inequality to be an opportunity for economic advancement, suggests that the health and social impacts of income inequality are partly context dependent being more likely to occur in regions and cities in China which are more economically developed. This is consistent with the findings of cross-national comparative research which suggests that the effects of rising income inequality are moderated by the level of economic development, with negative effects being most evident in richer countries (Ngamaba et al. 2018). Nevertheless there was still sufficient evidence of negative impacts of income inequality on smoking and other health outcomes, which were often most pronounced for lower-income groups, to suggest that there are certain similarities to the western experience.

In terms of the different pathways linking inequality and health, the evidence is less clear. With respect to psychosocial factors such as status comparisons, these are likely to become more important as urban development proceeds in China and residential differentiation increases. In the United States, Cheung and Lucas (2016) found that low-income individuals tended to report lower life satisfaction when they lived in richer counties with higher income inequality compared to high-income individuals. This is likely to occur at some point in China, depending upon the nature of urban development. In a similar way, the impact of income inequality on access to public resources, such as health care, is also likely to be context dependent. A growing body of evidence suggests that a strong health-care system, in particular improved access to primary care, may mitigate the adverse impacts of income inequality on population health (Detollenaere et al. 2018). Strong social capital will also play a role here, especially in rural areas where the adverse health effects of strong income differences may be countered by increased local support from friends and relatives. Much of the literature in China has emphasised the adverse impacts of poor access to health services on health (e.g. Zhang et al. 2017), but has not linked such outcomes to the presence of provincial or city policies or the extent to which these are an outcome of the income distribution in these areas.

The second main finding pertains to links between urbanisation and the social distribution of smoking. We have shown that when there is a greater community investment in urban social services, smoking prevalence tends to decline among more affluent groups, mimicking the pattern found in western countries. In the more urbanised regions of China, therefore, the social gradient of smoking has been inverted from that typical of rural areas, with consumption in urban areas being higher among more disadvantaged groups. This pattern is not surprising. It reflects not only the impacts of urban inequality upon more disadvantaged populations and the effects of *hukou* on restricting opportunities for poorer migrants but also the links between urbanisation and more progressive tobacco control policies.

There are also suggestions that the pattern in China is similar to that recently reported in South Korea where positive correlations between area deprivation and both male and female smoking were strongest at the metropolitan level and negative for rural areas (Kim et al. 2017, p. 107). The inversion of the social gradient in smoking between less and more urbanised areas deserves much more attention in different regions of China, especially those dominated by the tobacco industry.

There is also a need to explore the effects of both individual and area socio-economic status on smoking especially in urban China. As we have seen, in western cities it is common for area deprivation to affect smoking independent of individual SES. However, even though such trends (for administrative districts) have been reported for Korea (Yun et al. 2015), especially for males in urban areas, it is not clear whether this may also be true of China. The independent effect of area deprivation on smoking is most likely to occur in situations where structural forces, such as deindustrialisation, have affected community welfare and health behaviour. Within China it is likely that the higher smoking rate of some cities in the northeast 'rust belt' may reflect such trends, above and beyond individual social characteristics, especially since other evidence indicates that cancer mortality in northeast China is independently related to area deprivation (Kou et al. 2019).

Finally, urban development in China bears many similarities with other transition economies that have undergone economic and social reform. The pattern of urbanisation, once totally dominated by the state, now represents a quasi-market. State actors are still important in guiding urban growth, but development is now largely seen as the result of market processes. This is most evident in the housing market, and the reform of this sector has produced polarising property wealth gaps (Chen and Han 2014). Changes in the housing market are thus an increased source of stress reflected in unhealthy behaviour, such as smoking. In the future, spatial inequalities in housing segregation are likely to become more evident than inequalities based on a social hierarchy of different tenure categories. It is no accident that research on neighbourhood effects documenting the importance of residential context on health, independent of personal attributes, is largely absent in China. To some extent this reflects the past structure of *danwei* live-work communities embodying a certain degree of social mixing, but it also reflects the lack of neighbourhood social segregation and concentrated disadvantage. Nevertheless, even among more deprived migrant groups, neighbourhood level factors on self-rated health and psychological well-being tend to be weaker for migrants than more permanent residents. In Shanghai, for example, neighbourhood safety and social cohesion were health-promoting resources only for natives rather than migrants (Wen et al. 2010). Issues of neighbourhood stigma, for the present anyway, appear to be absent, and not a significant force affecting health behaviours as in western countries.

References

Abramovitz M, Strauss H, Littlechild B (2014) Economic crises, neoliberalism, and the US welfare state: trends, outcomes and political struggle. In: Noble C (ed) Global social work: crossing borders, blurring boundaries. Sydney University Press, Sydney, pp 225–240

Anderson GE (1990) The three worlds of welfare capitalism. Princeton University Press, Princeton, NJ

Anwar S, Hashmi F, Nasreen S (2017) Impact of environmental degradation and income inequality on health status in South Asian countries. J Appl Environ Biol Sci 7(6):178–190

Bakkeli NZ (2016) Income inequality and health in China: a panel data analysis. Soc Sci Med 157:39–47
Bambra C (2007) Going beyond: the three worlds of welfare capitalism: regime theory and public health research. J Epidemiol Community Health 61(12):1098–1102
Barnett R, Moon G, Pearce J, Thompson L, Twigg L (2017) Smoking geographies space, place and tobacco. Wiley Blackwell, Oxford, 301p
Bayer RB, Stuber J (2005) Tobacco control, stigma, and public health. Am J Public Health 96(1):47–50
Brenner N, Theodore N (2002) Cities and the geographies of "actually existing neoliberalism". Antipode 34(3):349–379
Buchmueller TC, Zuvekas SH (1998) Drug use, drug abuse, and labour market outcomes. Health Econ 7(3):229–245
Cevik S, Correa-Caro C (2019) Growing (un)equal: fiscal policy and income inequality in China and BRIC. J Asia Pac Econ. https://doi.org/10.1080/13547860.2019.1699985
Chan TW, Ermisch J, Gruijters R (2019) The dynamics of income inequality: the case of China in comparative perspective. Eur Sociol Rev 35(3):431–446
Chang Y, Kang HY, Lim D, Cho HJ, Khang YH (2019) Long-term trends in smoking prevalence and its socioeconomic inequalities in Korea, 1992–2016. Int J Equity Health 18:148
Chen G, Glasmeier AK, Shao Y (2016) Urbanisation and income inequality in post-reform China: a causal analysis based on time series data. PLoS One 11(7):e0158826
Chen H, Wang X, Chen G, Li Z (2018a) Upward social mobility in China: do cities and neighbourhoods matter? Habitat Int 82:94–103
Chen J, Han X (2014) The evolution of the housing market and its socioeconomic impacts in the post-reform People's Republic of China: a survey of the literature. J Econ Surv 28(4):652–670
Chen T, Lang W, Li X (2020) Exploring the impact of green space on residents' health in Guangzhou. J Urban Plan Dev 146(1):05019022
Chen WY, Hu FZY (2015) Producing nature for public: land-based urbanization and provision of public green spaces in China. Appl Geogr 58:32–40
Chen Y, Chen H, Li Z (2018b) Effect of characteristics and life in cities on residents' smoking behavior. J Int Med Res 46(1):4226–4334
Chen Z, Xu T, Liu XF (2012) Hukou identity, demonstration effects, and happiness: evidence from Shanghai and Shenzhen. (in Chinese). J World Econ 4:79–101
Cheung F (2016) Can income inequality be associated with positive outcomes? Hope mediates the positive inequality-happiness link in rural China. Soc Psychol Personal Sci 7:320–330
Cheung F, Lucas RE (2016) Income inequality is associated with stronger social comparison effects: the effect of relative income on life satisfaction. J Pers Soc Psychol 110(2):332–341
China Daily (2017) Walking tall in a world of their own. China Daily, 23 December. http://www.chinadaily.com.cn/a/201712/23/WS5a3d709ba31008cf16da2fbf.html
CGTN (China Global Television Network) (2019) China's Hangzhou City bans e-cigarettes in public. CGTN 1 February. https://news.cgtn.com/news/3d3d674e78557a4e31457a6333566d54/share_p.html
Cui X, Rockett TH, Yang T, Cao R (2012) Work stress, life stress, and smoking among rural-urban migrant workers in China. BMC Public Health 12:979
Dai T, Jiang S, Sun A, Wu S (2020) Inequality and social capital: how inequality in China's housing assets affects people' trust. Emerg Mark Financ Trade 56:562–575
Detollenaere J, Desmarest AS, Boeckxstaens P, Willems S (2018) The link between income inequality and health in Europe: adding strength dimensions of primary care to the equation. Soc Sci Med 201:103–110
Ding X, Billari FC, Gietel-Basten S (2017) Health of midlife and older adults in China: the role of regional economic development, inequality, and institutional setting. Int J Public Health 62(8):857–867
Du H, King RB, Chi P (2019) Income inequality is detrimental to long-tern Well-being: a large scale longitudinal investigation in China. Soc Sci Med 232:120–128

Factor R, Kawachi I, Williams DR (2011) Understanding high-risk behavior among non-dominant minorities: a social resistance framework. Soc Sci Med 73(9):1291–1301

Fan VS, Mhadevan R, Leung J (2020) Effect of income inequality, community infrastructure and individual stressors on adult depression. Heal Prom Int. https://doi.org/10.1093/heapro/daaa036

Fang H, Chen J, Rizzo JA (2009) Exploring urban-rural health disparities in China. Med Care 47(12):1209–1216

Fang H, Rizzo JA (2012) Does inequality in China affect health differently in high-versus low-income households. Appl Econ 44:1081–1090

Feng Z, Wang WW, Jones K, Li Y (2012) An exploratory multilevel analysis of income, income inequality and self-rated health of the elderly in China. Soc Sci Med 75(12):2481–2492

Festinger L (1954) A theory of social comparison processes. Hum Relat 7(2):117–140

Fu D, Wong K, Li CY, Song HX (2007) Rural migrant workers in urban China: living a marginalized life. Int J Soc Welf 16:32–40

Fu Q, Wu C, Liu H, Shi Z, Gu J (2018) Live like mosquitoes: Hukou, rural-urban disparity and depression. Chin J Sociol 4(1):56–78

Gilmore AB, McKee M (2004) Tobacco and transition: an overview of industry investments, impact and influence in the former Soviet Union. Tob Control 13(2):136–142

Gong P, Liang S, Carlton EJ, Jiang Q, Wu J, Wang L, Remais JV (2012) Urbanisation and health in China. Lancet 379:843–852

Graham H (1993) When life's a drag: women, smoking and disadvantage. Her Majesty's Stationery Office, Norwich

Grosfeld I, Senik C (2010) The emerging aversion to inequality. Econ Trans Inst Chang 18:1–26

Gugushvili A, Reeves A, Jarosz E (2020) How do perceived changes in inequality affect health? Health Place 62:102276

Hao P (2015) The effects of residential patterns and Chengzhongcun housing on segregation in Shenzhen. Eurasian Geogr Econ 56(3):308–330

Hao Y, Liu J, Lu ZN, Shi R, Wu H (2020) Impact of income inequality and fiscal decentralization on public health: evidence from China. Econ Model. https://doi.org/10.1016/j.econmod.2020.02.034

Harvey D (1989) From managerialism to entrepreneurialism: the transformation of urban governance in late capitalism. Geogr Ann 71B:3–17

He AJ, Qian J, Ratigan K (2020) Attitudes toward welfare spending in urban China: evidence from a survey in two provinces and policy implications. J Chin Govern. https://doi.org/10.1080/23812346.2019.1709324

He S (2013) Evolving enclave urbanism in China and its socio-spatial implications: the case of Guangzhou. Soc Cult Geogr 14(3):243–275

He S, Li SM, Chan KW (2015) Migration, communities, and segregation in Chinese cities: introducing the special issue. Eurasian Geogr Econ 56(3):223–230

He S, Wu F (2007) Socio-spatial impacts of property-led redevelopment on China's urban neighbourhoods. Cities 24(3):194–208

Hou B, Nazroo J, Banks J, Marshall J (2019) Are cities good for health? A study of the impacts of planned urbanization in China. Int J Epidemiol 48(4):1083–1090

Hu H, Lu Y (2012) Income inequality, health and subjective well-being among the elderly: empirical evidence on the aging of China. China Soft Sci 263(11):41–56. (In Chinese)

Huang J (2019) Income inequality, distributive justice beliefs, and happiness in China: evidence from a nationwide survey. Soc Indic Res 142:83–105

Inoue Y, Howard AG, Yazawa A, Kondo N, Gordon-Larsen P (2019) Relative deprivation of assets defined at multiple geographic scales, perceived stress and self-rated health in China. Health Place 58:102117

Jain-Chandra S, Khor N, Mano R, Schauer J, Wingender P, Zhuang J (2018) Inequality in China: trends, drivers and policy remedies. IMF Working Paper WP/18/127. International Monetary Fund, Washington, DC

Jessop B (1994) Post-Fordism and the state. In: Amin A (ed) Post-fordism: A reader. Blackwell, Oxford, pp 251–279
Jin F, Yangyang Y (2007) Income inequality and health in rural China. Yanjiu/Econ Res J 42(10):79–88
Kan HY, Forsyth A, Rowe P (2017) Redesigning China's superblock neighbourhoods: policies, opportunities and challenges. J Urban Des 22(6):757–777
Kanbur R, Zhang X (2005) Fifty years of regional inequality in China: a journey through central planning, reform, and openness. Rev Dev Econ 9:87–106
Kim I, Bahk J, Yoon TH, Yun SC, Khang YH (2017) Income differences in smoking prevalences in 245 districts in South Korea: patterns by area deprivation and urbanity, 2008-2014. J Prev Med Pub Heal 50:100–126
Knight J, Song L, Gunatilaka R (2009) Subjective well-being and its determinants in rural China. China Econ Rev 20:635–649
Kondo N, Kawachi I, Subramanian SV, Takeda Y, Yamagata Z (2008) Do social comparisons explain the association between income inequality and health?: relative deprivation and perceived health among male and female Japanese individuals. Soc Sci Med 67(6):982–987
Kou K, Baade PD, Guo X, Gatton M, Cramb S, Lu Z, Fu Z, Chu J, Xu A, Sun J (2019) Area socioeconomic status is independently associated with esophageal cancer mortality in Shandong, China. Sci Rep 9:6388
Kragten N, Rözer (2017) The income inequality hypothesis revisited: assessing the hypothesis using four methodological approaches. Soc Indic Res 131:1015–1033
Kuo CT, Chiang TL (2013) The association between relative deprivation and self-rated health, depressive symptoms, and smoking behavior in Taiwan. Soc Sci Med 89:39–44
Le Galès P (2016) Neoliberalism and urban change: stretching a good idea too far? Territory, politics. Governance 4(2):154–172
Leng H, Li S, Yan S, An X (2020) Exploring the relationship between green space in a neighbourhood and cardiovascular health in the Winter City of China: a study using a health survey for Harbin. Int J Environ Res Public Health 17:513
Li DX, Guindon GE (2012) Income, income inequality and youth smoking in low- and middle-income countries. Addiction 108(4):799–808
Li H, Zhu Y (2006) Income, income inequality and health: evidence from China. J Comp Econ 34(4):668–693
Li J, Wan G, Wang C, Zhang X (2019) Which indicator of income distribution explains crime better? Evidence from China. China Econ Rev 54:51–72
Li M, Wu YF (eds) (2016) Urbanization and public health in China. Imperial College Press, London
Li X, Song J, Lin T, Dixon J, Zhang G, Ye H (2016) Urbanization and health in China, think at the national, local and individual levels. Environ Health 15(Supp 1):32
Li Z, Wu F (2008) Tenure-based residential segregation in post-reform Chinese cities: a case study of Shanghai. Tran Inst British Geograph NS 33:404–419
Lin Y, Zhang Q, Chen W, Ling L (2017) The social income inequality, social integration and health status of internal migrants in China. Int J Equity Health 16:139
Ling DC (2009) Do the Chinese keep up with the Jones'? Implications of peer effects, growing economic disparities and relative deprivation on health outcomes among older adults in China. China Econ Rev 20(1):65–81
Liu Y, He S (2017) Unpacking the heterogeneity of poor neighbourhoods and neighbouring in large Chinese cities. Area Dev Pol 2(3):294–313
Liu Y, He S, Wu F (2008a) Urban pauperization under China's social exclusion: a case study of Nanjing. J Urban Aff 30(1):21–36
Liu Y, Wu F, He S (2008b) The making of the new poor in transitional China: market versus institutionally based exclusion. Urban Geogr 29(8):811–834
Liu Y, Wu F, He S (2008c) The making of the new urban poor in transitional China: market versus institutionally based exclusion. Urban Geogr 29(8):811–834.

Lu L, Kandilov IT, Zhu R (2020) The impact of internal migration on the health of rural migrants: evidence from longitudinal data in China. J Dev Stud 56(4):840–855
Lu M, Wan G (2014) Urbanization and urban systems in the People's republic of China: research findings and policy recommendations. J Econ Surv 28(4):671–685
Luo W, Xie Y (2020) Economic growth, income inequality and life expectancy in China. Soc Sci Med 256:113046
Luo H, Li J, Zhang Q, Cao P, Ren X, Fang A, Liao H, Liu L (2018) Obesity and the onset of depressive symptoms among middle-aged and older adults in China: evidence from the CHARLS. BMC Public Health 18:908
Lv J, Liu Q, Ren Y, Gong T, Wang S, Li L, Community Interventions for Health (CIH) (2011) Socio-demographic association of multiple modifiable lifestyle risk factors and their clustering in a representative urban population of adults: a cross-sectional study in Hangzhou, China. Int J Behav Nutr Phys Act 8:40
Mariapun J, Hairi NN, Ng CW (2019) Socioeconomic differences in smoking and cessation across a period of rapid economic growth in an upper-middle-income country. Nicotine Tob Res 21(11):1539–1546
Massa KHC, Pabayo R, Chiavegatto F (2018) Income inequality and self-reported health in a representative sample of 27017 residents of state capitals in Brazil. J Public Health 40(4):e440–e446
Meng R, Chen H (2014) A multilevel analysis of social capital and self-rated health: evidence from China. Health Place 27:38–44
Miao J, Wu X (2016) Urbanization, socioeconomic status and health disparity in China. Health Place 42:87–95
Mullins-Owens H (2015) Inequities in Chinese health services: An overview of the recent history of Chinese health care and recommendations for reform. SAGE Open 5(1):1–7
Ngamaba KH, Panagioti M, Armitage CJ (2018) Income inequality and subjective Well-being: a systematic review and meta analysis. Qual Life Res 27:577–596
Ngok K, Li D (2010) Tobacco control in China: process, actors and policy initiatives. J Asian Pub Pol 3(1):100–110
Ni Y, Tein JY, Zhang M, Yang Y, Wu G (2017) Changes in depression among older adults in China: a latent transition analysis. J Affect Disord 209:3–9
Odusanya IA, Agboola B (2017) Income, income inequality and health: evidence from Nigeria. Izvestiya J Varna Univ Econ 61(4):345–361
Oshio T, Nozaki K, Kobayashi M (2011) Relative income and happiness in Asia: evidence from nationwide surveys in China, Japan and Korea. Soc Indic Res 104:351–367
Pei X, Rodriguez E (2006) Provincial income inequality and self-reported health status in China during 1991-7. J Epidemiol Comm Heal 60:1065–1069
Perlman F, Bobak M, Gilmore A, McKee M (2007) Trends in the prevalence of smoking in Russia during the transition to a market economy. Tob Control 16(5):299–305
Piketty T, Yang L, Zucman G (2019) Capital accumulation, private property, an rising inequality in China, 1978-2015. Am Econ Rev 109(7):2469–2496
Pinson G, Journel CM (2016) The neo-liberal city – theory, evidence, debates. Terr Pol Govern 4(2):137–153
Rarick JRD, Dolan CT, Han WJ, Wen J (2018) Relations between socioeconomic status, subjective social status, and health in Shanghai, China. Soc Sci Q 99(1):390–405
Ravillion M, Chen S (2004) China's uneven progress against poverty. World Bank Policy Research Working Paper 3408. World Bank, Washington, DC
Ruhm CJ (2005) Healthy living in hard times. J Health Econ 24(2):341–363
Shi L, Starfield B (1993) Primary care, income inequality, and self-rated health in the United States: a mixed-level analysis. Int J Health Serv 30(3):541–555
Siahpush M, Borland R, Taylor J, Singh GK, Ansari Z, Serraglio A (2006) The association of smoking with perception of income inequality, relative material well-being, and social capital. Soc Sci Med 63(11):2801–2812
Smyth R, Qian X (2008) Inequality and happiness in urban China. Econ Bull 4(24):1–10

State Council of the People's Republic of China (2011) The twelfth five year plan for National economic and social development of the People's Republic of China (2011–2015). (in Chinese) http://news.sina.com.cn/c/2011-03-17/055622129864.shtml
Stead M, MacAskill S, MacKintosh AM, Reece J, Eadie D (2001) 'It's as if you're locked in': qualitative explanations for area effects on smoking in disadvantaged communities. Health Place 7(4):333–343
Stevens ARA, Caan W (2008) Gender, economics and culture: diversity and the international evolution of smoking prevalence. J R Soc Promot Heal 128(3):113–116
Storper M (2016) The neo-liberal city as idea and reality. Terr Polit Govern 4(2):241–263
Stubbs B, Veronese N, Vancamfort D, Prina AM, Pao-Yen L, Tseng PT, Evangelou E, Solmi M, Kohler C, Carvalho AF, Koyanagi A (2017) Perceived stress and smoking across 41 countries: a global perspective across Europe, Africa, Asia and the Americas. Sci Rep 7:7597
Stubbs B, Vancampfort D, Firth J, Solmie M, Siddiqi N, Smith L, Carvalho AF, Koyanagijk A (2018) Association between depression and smoking: a global perspective from 48 low-and middle-income countries. J Psychiatr Res 103:142–149
Sun P, Unger JB, Palmer P, Ma H, Xie B, Sussman S, Johnson CA (2012) Relative income inequality and selected health outcomes in urban Chinese youth. Soc Sci Med 74(1):84–91
Sun X, Yuan L (2019) Analysis of the effects of social capital on smoking behavior among older adults in China. Innov Aging 3(S1):900–901
Tan Z, Shi F, Zhang H, Li N, Xu Y, Liang Y (2018) Household income, income inequality, and health-related quality of life measured by the EQ-5D in Shaanxi, China: a cross-sectional study. Int J Equity Health 17:32
Tang K, Wang H, Liu Y, Tan SH (2019) Interplay of regional economic development, income, gender and type 2 diabetes: evidence from half a million Chinese. J Epidemiol Community Health 73:867–873
Tian G, Wang Z (2013) Socio-economic determinants of smoking and drinking in China: urban-rural difference. J Camb Stud 8(2):1–16
Townsend P, Davidson N (eds) (1992) Inequalities in health. The Black Report: The health divide. Penguin Books, London
Truesdale BC, Jencks C (2016) The health effects of income inequality: averages and disparities. Annu Rev Public Health 37:413–430
Viens AM (2019) Neo-liberalism, austerity and the political determinants of health. Health Care Anal 27:147–152
Vincens N, Emmelin M, Stafström M (2018) Social capital, income inequality and the social gradient in self-rated health in Latin America: a fixed effects analysis. Soc Sci Med 196:115–122
Wan G (2008) Understanding inequality and poverty in China: methods and applications. Palgrave Macmillan, London
Wan G, Zhang X (2006) Rising inequality in China. J Comp Econ 34:651–653
Wang C, Wan G, Yang D (2014) Income inequality in the People's Republic of China: trends, determinants and proposed remedies. J Econ Surv 28(4):686–708
Wang D, Lau KK, Yu R, Wong SYS, Kwok TTY, Woo U (2017) Neighbouring green space and mortality in community-dwelling elderly Hong Kong Chinese: a cohort study. BMJ Open 7:e015794
Wang H, Chen Z, Lu M (2009) Household registration, social segmentation and trust: the empirical studies from Shanghai. (in Chinese). J World Econ 10:81–96
Wang H, Yang XY, Yang T, Cottrell RR, Yu L, Feng X, Jiang S (2015a) Socioeconomic inequalities and mental stress in individual and regional level: a twenty one cities study in China. Int J Equity Health 14(1):25
Wang P, Pan J, Luo Z (2015b) The impact of income inequality on individual happiness: evidence from China. Soc Indic Res 121:413–435
Wang R, Xue D, Liu Y, Chen H, Qiu Y (2018) The relationship between urbanization and depression in China: the mediating role of neighbourhood social capital. Int J Equity Health 17:105
Wang YP, Murie A (2000) Social and spatial implications of housing reform in China. Int J Urban Reg Res 24(2):397–417

Wang YP, Shao L, Murie A, Cheng J (2012) The maturation of the neo-liberal housing market in urban China. Hous Stud 27(3):343–359
Wen M, Fan J, Jin L, Wang G (2010) Neighbourhood effects on health among migrants and natives in Shanghai, China. Health Place 16:452–460
Wilkinson RG, Pickett KE (2006) Income inequality and population health: a review and explanation of the evidence. Soc Sci Med 62(7):1768–1784
Wilkinson R, Pickett K (2011) The spirit level: why greater equality makes societies stronger. Bloomsbury Publishing, New York. 328 pp
Wilson WJ (1997) When work disappears: the world of the new urban poor. Vintage Books, New York
Wolch JR, Byrne J, Newell JP (2014) Urban green space, public health, and environmental justice: the challenge of making cities 'just green enough'. Landsc Urban Plan 125:234–244
Wong IOL, Cowling BJ, Lo SV, Leung GM (2009) A multilevel analysis of the effects of neighbourhood income inequality on individual self-rated health in Hong Kong. Soc Sci Med 68(1):124–132
Wu F (2016) Emerging Chinese cities: implications for global urban studies. Prof Geogr 68(2):338–348
Wu X, Li J (2017) Income inequality, economic growth, and subjective Well-being: evidence from China. Res Soc Stratif Mob 52:49–58
Xie Y, Zhou X (2014) Income inequality in today's China. Proc Natl Acad Sci 111(19):6928–6933
Xue X, Cheng M (2017) Social capital and health in China: exploring the mediating role of lifestyle. BMC Public Health 17:863
Yang J, Zhang C, Liu K (2019a) Income inequality and civil disorder: evidence from China. J Contemp China. https://doi.org/10.1080/10670564.2019.1705000
Yang M, Dijst M, Helbich M (2018) Mental health among migrants in Shenzhen, China: does it matter whether the migrant population is identified by hukou or birthplace? Int J Environ Res Public Health 15:2671
Yang T, Barnett R, Rockett IRH, Yang YY, Wu D, Zheng W, Li L (2015) The impact of regional economic reliance on the tobacco industry on current smoking in China. Health Place 33:159–171
Yang T, Barnett R, Fan Y, Li L (2019b) The effect of green space on uncertainty stress and life stress: a nationwide study of university students in China. Health Place 59:102199
Yang XY (2017) How community-level social and economic developments have changed the patterns of substance use in a transition economy? Health Place 46:91–100
Yang XY, Anderson JG, Yang T (2014) Impact of role models and policy exposure on support for tobacco control policies in Hangzhou, China. Am J Health Behav 38(2):275–283
Yang XY, Hu A, Schieman S (2019c) Relative deprivation in context: how contextual status homogeneity shapes the relationship between disadvantaged social status and health. Soc Sci Res 81:157–169
Yao J, Asiseh F (2019) An economic analysis of household income inequality and BMI in China. J Econ Dev 44(1):23–27
Yu K, Zhang Y, Zou H, Wang C (2019) Absolute income, income inequality and the subjective well-being of migrant workers in China: toward an understanding of the relationship and its psychological mechanisms. Int J Environ Res Public Health 16:2597
Yun WJ, Rheee JA, Kim SA, Kweon SS, Lee YH, Ryu SY, Park SW, Kim DH, Shin MH (2015) Household and area income levels are associated with smoking status in the Korean adult population. BMC Public Health 15:39
Zeng J, Jian W (2018) Income-related inequality and its effect on depression prevalence in China: a longitudinal population study. Lancet 392(Suppl 1):s22
Zeng S, Lin L (2000) Study on smoking pattern and related factors among residents aged over 15 years in Guangdong Province. Zhonghua Liu Xing Bing Xue Za Zhi 21(2):134–136. (In Chinese)
Zhang M, Zhou G, Fan G (2020) Political control and economic inequality: evidence from Chinese cities. China Econ Rev 61:101272

Zhang Q, Churchill SA (2020) Income inequality and subjective Well-being: panel data evidence from China. China Econ Rev 60:101392

Zhang X, Dupre ME, Qiu L, Zhou W, Gu D (2017) Urban-rural differences in the association between access to healthcare and health outcomes among older adults in China. BMC Geriatr 17:151

Zhang Y, Jiang J (2019) Social capital and health in China: evidence from the Chinese general social survey 2010. Soc Indic Res 142:411–430

Zhou M, Feng X, Yong J, Li Y, Zhang M, Page A, Astell-Burt T, Zhao W (2017) Lifting the lid on geographic complexity in the relationship between body mass index and education in China. Health Place 46:1–5

Zhou Y, Song L (2016) Income inequality in China: causes and policy responses. China Eco J 9(2):186–208

Zhu Y, Fu Q (2017) Deciphering the civic virtue of communal space: neighbourhood attachment, social capital, and neighbourhood participation in urban China. Environ Behav 49(2):161–191

Chapter 7
Policy Environments for Tobacco Control

Pauline Barnett, Weifang Zhang, and Shuhan Jiang

Abstract This chapter outlines the policy context for tobacco control and how this has changed in recent decades. The chapter identifies key policy-making organs of the state and the turning point around 2012–2013 that moved China from the reform era of 'fragmented authoritarianism' to a more strategic and centralised system of 'top-level policy design'. Historical and current challenges to tobacco control in China include community understanding of the harms of smoking and social norms surrounding tobacco consumption; national political leadership for tobacco control; the central and local institutional environments in which tobacco control occurs; the strategies used by the industry to resist tobacco control; and the experience of advocacy in civil society. A critical review of the tobacco control narrative since 1979 includes early efforts to promote tobacco control (1979–1999); the period (1999–2006) occupied negotiating the Framework Convention on Tobacco Control (FCTC); efforts towards implementation (2007–2012); and the period after 2012 when the new focus on central 'political steering' and 'top-level design' provided better prospects for action. The final section assesses current barriers to tobacco control and the changing positions of the 'players' in the shifting policy field.

Keywords China Tobacco Control Association · Corporate social responsibility · Healthy China 2030 · MPOWER · Non-government organisations · Reform era · State institutions · Top-level design · WHO Framework Convention on Tobacco Control (FCTC)

P. Barnett (✉)
School of Health Sciences, University of Canterbury, Christchurch, New Zealand
e-mail: pauline.barnett@canterbury.ac.nz

W. Zhang
Stomatology Hospital, School of Medicine, Zhejiang University, Hangzhou, China
e-mail: chzwf@zju.edu.cn

S. Jiang
School of Humanities and Management, Zhejiang Chinese Medical University, Hangzhou, China

R. Barnett et al. (eds.), *Smoking Environments in China*,
Global Perspectives on Health Geography,
https://doi.org/10.1007/978-3-030-76143-1_7

This chapter provides context for the next two chapters (Chaps. 8 and 9) which discuss tobacco control measures in some detail. The focus of this chapter is the environment in which China engaged with the WHO Framework Convention on Tobacco Control, recognised as the most significant antismoking policy initiative the world has seen (Chung-Hall et al. 2019). The chapter is divided into four sections. Section 7.1 provides a brief overview for non-Chinese readers on how the policy environment in China has changed over recent decades. It identifies key policy-making organs of the state, the turning point around 2012–2013 that moved China from what is called the 'reform era' of fragmented authoritarianism to a more strategic and centralised system of 'top-level policy design', and introduces the 'policy players' who influence tobacco control.

The second Section (7.2) sets out the historical and current challenges to tobacco control in China, relying on assessments in the literature from both Chinese and western sources. These challenges include community understanding of the harms of smoking and the social norms surrounding tobacco consumption; national political leadership for tobacco control; the central and local institutional environments in which tobacco control occurs, including the presence of the state-institutionalised tobacco industry; an outline of the strategies used by the industry to resist tobacco control; and the experience of advocacy in civil society.

This allows, in the third Section (7.3), a critical review of the tobacco control narrative since 1979. This includes early efforts to promote tobacco control (1979–1999), the period (1999–2006) occupied in negotiating the Framework Convention on Tobacco Control, the early difficult efforts towards implementation (2007–2012) and the period after 2012 when the new focus on central 'political steering' and 'top-level design' provided better prospects for action. The final section of the chapter (Sect. 7.4) assesses progress overall, in particular examining the status of the barriers to tobacco control and the changing positions of 'the players' and their contribution to the shifting policy field.

7.1 Policy Processes in China

It is not always helpful to try to fit policy-making in China into western theoretical frameworks, although some Chinese scholars have adapted elements of these to local situations (Ma and Lin 2012). Concepts of agenda setting (Wang 2008), policy transfer (Jin 2014), institutional contexts (Peters and Zhao 2017), policy networks (Zheng et al. 2010; Teets 2018) and implementation (Li 2010) have all found useful application in relation to policy-making in China, and are mentioned below where relevant to tobacco control.

7.1.1 Post-war Instability and the 'Open-Door' Policy (1949–2012)

Modern policy-making in China draws on its revolutionary past, with the institutional structures of the newly formed People's Republic in 1949 reflecting the centre-periphery balance of the guerrilla operations of the recent civil war (Heilemann and Perry 2011). The authoritarian control structures (1949–1978) allowed for centrally driven radical and experimental change, such as the Great Leap Forward (1958–1962) and the Cultural Revolution (1966–1976) (Jung and Chen 2019). The disruptions of the period, including the absence of effective central policy, failed to provide for the desired economic growth. The emergence, following Mao's death in 1976, of a new leadership group within the Chinese Communist Party, led by Deng Xiaoping, created circumstances for new approaches to policy and modernisation in critical areas such as the economy, agriculture, science and defence (Perkins 1986).

From approximately 1979 to 2012, 'open-door' policies prevailed, embracing a market economy, receptivity to outside influences (including foreign capital) and a pluralisation of contributions to decision-making, leading to increased productivity, a more diverse economy and higher personal and collective wealth (Huan 1986; Hu and Kahn 1997). The system for implementing these policies, described as 'fragmented authoritarianism' (Mertha 2009), included both central state leadership and local experimentation. While recognising the central authority of the state, the structure allowed 'spaces' for officials, new players or 'policy entrepreneurs' to 'wriggle their way' into policy formation or implementation (Mertha 2009 p. 996). Mertha (p. 1001) argues that these moves towards a less authoritarian policy approach were essential, with problems increasingly complex and requiring actors with a wider range of expertise than hitherto available to the bureaucracy.

Policy decisions, nevertheless, remain with the Chinese Communist Party (CCP) through its Central Politburo Standing Committee (CPSC). The Party works through its National People's Congress (NPC), which is elected for 5 years and meets annually to confirm national leadership appointments, membership of the Politburo and its Standing Committee and other Party groups (Lawrence and Martin 2013). The NPC is the legislative arm of the Chinese state and enacts legislation as required by the Party/Government. There is also a wider Chinese People's Political Consultative Conference (CPPCC) which represents a diverse range of interests, regions and ethnic groups and meets annually. The NPC and CPPCC have the opportunity to debate important matters, including tobacco control (Yang 2018a), but no action will be taken without consent of the Party. Implementation of policy is exercised through the top executive body in China, the State Council. The State Council is appointed by the Party and is responsible for all aspects of policy management and public administration.

Besides the central leadership, there has been a diverse group of other players engaged in both trying to set the policy agenda across many sectors and with responsibilities for policy implementation. This included a wider range of officials both in lower levels of the state bureaucracy and local government, state-owned enterprises (SOEs), the media, academics and non-government organisations (NGOs), all of which found greater access to the policy process under 'open-door' reforms (Singh 2007).

7.1.2 The Perfect Storm: Policy Process Under Pressure

The increased energy and flexibility of the 'open-door' years, when combined with centrally driven modernisation policies, created conditions for effective experimentation and vigorous advocacy leading to rapid economic growth and social development (Huan 1986; Hu and Kahn 1997). The economic outcomes of reform reinforced the acceptance of a policy model that has been described as 'central steering' and 'local rowing' (Schubert and Alpermann 2019), that is, with overall direction set from the centre but with flexibility in implementation. After the death of Deng Xiaoping in 1997, successive leaders continued with this policy model. Policy decisions continued to be made by top leadership, but the style was more collaborative with less strong central control. While this brought some benefits, it perhaps gave less clarity for implementers (Li 2012).

By the mid-2000s there was increasing concern that, despite significant economic growth, there were some negative outcomes from the reforms. There was evidence of increasing inequalities in Chinese society, between urban and rural areas, between different groups of workers and between coastal and other provinces (Bao et al. 2002; Pedroni and Yao 2006), and evidence of social dislocation (Quach and Anderson 2008). State companies were less amenable to being monitored (Leutert 2018), and cosy relationships between officials and companies spawned corruption (Li 2018; Schubert and Alpermann 2019). There were concerns about social issues, such as access to healthcare (Cheng 2014) and the environment (Teets 2018), with the more open discourse encouraging NGOs and the media to advocate for particular positions (Wu 2013). Li (2012, xiv) notes 'extraordinary openness and pluralism'.

Arguably the centre and the Party had let things drift, with weak national 'steering' undermining the perceived strengths of fragmented authoritarianism and exposing the policy system as inadequate for managing increased diversity and rapid economic and social change (Minzner 2015). Certainly, the Party leadership appeared weaker, with factionalism within the Standing Committee, divisions between 'populist' and 'elitist' groups and a lack of accountability for the implementation of decisions (Li 2012).

7.1.3 Steering and 'Top-Level Design'

The balance between 'central steering' and 'local rowing' has changed since November 2012 following the 18th meeting of the Central Committee of the Chinese Communist Party. That meeting both elected Xi Jinping as General Secretary of the Party and approved a 'Decision on Some Major Issues Concerning Comprehensively Deepening the Reform'. The 'Decision' marked a new focus on top-down political steering (Lee 2017). This reasserted the control that had always been available to Party leadership, and was seen as essential to address what were seen as flaws in a growth model that had led to wasteful investments, local government debt and unacceptable corruption (Pei 2016; Lee 2017). This was clarified at the 19th Party Congress in October 2017 where the Party's intention to engage in a higher level of policy steering, characterised as 'top-level design', was affirmed (Schubert and Alpermann 2019). The 'Decision' at the Party Congress in 2012 highlighted other changes in direction, recognising that the successful economic reforms had not been without cost. Future goals included 'deepening ecological environment management reform' as well as 'safeguarding and improving the people's wellbeing and promoting social fairness and justice', giving heart to tobacco control campaigners.

Top-level design addresses the integration of party and government organs, with party committees within government institutions strengthened (Teets and Hasmuth 2020). There is less scope for local leaders in policy initiation, with changes to reporting mechanisms, and requirements for sign-off at upper levels. Chen (2017) reports that local officials perceive that the 'spaces' for innovation have shrunk, with these officials becoming primarily policy implementers rather than developers. Nevertheless, there is still a strong institutional norm that encourages experimentation, with local officials having intrinsic motivations to serve their province and community (Gao 2017; Lu et al. 2020; He 2018).

In order to consider how tobacco control has fared under this evolving policy regime, it is important to identify briefly the participants in the tobacco control policy process.

7.1.4 Policy Actors in Tobacco Control

7.1.4.1 The Party and the Legislature

Effective tobacco control depends on a combination of community acceptance and strong government regulation. In China, Communist Party leadership is essential for both legislation and executive action. For example, the National People's Congress (NPC), the highest legislative body in the state, formally ratified the Framework Convention on Tobacco Control in August 2005. Implementation, however, relied on the executive arm of the State, led by the State Council. As will be shown, high-level decisions regarding the management of the implementation of the Framework Convention meant that, initially, policy transfer did not occur.

7.1.4.2 Government Bureaucracies

It is the task of the government bureaucracy to advise the state on actions for tobacco control and, within its own powers, to take action. The Ministry of Health has, for decades, been the primary advocate within the bureaucracy for tobacco control; has co-ordinated the work of other like-minded departments, such as the Ministry of Education and the Ministry of Civil Affairs; and worked closely with experts, academics and non-government organisations. It was the primary instigator for the government to enter into negotiations for the FCTC (Yang 2018a) but found itself in a minority position in implementation planning which was dominated by state economic interests and the tobacco industry.

7.1.4.3 Local Governments

Within China, with its great size and diversity, there is a natural tension between the centre and the provincial and other levels of government (Zheng 2016). As part of the economic reforms of the 1980s, central government granted substantial fiscal and administrative autonomy to local governments, providing significant opportunities for economic gains, particularly in the favoured eastern and coastal provinces, but reinforcing some regional inequalities across China (Kanbur and Zhang 2005). 'Central steering' and 'top-level design' involve closer management by and accountability to the centre, with scholars confirming shifting relationships between central and provincial governments (Wu 2019). Nevertheless, local officials, despite being embedded in an authoritarian bureaucracy, continue to engage in policy experimentation (Gao 2017; Zhu and Zhang 2016). Teets and Hasmuth (2020) indicate that this is in fact an institutional norm in China, and particularly important when the central state leaves a policy vacuum, as we shall see in the case of tobacco control.

7.1.4.4 The State-Owned Tobacco Monopoly

The State Tobacco Monopoly Administration/National China Tobacco Company (STMA/CNTC) (see Chap. 3) is one of over 100,000 state-owned enterprises (SOEs) in China. All are participants in the market economy but subject to the necessary monitoring of state assets (Miller 2014). The STMA/CNTC is of significant size and importance and nationally accountable, reporting to the Ministry of Finance. Larger SOEs, such as the STMA/CNTC, are not passive actors but often have the resources to shape government policy within the context of continuing competition between the various players, including bureaucrats, party officials and others (Yi-chong 2018). Since 2013 SOEs have been brought under closer government scrutiny, particularly to control corruption (Leutert 2018; Schubert and Alpermann 2019).

7.1.4.5 Intergovernmental Agencies

The most important intergovernmental authority contributing to tobacco control in China is the World Health Organization. In technical areas the WHO engages in research and capacity building within countries and regions. As an agency of the United Nations, it has taken leadership of tobacco control internationally, using its scientific expertise and moral authority to ensure that most countries worldwide, including China, signed the voluntary International Framework Convention on Tobacco Control (WHO 2003). The WHO now uses the FCTC to monitor the implementation of tobacco control internationally, publishing country comparisons and further guidance (Chung-Hall et al. 2019).

7.1.4.6 Non-government Organisations (NGOs) and Expert Advocates

Tobacco control, as a health issue, has not been considered politically sensitive in China, with NGOs operating freely and in close association with central agencies and local governments. The foremost domestic NGO in the field, the China Tobacco Control Association (CTCA), was formed in 1990 under the supervision of, and a close partner with, the Ministry of Health with which it shared common tobacco control goals. Over three decades it has maintained an outstanding record of research and education, recruiting senior officials, academics and high-profile personalities to its cause, providing essential technical and advocacy contributions to the policy process and establishing partnerships with central and local governments and other NGOs (Mackay 2016). Partners included international NGOs which, as philanthropic funders and experts, entered China in the 1980s (Redmon et al. 2013). Recent regulatory change has created uncertainty for both domestic and foreign NGOs (Gan 2017).

7.2 Challenges to Policy Action for Tobacco Control

In this section we discuss five major challenges which have faced tobacco control in China: (1) lack of community knowledge of tobacco harms and importance of community norms; (2) lack of political leadership in tobacco control; (3) institutional barriers at national and provincial levels and the institutionalisation of 'Big Tobacco'; (4) tobacco industry resistance to regulation; and (5) barriers facing advocacy and policy networks.

7.2.1 *Community Knowledge and Smoking Norms*

High levels of smoking in China were documented in Chap. 2. A poor understanding of the harms from smoking and the embeddedness of smoking in Chinese society are among the primary barriers to tobacco control.

7.2.1.1 Community Understanding of Smoking

Assessments of the knowledge of smoking harms have been undertaken over more than two decades in China through population studies, local surveys and qualitative research. Wang (2006) notes that the China Tobacco Prevalence Survey in 1996 reported that 60% of smokers assessed the effects of tobacco to be 'limited or no harm'. Later studies investigated knowledge of specific conditions rather than general harm, with a population study in 2006 reporting a reasonable knowledge of some harms, with smokers reporting significantly less knowledge than former or non-smokers (Yang et al. 2010b). There were particularly low levels of knowledge of important diseases such as stroke (20%) and CHD (40.2%) and with lower levels of knowledge among rural dwellers. From the same survey, Elton-Marshall et al. (2010) reported that 71% believed that low-tar/low-nicotine cigarettes to be less addictive than other types of cigarettes.

By 2010 there had been a no real improvement, with WHO's Global Adult Tobacco Survey (GATS) reporting that while 81.8% of Chinese were aware that smoking causes disease, with 77.5% aware of the impact on lung cancer, only 27.2% were aware that stroke and 38.7% aware that heart attack can also be caused by smoking (Yang et al. 2010c). Most of these indicators were worse than in countries at comparable stages of development within the Asia-Pacific Region (WHO 2015, pp. 24–25). In a national population survey (2013–2014) of chronic disease, 75.9% of the population reported smoking as hazardous, but the knowledge of specific conditions remained low (lung cancer 67%, heart attack 33.2% and stroke 29.5%). Lesser levels of knowledge were reported among people living in less economically developed areas and rural communities (Yang et al. 2019). Local surveys (e.g. Au et al. 2016) reported similar levels of knowledge, and qualitative study (Wang et al. 2014) revealed the lack of smoking knowledge in rural areas among local residents, village leaders and even doctors. Concerningly, adolescents appear to have a poor perception of risk, with the Chinese CDC Youth Tobacco Survey (Chinese CDC 2015) indicating that 68.4% of young people did not have a correct understanding of nicotine addiction. A local survey of secondary school students in Chongqing (Xu et al. 2016) reported that less than 20% of students understood the impact of smoking on heart disease, with much lower levels for other conditions. Even among doctors, where smoking rates have been high (Ceraso et al. 2009), studies indicate that smoking was associated with less knowledge of smoking-related risks, with the exception of lung cancer, but especially in the risks of exposing children to second-hand smoke.

7.2.1.2 Community Smoking Norms

The lack of knowledge among Chinese communities of the harms from smoking is reinforced by the historical social and cultural support for smoking in Chinese society (Pan 2004; Hu et al. 2013). Ma et al. (2008) identified two broad groups of 'myths and misperceptions' about smoking. The first is the misunderstanding that the dangers from smoking are minimal and can be controlled, that using good-quality tobacco can reduce harm and that the risks of second-hand smoke can be reduced by 'good air circulation' (Ma et al. 2008). Second are the perceptions that tobacco is 'normal' in China, including the view that as tobacco is legal, then smoking is a 'lifestyle choice'. Historically smoking has been practised by respected community leaders and professionals. Doctors, for example, have had relatively high rates of smoking in China (see Chap. 8). Smith et al. (2012) reviewed 17 studies of doctors smoking in China from 1987 to 2011. There was a great diversity in smoking rates reflecting the variety of studies, settings and samples, but overall rates were high compared with other countries. The late President Deng Xiaoping is cited as an endorsement of smoking: he lived to be 90 years old and smoked his own blue-wrapped brand of Panda cigarettes made for him by Shanghai Tobacco (Kohrman 2018a, p. 142). The China National Tobacco Corporation itself is seen positively by the community; it is acknowledged as an important employer, a provider of state and local revenue and the sponsor of local or disadvantaged communities (Wang 2006).

The social acceptability of smoking in China is reinforced by the sharing and gifting of cigarettes, established cultural traditions that are important in making and maintaining social relationships (*guanxi*) and expressing courtesy and regard (Ma et al. 2008; Pan 2004) (see Chap. 5). 'Sharing' (i.e. the offering and accepting individual cigarettes) is often the means by which people, particularly young males, start smoking and are unable to quit (Sheer and Mao 2018), feeling under social pressure to accept a cigarette and not give offence. Among rural adults in Hunan province, 92% of men and 19% of women reported being offered a cigarette in the previous week, and 90% of respondents reported that their friends had tried to prevent them from quitting by offering cigarettes (Rich et al. 2014).

'Gifting' (i.e. giving packets or cartons of cigarettes, sometimes extravagantly presented) is an important cultural phenomenon in China (Rich and Xiao 2012) and implicitly endorsed by both opinion leaders (such as respected professionals) and political leaders (Wang et al. 2014). Described as 'social currency' (Rich and Xiao 2012), gifting is a relatively formal process designed to show and seek respect (Rich et al. 2014). Gifting may be used to mark an event or holiday, or to show respect in social and business relationships (Chu et al. 2011). Ding and Hovell (2012) argue that the excesses of gifting were actually created by the tobacco industry. Expensive cigarettes, initially foreign but later including Chinese 'premium' brands, were seen as 'badges of affluence' (Chu et al. 2011). Zhang et al. (2016), reporting from a national survey of both urban and rural areas, across 31 provinces, found that over 23% of smokers reported either giving or receiving cigarettes as gifts. The highest rates of gifting were among smokers working in government agencies.

7.2.2 Political Leadership in Tobacco Control Policy

Over much of the past four decades, observers have regarded the Chinese government itself as an important barrier to tobacco control policy, even after ratifying the Framework Convention on Tobacco Control in late 2005. Wang (2006, p. s144) points to the peculiar situation where the Chinese Government 'has made significant efforts to discourage cigarette smoking through public campaigns, yet itself runs the largest tobacco enterprise in the world'.

During World War II, the Red Army, later to become the People's Liberation Army, held several base areas in Central China surrounded by territory of both the Nationalist forces and the Japanese invaders. The Mao-led government in these enclaves came to see tobacco production as both an opportunity for self-reliance and an area of potential industrial growth offering significant returns. After 1949, in the early years of the PRC, this translated into a strong government commitment to assist industrialised tobacco production. This was demonstrated in 1953, when Mao and the Central Committee of the CCP explicitly rejected the Ministry of Health proposed ban on smoking as a part of its public health approach, because it would 'harm production' (Liu 2018, p. 52). This was the beginning of the Chinese government's overt prioritising of production over health, and seeing a strong tobacco industry as a way of ensuring economic stability and revenue.

Following the death of Mao, the opening up of China after 1979, under the leadership of Deng Xiaoping's 'open-door' policy, exposed China to the research and debate about smoking and tobacco control that had challenged governments around the world. Deng himself was a heavy smoker and there are numerous reports of him endorsing smoking (Li 2012). He was also deliberate in his efforts in the 1980s to revive the tobacco industry after the damage of the Cultural Revolution, seeing it as a source of stability (Li 2012). Deng's immediate successor, Jiang Zemin (1989–2000), while himself a non-smoker, seemed uninfluenced by the international discourse on smoking, and more by the economic imperatives of the 1980s and 1990s, complemented by an entrenched 'tobacco culture' reinforced by the China National Tobacco Corporation.

Nevertheless, in line with the openness of the times, the government was quite permissive about tobacco control efforts. International philanthropy became involved in China in 1986 (Redmon et al. 2013), and in 1990 the China Association of Tobacco Control, with strong community and academic support, was established (Mackay 2016). Despite these initiatives, the tobacco industry maintained its influence over tobacco control, with the Standing Committee of the seventh National People's Congress ratifying the *Law on the State Tobacco Monopoly*, 1991. This gave responsibility for public health education and smoking bans to the China National Tobacco Corporation. This was typical of the government stance during this period and under the next president, Hu Jintao (2000–2012). Government took a few actions, leaving the initiative with experts, advocates and interested government departments such as the Ministry of Health and the Ministry of Education, both of which developed concrete measures in their areas of control, but with limited results.

Some changes were in the air. China's new exposure in the international arena encouraged alignment with international practice on tobacco control, with increasing concern to be seen as a good 'international citizen'. Examples of this international alignment included China's engagement with the WHO and its ratification in 2005 of the Framework Convention on Tobacco Control (FCTC), and its commitment to a smoke-free Beijing Olympics (2008). These have both been described as relatively weak government responses (Yang et al. 2010a). The greatest leadership failure of this period for tobacco control was the decision in 2007 of the State Council of the CPC, the highest executive organ of the state, to appoint the STMA to the Inter-Ministry Coordinating and Steering Committee (hereafter 'Co-ordinating Committee') for Implementation of the FCTC. Soon after, the State Council also allocated the Chair of the committee to the tobacco industry's controlling ministry (Yang 2018a, p. 52). (See Sect. 7.3.3 for details.) Giving priority to economic development over tobacco control significantly undermined progress on implementing the Framework Convention and demonstrated the lack of political leadership required under the FCTC, Article 4. Hu et al. (2013) reported conversations with officials from many sources over the ensuing years, all reluctant to upset the tobacco industry or become unpopular with a population in which tobacco consumption was still highly normalised. Leadership change and progress on tobacco control is reported in Sect. 7.4.

7.2.3 *Institutional Challenges to Tobacco Control*

In the previous section, it was shown that there had been no real political will to promote tobacco control in China, and it could be argued that even the signing of the FCTC was a cynical act to enhance the country's international reputation rather than a commitment to policy transfer (Jin 2014). Nevertheless, ratifying the Convention imposed policy transfer obligations on China, and it is important to examine the receptivity of national and local institutional arrangements to the introduction and enforcement of measures aligned with the FCTC.

In China, policy regulation and implementation are the responsibilities of officials (Qian 2017). Policy transfer for tobacco control plays out through both national bureaucracy co-ordination and leadership and provincial and local action on tobacco control (Li 2012), including specific relationships with the state-institutionalised tobacco industry. This section will examine the institutional arrangements affecting tobacco control from the national and local/provincial perspectives, but also show how the state tobacco company has acted not just as a commercial enterprise but as an institutionalised agency of the state.

7.2.3.1 National Policy and Co-Ordination

National institutional arrangements determine where responsibility lies for policy transfer (Lodge 2003), and under FCTC Article 5, the government of China was expected to establish national co-ordinating mechanisms for the implementation of tobacco control. In China, there were difficulties within the national Co-ordinating Committee for FCTC implementation, where the leadership and membership from 2007 onwards was heavily weighted towards economic and tobacco industry interests. This undermining of health initiatives was not confined to tobacco control but was experienced in matters such as drug safety, food handling and the management of outbreaks (Huang 2013). This reflected what Jin (2014, p. 637) has referred to as 'GDP-ism', where economic considerations always 'trumped' the social good. Nevertheless, the problem with national institutional arrangements was most obviously present in the case of implementing the FCTC, elsewhere internationally an outstanding example of worldwide policy transfer (Jin 2014).

In China, the fragmented authoritarianism model of co-ordination and bargaining should theoretically achieve suitable negotiated regulation and outcomes on tobacco control, but this was precluded by the dominance of economic interests on the Co-ordinating Committee for FCTC implementation. When it was established in 2007, the Co-ordinating Committee was led by the vice-chairman of the National Development and Reform Commission (NDRC), a macro-management agency under the State Council and the supervising body for the STMA. Membership of the Co-ordinating Committee included the Ministry of Health, five other economic ministries or agencies and the STMA itself, its presence a clear contravention of FCTC Article 5.3. Although the NDRC also had social development policies in its responsibilities, it was reported that in relation to tobacco control, it 'did not stand as firm as the MoH' because of financial considerations (Ngok and Li 2010). (See Sect. 7.3.3).

The resources available to the STMA were far in excess of those available to the Ministry of Health, the lone tobacco control advocate on the Co-ordinating Committee. The STMA was involved in drafting national guidelines and legislation for tobacco control, with deadlines for regulations routinely missed, including the January 2011 WHO deadline for regulations to enforce FCTC policies. In late 2011 the Ministry of Health issued *Implementation Rules on the Regulations of Public Space Sanitation (MOH Order No. 80)*. Enforcement of regulations is the role of provincial and local authorities, but without action from the National People's Congress of the People's Republic of China, the national law-making body, these were largely unenforceable (Jin 2014). The 2012–2015 China Tobacco Control Action Plan was published in December 2012. Although the planning goals for reducing smoking prevalence were ambitious, the Plan lacked relevant policy measures to support these. By the end of 2015, goals had not been achieved (Yang 2018a). The WHO (2015, p. 10) reported, somewhat pointedly, that China was 'lagging behind' other rapidly developing BRICS countries (Brazil, Russia, India and China) and that a national-level law was yet to be enacted (see Chap. 8).

7.2.3.2 Provincial and Local Institutions and Tobacco Control

As noted above, following the FCTC there was no national tobacco legislation that could require provincial or other jurisdictions to implement and enforce tobacco control measures. Without this mandate local authorities had neither incentive nor strong legal mechanisms for adopting tobacco control measures, so the well-recognised 'crossing the river by feeling the stones' experimentation (Teets and Hasmuth 2020) was unlikely to occur. Qian argues that the fragmented authoritarianism model can be used to understand the nonengagement of provinces, through the reliance on bureaucrats to develop and enforce rules (Qian 2017). With high levels of bargaining at provincial level where there are many competing claims between sectors, officials were likely to be overly influenced, or certainly until recently, by the need for economic growth. Their own performance measures had been almost exclusively linked to economic performance, with promotions based on such criteria (Shih et al. 2012). In these circumstances it is unlikely that provincial priorities would be aligned with the social good. For example, there is evidence of provincial officials diverting funds earmarked for poverty reduction to payroll or other commitments in order to meet performance criteria (Liu et al. 2009) or local cadres' efforts in air pollution control being negatively associated with promotion opportunities (Wu et al. 2014).

Financial imperatives and official self-interest tend to work in tandem. Some financially constrained provinces gave priority to maximising revenue, including from tobacco production, a situation fully exploited by the tobacco industry. By 2009 the industry was consolidated into 31 manufacturing companies across the country, but with many more local bureaus, branches and sales offices distributed across all provinces (Yang 2018a). The industry has been described as 'deeply enmeshed in regional government' and crucial to local prosperity (Kohrman et al. 2018, p.157). Li (2018) reports that, in many provinces, local officials (e.g. governors, mayors, party secretaries) engage personally in negotiations with the STMA on matters of production, distribution and revenue sharing. In an extremely complex tax system, most tax, although collected by the centre, is paid on local production and a portion returned to provinces. There are, therefore, strong incentives for tobacco companies and provinces to work together and examples in the literature of local efforts to boost production, such as Hubei province ordering all public sector workers to smoke locally produced brands (Li 2012) (see Chap. 3). Thus, it is not surprising that there is a high degree of variability in provincial performance in tobacco control, and in that there is little evidence of committed action on the part of provinces themselves.

Individual cities are the places where local tobacco control initiatives have been more apparent, and where different incentives may apply (see Chaps. 8 and 9). Cities most likely to engage in smoke-free initiatives have been either provincial capitals or nationally managed cities such as Beijing or Shanghai (Lin et al. 2019). These tend to be important economic centres, with international exposure and access to resources (including for enforcement), a modern outlook and concern about their perceived status. They are also likely to have more professional scientific and academic expertise. Where smaller and perhaps less obviously well-equipped cities move on tobacco control, there is generally a powerful local incentive. For example, Haikou City, one of the least developed provincial capitals with a population of only 0.54m and few resources, was motivated, despite the lack of legislative mandate, to embark on its own tobacco control initiative primarily to secure its image as a tourist destination and promote itself as a 'healthy city' (Su and Deng 2019). Even with the important cities in China, there is usually a local story about how tobacco control came about and progressed, suggesting ad hoc local responses, responding to local cultures but often supported by local advocacy or international philanthropic efforts (Jiang et al. 2019; Yang 2018b).

7.2.3.3 The Institutionalisation of 'Big Tobacco'

The formation and emergence of the CNTC and the STMA as 'Big Tobacco' was discussed in Chap. 3. It is well-known that the Chinese tobacco industry is an enormous entity of great economic power, a great contributor to tax revenue, employment, foreign exchange and 'economic stability' (Kohrman 2018b). It was an industry favoured by former leaders such as Mao Zedong and Deng Xiaoping, and the producer of a consumer product highly valued in the community (Kohrman 2018b). But this cultural and economic success does not fully explain how its present position and upward commercial trajectory have been assured. As a producer of consumer goods, it would not normally be considered a 'strategic industry' in the sense that banks, infrastructure and telecommunications might be, although it is as large as, or larger than, some other state enterprises, and more profitable (Yang 2018a). Similarly, the production of a 'non-essential', even dangerous, product would not normally be seen as a desirable commodity for state stewardship (Cohen and Lee 2019).

How then did 'Big Tobacco', the purveyor of death and illness, resist the overwhelming scientific evidence of its harm, and counter and even dominate for so long both the local and international forces of tobacco control? These questions can only be answered by recognising that the tobacco industry in China is not merely a commercial enterprise but an institutionalised agency of the state.

There are two important turning points in the modern history of tobacco in China that help explain its institutional 'exceptionalism'. The first was in the 1950s when the new government of China recognised that tobacco manufacturing could become an important state asset (Sha 2018), and executed plans to bring all tobacco manufacturing, public and private, into a state-owned company. The record of this process has only recently come to light in the Shanghai Museum, and has been analysed by Sha (2018). Despite opposition, government action, through intervention to disrupt both the tobacco leaf and paper supply chains, weakened the position of the companies with the takeover given moral authority by the chairman who placed a prohibition on officials smoking cigarettes from private or foreign companies (Sha 2018). This affirmed the patriotic choice of local tobacco, and by 1955 the process of consolidation and state ownership was complete.

At this time the government's lead agencies for tobacco manufacture were the Central Finance and Economic Commission and the Ministry of Food – the latter providing an indication of how cigarettes were seen as one of the essentials of daily living. Liu (2018, p. 52) points out that not only was producing cigarettes 'a responsibility of CCP governance' but that 'all impediments to cigarette manufacturing...should be overcome'. This was a critical moment for the industry; the normalisation of integrated tobacco production and sales as a state interest inevitably confirmed the normalisation of cigarette consumption itself. Despite some stumbling during the Great Leap Forward, it was clear by the 1960s that the population's 'right' to cigarettes was to be enhanced by the production of cheaper brands (Huangfu and Korhman 2018). Even with fragmentation of enterprises following the Cultural Revolution (Fang et al. 2017), tobacco had become, against all logic, a strategic industry in which the government would intervene when necessary.

The second important event that institutionalised Chinese tobacco manufacturing as an agency of the state was the State Tobacco Monopoly Law of 1982 (see Chap. 3). This set up the China National Tobacco Corporation to manage the 28 provincial companies and later in 1985 the State Tobacco Monopoly Administration (STMA) to both administer and regulate the industry (He et al. 2013). The economic reforms from the 'open-door' policy had brought new technology, expertise and energy to the industry (Fang et al. 2017). From the 1980s a highly regionalised structure evolved, with intimate relationships between companies and local governments, and evidence of corruption, especially in high producing areas such as Yunnan (Li 2018). Despite intermittent scandals, over a quarter of a century of operating under the State Tobacco Monopoly Law has given the industry the opportunity to build its commercial strength and compile political capital, and cement its position as a large state-owned enterprise. Yi-chong (2018) claims that despite 'top-level design', large SOEs in China can be effective in shaping government decisions.

7.2.4 Tobacco Industry Resistance to Regulation

It is well-documented in low- and middle-income countries that transnational tobacco companies actively obstruct public health policies designed to reduce smoking (Lee et al. 2012; Gilmore et al. 2015; Bhatta et al. 2020). The STMA/CNTC has certain advantages from the structural arrangements and its favoured industry standing within the state (Jin 2014), building on these advantages to undermine tobacco control and expand its commercial reach. It has done so via four main strategies: (1) maintaining a positive 'patriotic' image; (2) marketing misinformation about tobacco harms; (3) targeting vulnerable segments in the market; and (4) being a 'good' citizen by emphasising corporate social responsibility.

7.2.4.1 Maintaining a Positive 'Patriotic' Image

The tobacco industry continues to promote its role in supporting the economy of China by trumpeting its contribution to government taxation and employment, and therefore to social stability (Jin 2014), a tactic typical of tobacco companies internationally (Salooji and Dagli 2000). In 2007 an administrator of the STMA reportedly claimed that 'Smoking harms people's health, but restraining smoking threatens social stability' (Reuters 2007). Research indicates, however, that these benefits are overplayed and health costs and social harms are significant (WHO 2017; Hu et al. 2013). The industry-funded Tobacco Museum in Shanghai is a comprehensive approach to promoting a patriotic public image of the tobacco industry. According to Kohrman (2018a), p. 137), an important aim of the museum is to shore up the confidence of industry employees so that that they see themselves not as engaged in an increasingly socially undesirable activity but as working in the interests of the community 'building a stronger and more modern nation, and helping to finance strategic priorities of the party-state, both local and national'.

Maintaining a confident public image permits tobacco industry representatives to challenge or even intimidate tobacco control advocates. At the negotiations for the Framework Convention on Tobacco Control (1999–2003), a delegate from the tobacco industry accused a Ministry of Health tobacco official of 'betraying China' (Yang 2018a, p. 69). Other charges included that western tobacco control NGOs aimed to 'spread western ideological thought' and were 'under the manipulation of foreign investors and transnational tobacco companies' (Yang 2018a, p. 69). Redmon et al. (2013) report that international partners who have worked with Chinese tobacco control advocates have been criticised on the basis that outside investment will 'bring down' Chinese tobacco companies in the interest of American industry.

7.2.4.2 Research and Marketing Misinformation About Tobacco Harms

Like tobacco companies worldwide, the CNTC spent several decades first denying the harm caused by tobacco and then explaining the harm mitigation measures they were taking. Most recently this has been the promotion of 'low-tar/low-nicotine' products and 'higher-quality' cigarettes as less harmful (Yang 2014), a claim rejected internationally (Guo 2018) and specifically by independent research in China (Gan et al. 2010). Despite the evidence, the CNTC has continued to market tar/nicotine 'light' cigarettes and used spurious research to support this, causing embarrassment within Chinese scientific circles (Yang 2014) and serious misunderstanding about the status of 'low-tar/low-nicotine' cigarettes. The Chinese CDC (2011) reported that in the GATS 2010 survey, 86% of respondents believed that 'low tar' meant 'less harmful' and that those with higher education (45%) and health workers (54.7%) were more likely to believe the industry's 'science-backed' claims. Of growing concern are the unsubstantiated claims that e-cigarettes are healthy alternatives (Yao et al. 2016). The lack of product disclosure of these and other new products, such as heated products and flavoured capsule sticks, contravenes Articles 9 and 10 of the FCTC (Guo 2018).

7.2.4.3 Targeting Vulnerable Segments in the Market

Despite tobacco company claims that marketing is designed to secure market share rather than recruit new smokers, it is well-known that transnational tobacco companies target vulnerable populations to expand sales (Gilmore et al. 2015; Lee et al. 2012), with the CNTC no exception (see Chap. 4). In spite of strong norms against female smoking in China, there are indications that there is increasing acceptance of women smoking (Sansone et al. 2015). In an early study of both rural and urban young women (aged 14–24 years), both smoking and the intention to smoke were higher among urban females. 'Ever smoking' was associated with awareness of 'cigarettes made for women' and thinking that they were 'less harmful' than other cigarettes (Ho et al. 2010). The authors warn of the risks from, and receptivity of this group to, directed advertising.

Targeting the young is a particular tactic, with national surveys reporting smoking prevalence among male adolescents increasing from 16·0% in 2003 to 23·5% in 2013, with female prevalence increasing from 0·4% to 1·1% during the same period. The restrictive Advertising Law of 2015, passed to align with Article 15 of the FCTC, has been flouted with little enforcement of advertising in proximity to schools (Wang et al. 2017). In Hong Kong, nevertheless, research indicates a high level of awareness among adolescents of the positioning of the tobacco industry, with 77.6% of respondents considering the tobacco industry 'not respectable', and 56.6% believing the industry tries to 'get youth to smoke' (Chen et al. 2019).

There is growing concern that both women and the young are the targets of e-cigarette marketing, although rates of e-cigarette use in China remain low internationally (Wang et al. 2019). Among young people awareness is high, with around 90% of 12–18 year olds knowing about e-cigarettes and 36% indicating that they used e-cigarettes to help quit smoking (Wang et al. 2018). Reports from the Global Youth Tobacco Survey (2014) in China indicated a high level of awareness (45%) but a low level of use (1.2% of students in the previous 30 days), but with users more likely to report an intention to use tobacco products in the next 12 months (Xiao et al. 2019). As early as 2015, the State Tobacco Monopoly Administration in China indicated that it supported regulation of e-cigarettes for the purposes of 'quality control'; other agencies such as the WHO preferred bans on e-cigarettes wherever smoking is banned (Shan 2015; WHO 2020). In China there has been a ban on Internet sales and advertising. The former has been effective but only addresses 15% of the market and protects China's products from imports, and the latter is relatively ineffective with 'workaround' opportunities (Zhuang and Owermohle 2019). In China, though, clearly e-cigarette control advocates have had some early successes compared with the long struggle for tobacco control, but much remains to be resolved.

7.2.4.4 Corporate Social Responsibility

The practice of 'corporate social responsibility' (see Chap. 4) for tobacco companies has been described by Barraclough and Morrow (2008) as a 'grim contradiction', or an 'oxymoron'. The WHO (2004) warned particularly that this can be used to disguise promotional activities. A rudimentary advertising ban was put in place in China in 1994, but even since the more restrictive Advertising Law (2015) and Charity Law (2016), the CNTC has continued to use a sophisticated repertoire of 'tobacco advertising, promotion and sponsorship' (TAPS). It is the 'sponsorship' element of this that claims to be corporate social responsibility, but has been more unflatteringly described as 'political activity' (along with lobbying and self-regulation) as a means of improving its public image with government and the community and promoting tobacco brands (Lee et al. 2012).

An overview of CNTC 'social responsibility' projects has been drawn up by Zheng et al. (2018). They note a report from the Beijing Morning Post (2015) that summarises the work of the China Tobacco Control Association (CTCA) in monitoring tobacco sponsorship. The CTCA identified 89 sponsorships over a 3-month period, relating to poverty alleviation, disaster relief, educational funds, social activities and social construction.

Education sponsorship has been singled out for particular criticism because, when associated with tobacco branding, it creates a positive image of smoking among both parents and children. Fang et al. (2020) draw attention to critical reports of school sponsorship in China, specifically referencing the government initiative for rural schools, Project Hope, over 100 of which are supported by the CNTC. Some have had tobacco logos and slogans incorporated, the most blatant being 'Genius is from hard work; tobacco makes you excel'. Some of these schools were built following the Sichuan earthquake in 2008, and, while sorely needed, the overt sponsorship drew significant criticism (Sohu News 2009). Fang et al. (2020) recognise that educational sponsorship reaches out specifically to the young. They report research from Yunnan province, where tobacco production and cigarette manufacture are prominent, and note positive attitudes from parents and local communities to tobacco company sponsorship.

7.2.5 Advocacy and Policy Networks: Barriers to Achievement

Where do NGOs and advocacy organisations fit into the tobacco control project in China? Conceptually the literature from pluralistic democracies sees advocacy organisations as part of a policy network and as agents of policy transfer (Berg 2019; Jin 2014). Policy networks have been vital in tobacco control internationally, mainly high-income countries, developing formal and informal coalitions to push for legislation and take a leading role in community education (Berg 2019). There are recent initiatives to promote tobacco control coalitions in low- and middle-income countries, supported by international NGOs (Koplan and Redmon 2014), but how does this apply in China where there is strong state-centred responsibility for policy-making as well as an element of mistrust of 'civil society' models (Jin 2014) and concern about foreign 'interference'? (Redmon et al. 2013).

Popovic (2017) identifies two possible roles for advocacy networks in authoritarian states: engagement with policy-makers in consultations about policy decisions and working to inform and educate the community and policy-makers about the issues and possible strategies for change. In the case of tobacco control in China, advocacy organisations have had limited formal engagement with the policy process, but have provided expert support to state agencies. For example, since 1990 the China Association of Tobacco Control (CATC) has been under the sponsorship of, and a close partner with, the Ministry of Health. It shares common tobacco control goals with the Ministry and the China CDC. In relation to the FCTC, the policy role of CATC and other advocacy groups was to support the Ministry of Health as it engaged as part of the China delegation with negotiations for the FCTC and sought legislative implementation of its provisions.

Despite the lack of direct engagement with policy, the Association has worked successfully over three decades to fulfil the second role identified by Popovic (2017), that is, to challenge community attitudes to smoking, to engage in research and to set up local groups to advocate actively at provincial and city levels. The CATC is a well-connected organisation with former senior officials serving as presidents of the Association, although since 2015 this is no longer allowed by the government. It has been very successful in recruiting elites (Ngok and Li 2010) to its cause, and its supporters are members of legislative bodies, such as the National People's Congress and the CPPCC (Yang 2018a). A number of high-profile advocacy successes include, in partnership with other public health NGOs, influencing the City of Shanghai to refuse tobacco sponsorship for the 2010 World Expo (Redmon et al. 2013).

Besides organisations specifically directed towards tobacco control, there are a range of other local NGOs and expert societies involved in tobacco control advocacy, such as the Cancer Foundation of China, Chinese cardiologists and the Chinese Medical Doctor Association (Yang 2018a). Legal experts, too, joined the tobacco control network, in 2013 proposing national tobacco control legislation (Beijing News 2013).

Popovic's research (Popovic 2017) reports that that major environmental and social NGOs continue to maintain positive informal relationships with policy agencies in China, but there is nevertheless ambivalence over future status. The new Charity Law (2016) relaxes registration requirements for the majority of domestic NGOs, estimated in 2017 to number approximately 816,000. Nevertheless the government has strengthened scrutiny and regulation to ensure that agencies to not 'endanger national security or violate public morality' (China File 2017). Measures include increased information disclosure and protection against corruption and strengthened CCP presence in organisations. International NGOs, under a 2017 Law on the Management of Overseas Non-governmental Organisations (2017), must register with a Bureau of Public Security and may be subject to closer scrutiny, including by the police (Gan 2017). Some international NGOs have withdrawn from China, but tobacco control NGOs have a history of engagement since the 1980s and established partnerships with central and provincial agencies and academic institutions. Huang (2018) nevertheless sees risk in the more regulated environment for NGOs, such as Gates and Bloomberg Foundations, which fund more 'activist' public interest litigation (Boseley 2015).

7.3 The Progress of Tobacco Control (1979–2012)

As Chap. 8 provides a detailed evaluation of tobacco control measures, this section just sets out key events on the pathway to tobacco control.

7.3.1 Tobacco Control During the Reform Period 1979–1999

During this period there were few tobacco control measures. The only national law enacted in relation to tobacco in this period was the 1992 Law on the Tobacco Monopoly which further institutionalised the tobacco industry, giving it monopoly jurisdiction over advertising bans, warning labels and levels of tar in cigarettes and undermining other efforts to campaign for smoking reduction or introduce specific measures. The 'fox was now in charge of the henhouse', and from this point on, the tobacco industry actively promoted the social and economic benefits of smoking and, as in western countries, questioned the evidence of smoking harm (Mackay 2016).

This two-decade period was characterised by a lack of political will to support tobacco control, largely because of the priority given to economic development, but is nevertheless remarkable for the indications of growing professional and community advocacy, including through research, bureaucratic support and the formation of CATC. With respect to research, there had been little previous academic interest or research funding for tobacco control in China (Koplan et al. 2013), but from 1986 onwards major epidemiological research initiatives were undertaken with the financial and technical support of the WHO Collaborating Centre, set up in Beijing in1986. In addition, support was provided through a World Bank loan (1996–2004) to set up and evaluate model tobacco control projects in seven cities (Redmon et al. 2013).

The Ministry of Health, with other government partners, engaged in health education and promotion, and introduced some regulations within their scope, such as the banning of smoking on domestic and international flights and establishing smoke-free carriages on some trains. The Ministry encouraged smoke-free hospitals and schools but with few opportunities for enforcement. The establishment of CATC in 1990 was an important development, and the beginning of an outstanding three-decade commitment to tobacco control.

7.3.2 *Negotiating the Framework Convention on Tobacco Control (1999–2006)*

The WHO launched its invitation for all governments to participate in the Framework Convention on Tobacco Control in 1996, and negotiations began in 1999. In China, the Ministry of Health strongly endorsed this initiative with support, from 2002, from the newly established Chinese Center for Disease Control (Walgate 2002). There was some ambivalence on the part of the government, with Huang (2014) reporting that China, while wishing to be a 'good international citizen' and supporting the public health goals of the Convention, was concerned about any challenge to its sovereignty and the position of tobacco in the economy.

China's part in the negotiations process has been well-documented (Yang 2018c; Wipfli 2015), confirming the ambivalence within the Chinese delegation. China was the only delegation that included representatives of the tobacco industry (WHO 2000), and at various times during the negotiations delegation members expressed opposing views. On the one hand, China's spokesperson repeatedly insisted that 'Chinese Government supported the formulation of the proposed framework convention on tobacco control, appreciating the great significance of the WHO FCTC' (Yang 2018c, p. 6). On the other hand, the CNTC representative explicitly expressed the view that 'Firstly, tobacco as a legal product should have its existence in the market place and should not be 'eliminated'; … Secondly, individual countries are different in many aspects and can have their own choices and therefore tobacco control should respect these differences' (Yang 2018c, p. 6; Wipfli 2015). The contradictions revealed in the negotiations highlighted the historic tensions over who was in charge of tobacco policy in China: the industry or public health interests.

The China delegation had only one Ministry of Health representative with limited leverage, whereas the resources available to the SMTA/CNTC members were substantial, particularly the work of the 'Counterproposals Research Work Group'. The 'Counterproposals Group' was set up by the STMA to 'conduct proactive research, facilitating decision-making, [and] servicing the needs of the industry' (Wu et al. 2018, p. 255). The Group had the benefit of experts seconded from the tobacco industry and from such influential groups as the STMA sponsored Yunnan Academy of Agricultural Science and the Zhengzhou Tobacco Research Institute as well as representatives from relevant departments of the STMA.

Members of the Counterproposals Group travelled with the China delegation and studied draft proposals from the WHO's Negotiating Body, making suggestions to the delegation for changes to the WHO text (Wu et al. 2018). The effect was to weaken the provisions of the Treaty in relation to China, a claim made proudly ('some strict provisions of the FCTC were modified') in the final report of the Counterproposals Group. The report said: 'we closely track…the progress of the negotiations,…carefully reading each Chair's text released by WHO, and form a draft proposal for the revision of the next Chair's text in the viewpoint of STMA to submit to the Chinese government delegation… Of 128 recommendations made by the working group, 51 were adopted by the Chinese government delegation and included in the Chinese negotiating documents [for] the meetings of Intergovernmental Negotiating Body' (Cheng 2004; Zhou and Cheng 2006, reported by Wu et al. 2018, p. 256). The FCTC text was weakened by the intervention of the Counterproposals Group which, through STMA's Resource and Translation subgroup, was able to convert the direct and affirmative language of the WHO into more tentative Chinese terms. For example, Chinese language terms such as 'recommended that' and 'when appropriate' were used instead of stronger English words such as 'should' and 'shall' (Wu et al. 2018, p. 257).

The FCTC was approved by the World Health Assembly in 2003 and ratified by the Chinese government in 2005, coming into force on 1 January 2006. Throughout the negotiations and afterwards, the Counterproposals Group engaged in a research, communications and information strategy designed to undermine the implementation of the Convention in China. This was based on the strategy developed by the Yunnan Academy of Tobacco Science to address challenges to tobacco use, with the aim of obtaining political and social support for the tobacco industry in its resistance to tobacco control. The output of this activity included publishing more than 30 research papers, funding a journal specifically related to the FCTC (132 issues) and publishing 4 research monographs. This work presented the STMA perspective on the FCTC, in which the economic importance of the tobacco industry was exaggerated and excuses are made not to place graphic warnings on cigarette packs, the benefits of smoking re-emphasised and 'low-tar, low-risk' myths propounded. The final report of the Counterproposals Group, approved and registered by STMA as a major science research project, was published as *The Study on Counter-proposals to the WHO FCTC and Countermeasures to Address Its Impacts on the Chinese Tobacco Industry in China* (Zhou and Cheng 2006). This brought accolades to members of the Counterproposals Group: eight people gained promotion to higher administrative posts and five received other significant promotions, with others promoted within academic and scientific spheres (Wu et al. 2018).

Even after ratification, the CNTC, along with other TTCs, continued to undermine the FCTC. In November 2010 the FCTC Parties met to approve Guidelines for Articles 9 and 10 which dealt with measuring emissions and the disclosure of the contents of tobacco products (Assunta 2012). The TTCs mobilised their 'front' organisation of international tobacco-growing countries to oppose the Guidelines which had no particular relevance to tobacco leaf growers, but posed important challenges to manufacturers. The industry members of the Chinese delegation (expanded from the usual two delegates to six) strongly opposed the provisions which were finally approved only as 'partial' guidelines.

7.3.3 Implementing the Framework Convention on Tobacco Control 2006–2012

The FCTC contained 24 Articles, some of which were purely administrative, but a number (Articles 6–17) outlined specific measures to reduce the impact of tobacco on society (WHO 2003). These have been helpfully summarised using the MPOWER acronym, launched in 2008 (WHO 2008), to reflect the main areas of intervention:

- **M**onitoring tobacco use and prevention policies.
- **P**rotecting people from tobacco smoke.
- **O**ffering to help quit tobacco use.
- **W**arning about the dangers of tobacco.
- **E**nforce bans on tobacco advertising, promotion and sponsorship.
- **R**aise taxes on tobacco.

7.3.3.1 Failure of Policy Transfer

The FCTC legitimised tobacco control in China. Nevertheless, it has been argued that the 5 years following ratification represented a failure of policy transfer (Jin 2014) with previous sections of this chapter noting how barriers were posed by lack of political will (Sect. 7.2.2) and the institutional environment (Sect. 7.2.3). Two specific articles of the Convention, Articles 4 and 5, gave guidance on how governments should act, but received little action after ratification (WHO 2003).

Article 4 noted that 'Strong political commitment is necessary to support, at the national, regional and international levels, comprehensive multi-sectional measures and coordinated responses'. In China the presence of high-level national laws, confirmed by meetings of the National People's Congress, is essential for the effective development and enforcement of provincial and local initiatives. For several years there were no new laws to support the FCTC. By 2012 the only national law was the updating of national legislation related to young people, the Law of the People's Republic of China on the Protection of Minors, promulgated on 26 October 2012 by the National People's Congress.

Article 5 expected governments to 'establish or reinforce or finance a national coordinating mechanism or focal points for tobacco control, such as working groups or steering committees'. This supportive institutional structure, required by Article 5, was denied to FCTC implementation in China. In April 2007 the State Council approved the establishment of the Inter-Ministry Coordination and Steering Committee for Tobacco Control (State Council 2007). This included eight major government departments, with the Ministry of Health the lone tobacco control advocate and with the STMA a member. Such composition had two fatal weaknesses. Firstly, naming the STMA as a participant in tobacco control has been described as 'asking a tiger for its skin'. Secondly, the lead unit of the Co-ordinating Committee was the National Development and Reform Commission (NDRC), rather than the MOH. The NDRC's responsibilities included formulating and organising the implementation of national economic and social development strategies, with 15 of its 17 functions directly related to economic matters. It was also the overseeing authority for the STMA. When, in 2008, the authority overseeing SMTA changed to the Ministry of Industry and Information Technology (MIIT), this organisation became head of the Co-ordinating Committee. The MIIT was regarded as market oriented, focused on short-term gains and resistant to regulation (Wipfli 2015). Just as the tobacco industry had undermined and obfuscated the terms of China's joining the FCTC, so it could now undermine the implementation of the framework.

7.3.3.2 Alternative Action on Tobacco Control

Despite the lack of legislation or institutional endorsement, the Ministry of Health, with support of the Tobacco Control Office (set up in 2009) of the Chinese CDC and other network allies, challenged the barriers to tobacco control in a variety of ways. Setting guidelines and pushing legislation and regulation for tobacco control were important, if sometimes thankless, tasks of the Ministry of Health. Despite no enforcement powers, by the late 2000s it had developed guidelines for smoke-free hospitals (2009), healthcare facilities (2011) and schools (2010) and proposals for 'Full Implementation of Smoking Ban in Public Places' (2011) (Wipfli 2015). There was a statement of commitment to tobacco control in the 12th Five-Year Plan approved by the National People's Council (Redmon et al. 2013).

Work by the tobacco control networks during this period included undertaking the routine work of health promotion and education through local offices of the Ministry and CDC, and by CATC and other local and international NGOs. In the absence of national policy, these bodies also supported city- and provincial-level efforts to become smoke-free. A number of localities in China had attempted their own non-smoking regulations in the late 1990s, but only after the FCTC was there a more widespread action. By 2007 88 cities in China had some local tobacco control regulations (Ngok and Li 2010) and work in cities intensified after 2008, despite lack of legislative mandate and limited enforcement (Lin et al. 2019) (see Chap. 9).

7.3.3.3 Research and Monitoring

The findings of the 2010 GATS survey indicated that China had performed very poorly on the implementation of the MPOWER programme (Yang et al. 2010a). The lack of institutional support meant that official research and monitoring systems were limited, with the bureaucracy again supported by local and national non-government organisations. The Ministry and China CDC engaged in major population-level monitoring projects with the continuing support of international partners such as the WHO, Bloomberg Philanthropies and the Bill and Melinda Gates Foundation (Yang 2018b).

As in other countries (Fry et al. 2017; Egbe et al. 2019), there have been deficiencies in enforcement and monitoring the implementation of tobacco control (Wang et al. 2017) with NGOs moving to fill this vacuum. Monitoring tobacco control performance was taken up by the CATC and the ThinkTank Research Center for Health Development (TTH 2015). The CATC reported, over 100 days in 2014–2015, a total of 89 sponsorships and a number of prohibited Internet and point-of-sale promotions (Beijing Morning Post 2015). The ThinkTank, since 2010, has monitored lack of enforcement and improper advertising, with achievements including establishing a monitoring database, publishing a Monitoring and Tracking Briefing, publicity about tobacco industry behaviour and numerous successful complaints about violations of advertising regulations (Jinkui et al. 2018).

Despite the absence of national legislation and non-compliance with Articles 4 and 5, this short period saw positive bureaucratic initiatives at national and local level and an energetic civil society sector working for tobacco control. This progress was ad hoc, led by community, officials and academics, but nevertheless notable. There was some disquiet about the effectiveness of FCTC implementation (Yang et al. 2010a), and in 2011 two respected tobacco control advocates assisted by colleagues published a critical assessment of the government's performance on tobacco control in China (Yang and Hu 2011). This generated significant public debate, with Li (2012) reporting that more than 900 news articles appeared within 2 weeks of its release. The nature of discourse on smoking appeared to be changing.

7.4 The Turning Tide? Action on Tobacco Control (2013–2019)

The 'Decision' on the deepening of reform at the Party Conference in 2012, and the specific indication that social, health and equity issues would be on the political agenda (see Sect. 7.1.3), may have given advocates hope that tobacco control might receive more attention. This view was reinforced by the 2013 Report of the Party School of the Central Committee of the CPC, entitled *Tobacco Control: International Experience and China's Strategy* (Mackay 2016). This documented the most progressive and high-level thinking on tobacco control by the Party and had been initiated when Xi Jinping was Director of the School. This publication has been rightly regarded by Mackay (2016) as the political 'tipping point' on tobacco control. It sets out the 'unthinkable:' that China's tobacco control efforts were 'weak' with regulations unenforced, that the tobacco industry 'may mislead the public' and that its presence on the Co-ordinating Committee was a conflict of interest (Mackay 2016). The Report of the Party School signalled a change in Party and government policy.

During this period there were a number of nationally led measures for the removal of barriers to tobacco control. This included statements and guidance issued from official Party sources, such as instructions from the State Council to officials not to smoke in public or to be involved in gifting tobacco products (2013), an effort to denormalise smoking and create positive role models. Smoking was banned in schools and the People's Liberation Army (2014). Actions included legislation, such as the National Advertising Law (2015), regulations for a tax increase on tobacco (2015) and preliminary drafting of a national smoke-free law.

In the second half of the decade, there was evidence of a broader change in the policy climate, highlighting how new priorities were becoming institutionalised within government. These included the *Healthy China 2030* policy promulgated in 2016 (Tan et al. 2018) and China's *Medium-Long-Term Plan for the Prevention and Treatment of Chronic Disease 2017–2025* (Kong 2017; State Council 2019). These demonstrate a strategic 'top-level design' rather than reactive approach to major issues, and commit to genuine co-ordination rather than the 'bargaining' style power relationships more generally characteristic of policy-making in China. They also provide a supportive framework for tobacco control and an alternative approach to health issues. For example, the Healthy China strategy emphasised a collaborative approach and the shift from the previous pursuit of rapid economic growth to a health-based economy (Tan et al. 2018) and an alignment with the UN Sustainable Development Goals. Tobacco control was specifically included as Action 4 in *Healthy China Initiatives* launched in 2019 (Lin et al. 2019) (Fig. 7.1).

An even greater impact on the institutional environment was demonstrated by the State Council appointment in 2018 of the National Health Commission (previously the Ministry of Health/National Health and Family Planning Commission) to chair the Co-ordinating Committee for the Implementation of the FCTC (Leng 2018). By removing the Ministry of Industry and Information Technology from its leadership role, China was now compliant with Articles 4 and 5 of the Convention and, for the

Fig. 7.1 Poster linking tobacco control with *Healthy China 2030*. (Credit: WHO Western Pacific. https://open.who.int/2018-19/country/CHN Public domain)

first time, had the potential to enact more effective national policies of tobacco control, including greater WHO advocacy of smoke-free policies in major cities.

Besides the changes driven by national leadership, throughout the decade there was a continuation of leadership by officials and advocacy groups and an increased range of regulation and action by local authorities. The growth in city-level smoke-free projects, especially in large cities, has been an outstanding example of combined local, central and community initiatives (see Chap. 9). A notable feature of city-level regulation has been the effort to enforce where necessary and provide sanctions. Redmon et al. (2013) note heavy investment by foundations not only in research but broader aspects of tobacco control, including cessation and health education. For example, the Bloomberg Initiative and the Campaign for Tobacco-Free Kids (CTFK) have cooperated with domestic NGOs and local government as well as US and Chinese universities to educate the public and alter attitudes towards smoking while also working to strengthen China's smoking control laws (Koplan and Redmon 2014). Similarly, the Emory Global Health Institute Tobacco-Free Programs focused on evaluation of city-level projects (Redmon et al. 2014). Huang (2018), however, sees some risk in the more centralised, regulated environment for NGOs, such as the Gates and Bloomberg Foundations, which fund more 'activist' projects (Boseley 2015).

While this implicit endorsement of the need for action on the FCTC was heartening for tobacco control advocates, Lin et al. (2019) also noted the deficits in enforcement, even in some of the best city-based tobacco control initiatives. The 'top-level design' framework of the 19th Party Congress (2017), nevertheless, had confirmed the importance of follow-through on implementation and monitoring of performance, which can be seen as another opportunity to support tobacco control actions (Yang and Yan 2018). Besides the monitoring by CATC and the ThinkTank, already noted, Huang (2018) reports on a growing trend to challenge non-compliance through public interest litigation lawsuits, usually with limited success other than valuable media coverage. Nevertheless, in 2018 the Railway Court upheld that the 'right to health' is greater than the 'right to smoke' (Li and Galea 2019). Administrative complaints have been more successful, with some retreat on sponsorship by offending tobacco companies (Huang 2018) and particular success in ensuring that Internet advertising is subject to pre-approval.

7.5 Conclusion

In China the barriers to tobacco control have historically been 'top-down' environmental forces, such as the lack of political will and the institutional arrangements of the state. These were reinforced by 'bottom-up' factors such as the normalisation of smoking in the community over many decades. This combination, despite the open-door policy exposing China to external forces and giving more flexibility to academic and civil society, ensured that, until the late 1990s, the prospects for tobacco control were negligible. So, what has changed?

First, having opened up to the world, the Chinese government found it politically advantageous to engage with the FCTC through the WHO, although there is no indication that there was a whole-hearted embracing of the tobacco control cause nor any intention to push on with implementation of the provisions of the Convention. Despite this, by ratifying the Convention the government provided legitimacy and an institutional 'space' in which tobacco control networks could extend their work, and particularly engage with city leadership. This allowed the classic approach to policy-making and implementation in China, 'feeling the stepping stones', advancing cautiously but nevertheless, moving forwards with experimentation.

Second, there was, from 2013 onwards, a significant change in the policy approach and priorities of the Party leadership. 'Top-level design' meant that any 'feeling the stepping stones' would take place within the framework of strategic policy decisions, set out and monitored by the government and Party. Policy priorities were restated to include important social and equity dimensions as well as economic growth; in the case of tobacco control, the institutionalised power of the industry was reduced by removing it from the key role in implementing the FCTC and indicating clear intention to legislate for change.

Besides this institutional and policy leadership change, it is important to assess the earlier and ongoing contribution of the tobacco control networks. In the 1980s and 1990s, 'open-door' policies had provided early 'space' for the networks to develop, relatively unfettered and with opportunities to take advantage of the tobacco control research, action and policy vacuum. After 2006, the FCTC ensured some legitimacy, but little official support, for their activities which nevertheless expanded in both volume and scope, and allowed engagement with important groups such as the NPC and the CPPCC. After 2013 the tobacco control networks worked strongly and effectively to advocate for legislation and to monitor the behaviour of the industry. The advocacy groups now certainly occupy the 'moral high ground' of tobacco control but at the same time are themselves subject to additional regulation.

There has been some shift in power relationships over this 40-year period. The tobacco industry is a less powerful an actor in the tobacco control environment with its institutional protection removed and being subject to greater regulation. The tobacco control advocacy networks now have significant 'standing', i.e. 'soft' power, based on their capability, expertise, history and widespread support. The government has asserted its power and chosen to take leadership in tobacco control, including both legislation and interventions such as role modelling non-smoking. While there may be some indications of 'denormalisation' of smoking, as of mid-2020 there is still no national tobacco legislation, as anticipated by the WHO under the provisions of the FCTC (Galea 2018), and much work left to be done.

This changing environment allowed particular policy initiatives, and it is to the outcomes of these that we now turn in Chaps. 8 and 9.

References

Assunta M (2012) Tobacco industry's ITGA fights FCTC implementation in the Uruguay negotiations. Tob Control 21(6):563–568

Au W, Ma W, Zhu Q, Chen H, Tang L (2016) Problems with cigarette smoking and attitudes towards the ban on smoking in Shantou, China. Public Health 134:46–53

Bao S, Chang G, Sachs J (2002) Geographic factors and China's regional development under market reforms 1979-98. China Econ Rev 13(1):89–111

Barraclough S, Morrow M (2008) A grim contradiction: the practice and consequences of corporate social responsibility by British American Tobacco in Malaysia. Soc Sci Med 66(8):1784–1796

Beijing Morning Post (2015) China Tobacco Control Association: Tobacco propaganda "incognito" aids students in disaster relief. (15 April; original in Chinese). http://shipin.people.com.cn/n/2015/0417/c85914-26858847.html

Beijing News (2013) Dozens of jurists in China jointly proposed legislation for tobacco control. January 28. http://health.sina.com.cn/news/2013-01-28/091670787.shtml

Berg C (2019) Local coalitions as an underutilised and understudied approach to promoting tobacco control in low and middle-income countries. J Glob Health 9(1):010301

Bhatta D, Crosbie E, Bilous S, Glantz S (2020) Defending comprehensive tobacco control policy implementation in Nepal from tobacco industry interference (2011-2018). Nicotine Tob Res 22(12):ntaa067. https://doi.org/10.1093/ntr/ntaa067

Boseley S (2015) Bloomberg and gates launch legal fund to help countries fight big tobacco. The Guardian, 18th March. https://www.theguardian.com/society/2015/mar/18/bloomberg-gates-foundation-fund-nations-legal-fight-big-tobacco-courts

Ceraso M, McElroy J, Kuang X, Vila P, Du X, Lu L, Ren H, Qian N, Jorenby D, Fiore M (2009) Smoking, barriers to quitting, and smoking related knowledge, attitudes and patient practices among male physicians in China. Prev Chronic Dis 6(1):1–6

Chen J, Ho S, Leung L, Wang M, Lam T (2019) Tobacco industry denormalization beliefs in Hong Kong adolescents. Nicotine Tob Res 21(7):949–954

Chen X (2017) A U-turn or just pendulum swing: tides of bottom-up and top-down reforms in contemporary China. J Chin Polit Sci 22:651–673

Cheng J (2014) Institutions, perceptions and social policy-making of Chinese local governments: a case study of medical insurance policy reforms in Dongguan. J Asian Public Policy 7(1):58–70

Cheng Y (2004) STMA's report on the research on counterproposals and countermeasures of the WHO framework convention on tobacco control. http://www.tobacco.gov.cn/html/15/1503/63947_n.html. Accessed 10 Feb 2020

China File (2017) How are NGOs in China faring under the new law. The China File 27 September 2017. https://www.chinafile.com/conversation/how-are-ngos-china-faring-under-new-law

Chinese CDC (Center for Disease Control and Prevention) (2011) Global adults tobacco survey (GATS) China 2010 country report. San Xia Press, Beijing

Chinese CDC (Center for Disease Control and Prevention) (2015) Global youth tobacco survey. Fact sheet: China 2014. Beijing, Chinese Center for Disease Control and Prevention, Beijing. http://www.wpro.who.int/china/gyts_china_fs_en_20140528.pdf?ua=1

Chu A, Jiang N, Glantz S (2011) Transnational tobacco industry promotion of the cigarette gifting custom in China. Tob Control 20(4):e3

Chung-Hall J, Craig L, Gravely S, Sansone N, Fong G (2019) Impact of the WHO FCTC over the first decade: a global evidence review prepared for the impact assessment expert group. Tob Control 28(Suppl 2):119–s128

Cohen J, Lee K (2019) State versus private ownership of tobacco companies. Br Med J 365:l4056

Ding D, Hovell M (2012) Cigarettes, social reinforcement, and culture: a commentary on "Tobacco as a social currency: gifting and sharing in China". Nicotine Tob Res 14(3):255–257

Egbe C, Bialous S, Glantz S (2019) Framework convention on tobacco control implementation in Nigeria: lessons for low- and middle-income countries. Nicotine Tob Res 21(8):1122–1130

Elton-Marshall T, Fong G, Zanna M, Jiang Y, Hammond D, O'Connor R, Yong H, Li L, King B, LiQ BR, Cummings K, Driezen P (2010) Beliefs about the relative harm of "light" and "low tar" cigarettes: findings from the international tobacco control (ITC) China survey. Tob Control 19:i54–i62

Fang J, Lee K, Sejpal N (2017) The China National Tobacco Corporation: from domestic to global dragon? Glob Public Health 12(3):315–334

Fang J, Wang G, Wan X (2020) 'Pro-tobacco propaganda': a case study of tobacco industry -sponsored elementary schools in China. Tob Control 29:477–451

Fry R, Burton S, Williams K, Walsberger S, Tang A, Chapman K, Egger S (2017) Retailer licensing and tobacco display compliance: are some retailers more likely to flout regulations? Tob Control 26:181–187

Galea G (2018) Law must ban smoking in public places without exemptions. WHO, Geneva. https://www.who.int/china/news/commentaries/detail/law-must-ban-smoking-in-public-places-without-exemptions

Gan N (2017) Why foreign NGOs are struggling with new Chinese law. South China Morning Post, 13 June. https://www.scmp.com/news/china/policies-politics/article/2097923/why-foreign-ngos-are-struggling-new-chinese-law

Gan Q, Lu W, Xu J, Li X, Goniewicz M, Benowitz N, Glantz S (2010) Chinese 'low tar' cigarettes do not deliver lower levels of nicotine and carcinogens. Tob Control 19(5):374–379

Gao X (2017) Promotion prospects and career paths of local party-government leaders in contemporary China. J Chin Gov 2(2):223–234

Gilmore A, Fooks G, Drope J, Bialous S, Jackson R (2015) Exposing and addressing tobacco industry conduct in low-income and middle-income countries. Lancet 385:1029–1043

Guo H (2018) Tobacco product regulation and tobacco industry interference. In: Yang G (ed) Tobacco control in China. Springer, Singapore, pp 201–220

He A (2018) Manoeuvring within a fragmented bureaucracy: policy entrepreneurship in China's local healthcare reform. China Q 236:1088–1110

He P, Takeaki T, Yano E (2013) An overview of the China National Tobacco Corporation and state tobacco monopoly administration. Environ Health Prev Med 18:85–90

Heilemann S, Perry E (2011) Mao's invisible hand: the political foundations of adaptive governance in China. Harvard University Press, Cambridge, MA

Ho M, Ma S, Chai W, Xia W, Yang G, Novotny T (2010) Smoking among rural and urban young women in China. Tob Control 19(1):13–18

Hu T, Lee A, Mao Z (2013) WHO framework convention on tobacco control in China: barriers, challenges and recommendations. Glob Health Promot 20(4):13–22

Hu Z, Kahn M (1997) Why is China growing so fast? Staff Papers 44(1):103–131

Huan G (1986) China's open door policy, 1978-1984. J Int Aff 39(2):1–18

Huang J (2018) Public interest litigation and tobacco control in China. In: Yang G (ed) Tobacco control in China. Springer, Singapore, pp 119–140

Huang Y (2013) Governing health in contemporary China. Routledge, New York

Huang Y (2014) China's position negotiating the FCTC and the international health regulations. Public Health 128(2):161–166. http://theory.people.com.cn/GB/40557/350432/350790/index.html

Huangfu Q, Korhman M (2018) The Chinese cigarette industry during the 'Great Leap Forward'. In: Kohrman M, Gan Q, Liu W, Proctor R (eds) Poisonous pandas: Chinese cigarette manufacturing in critical historical perspectives. Stanford University Press, Stanford, pp 73–94

Jiang S, Yang Y, Yang T (2019) China: City smokefree policy challenge highlights China's FCTC quandary. Tob Control 28(3):241

Jin J (2014) Why FTC policies have not been implemented in China: domestic dynamics and tobacco governance. J Health Polit Policy Law 39(3):632–666

Jinkui L, Tong L, Xiao L, Jiarui W, Yu X (2018) Exploration into and practice in tracking and monitoring the tobacco industry. Sci J Public Health 6(1):15–20

Jung H, Chen J (2019) Causes, consequences and impact of the great leap forward in China. Asian Cult Hist 11(2):58–65
Kanbur R, Zhang X (2005) Fifty years of regional inequality in China: a journey through central planning, reform, and openness. Rev Dev Econ 9(1):87–106
Kohrman M (2018a) Curating employee ethics: self-glory and slow violence at the China Tobacco Museum. In: Kohrman M, Gan Q, Li W, Proctor R (eds) Poisonous pandas: Chinese cigarette manufacturing in critical historical perspectives. Stanford University Press, Stanford, pp 133–136
Kohrman M (2018b) Introduction. In: Kohrman M, Gan Q, Li W, Proctor R (eds) Poisonous pandas: Chinese cigarette manufacturing in critical historical perspectives. Stanford University Press, Stanford, pp 1–33
Kohrman M, Gan Q, Hu T (2018) Wrangling the cash cow: reforming tobacco taxation since Mao. In: Kohrman M, Gan Q, Li W, Proctor R (eds) Poisonous pandas: Chinese cigarette manufacturing in critical historical perspectives. Stanford University Press, Stanford, pp 157–178
Kong L (2017) China's medium-long-term plan for the prevention and treatment of chronic diseases under the Healthy China initiative. Chronic Dis Transl Med 3(3):135–137
Koplan J, Erikson M, Chen L, Yang G (2013) The value of research as a successful component of tobacco control in China. Tob Control 22(Suppl 2):ii1–ii3
Koplan J, Redmon P (2014) Foreign philanthropic initiatives for tobacco control in China. In: Ryan J, Chen L, Saich T (eds) Philanthropy for health in China. Indiana University Press, Bloomington, pp 175–194
Lawrence M, Martin S (2013) Understanding China's political system. Congressional Research Service, Washington, DC
Lee S (2017) Institutional analysis of Xi Jinping's centralization of power. J Contemp China 25(105):325–336
Lee S, Ling P, Glantz S (2012) The vector of the tobacco epidemic: tobacco industry practices in low and middle-income countries. Cancer Causes Control 23(Suppl 1):117–129
Leng S (2018) China's ministry in charge of tobacco control had ties to the tobacco industry. Not any more South China Morning Post, 15 March. https://www.scmp.com/news/china/article/2137208/how-new-health-commission-could-help-reduce-smoking-chin
Leutert W (2018) Firm control: Governing the state-owned economy in China under Xi Jinping. China Perspect 2018(1–2):27–36
Li C (2012) The political mapping of China's tobacco industry and anti-smoking campaign. John L Thornton Centre, Brookings Institution, Washington, DC
Li C (2018) Tobacco governance: elite politics, subnational stakeholders and historical context. In: Kohrman M, Gan Q, Lui W, Proctor R (eds) Poisonous pandas: Chinese cigarette manufacturing in critical historical perspectives. Stanford University Press, Stanford, pp 179–204
Li L (2010) Central-local relations in the People's Republic of China: Trends, processes and impacts for policy implementation. Public Adm Dev 30(3):177–190
Li X, Galea G (2019) Healthy China 2030: an opportunity for tobacco control. Lancet 394(10204):1123–1125
Lin H, Chang C, Liu Z, Zheng Y (2019) Subnational smoke-free laws in China. Tob Induc Dis 17(Nov):78
Liu M, Wang J, Tao R, Murphy R (2009) The political economy of earmarked transfers in a state designated poor county in Western China: central polices and local responses. China Q 200:973–974
Liu W (2018) Experimentation: cigarettes in the communist base areas during World War II. In: Kohrman M, Gan Q, Liu W, Proctor R (eds) Poisonous pandas: Chinese cigarette manufacturing in critical historical perspectives. Stanford University Press, Stanford, pp 35–55
Lodge M (2003) Institutional choice and policy transfer: reforming British and German railway regulation. Governance 16(2):159–178

Lu H, Mayer A, Wellstrud A, Zhou S (2020) Can the dual identity of policy entrepreneur and policy implementer promote successful policy adoption? Vertical greening policymaking in Shanghai. J Asian Public Policy 13(1):1–16
Ma J, Lin M (2012) Policy-making in China: a review of Chinese scholarship. China Rev 12(1):95–122
Ma S, Hoang M, Samet J, Wang J, Mei C, Xu X, Stillman F (2008) Myths and attitudes that maintain smoking in China. J Health Commun 13(7):654–666
Mackay J (2016) China: the 'tipping point' in tobacco control. Br Med Bull 120(1):15–25
Mertha A (2009) "Fragmented authoritarianism 2.0": political pluralization in the Chinese policy process. China Q 200(December):995–1012
Miller A (2014) More already on the Central Committee's leading small groups. China Leadersh Monit 44:1–8
Minzner C (2015) China after the reform era. J Democr 26(3):129–143
Ngok K, Li D (2010) Tobacco control in China: process, actors and policy initiatives. J Asian Public Policy 3(1):100–110
Pan Z (2004) Socioeconomic predictors of smoking and smoking frequency in urban China: evidence of smoking as a social function. Health Promot Int 19(3):309–315
Pedroni P, Yao J (2006) Regional income divergence in China. J Asian Econ 17(2):294–315
Pei M (2016) China's crony capitalism. Harvard University Press, Cambridge, Ma
Perkins P (1986) Chapter 2: Prospects for China's economy. In: Barnett AD, Clough R (eds) Modernizing China: post-Mao reform and development. Routledge, New York. (Electronic version reprinted 2019)
Peters B, Zhao Y (2017) Local policy-making process in China: a case study. J Chin Gov 2(2):127–148
Popovic E (2017) Lobbying practices of citizens' groups in China. SAGE Open 7(2):1–9
Qian J (2017) Tobacco control in China: institutions, bureaucratic noncompliance and policy ineffectiveness. In: Broedsgaard KE (ed) Fragmented authoritarianism: earthquakes, energy and environment. Routledge, Abingdon/Oxford, pp 56–76
Quach A, Anderson E (2008) Implications of China's open door policy on families: an impact analysis. J Fam Issues 29(8):1089–1103
Redmon P, Chen L, Wood J, Li S, Koplan J (2013) Challenges for philanthropy and tobacco control in China. Tob Control 22(Suppl 2):ii4–ii8
Redmon P, Koplan J, Eriksen M, Li S, Kean W (2014) Role of cities in reducing smoking in China. Int J Environ Res Public Health 11(10):10062–10075
Reuters (2007) Smoking curb could 'Upset China stability.' http://uk.reuters.com/article/2007/03/07/uk-china-parliament-smoking-idUKPEK30818820070307
Rich Z, Hu M, Xiao S (2014) Gifting and sharing cigarettes in a rural Chinese village: a cross-sectional study. Tob Control 23(6):496–500
Rich Z, Xiao M (2012) Tobacco as social currency: cigarette gifting and sharing. Nicotine Tob Res 14(3):258–263
Salooji Y, Dagli E (2000) Tobacco industry tactics for resisting public policy on health. Bull World Health Organ 78(7):902–910
Sansone N, Yong H, Li L, Jiang Y, Fong G (2015) Perceived acceptability of female smoking in China. Tob Control 24(Suppl 4):iv48–iv54
Schubert G, Alpermann B (2019) Studying the Chinese policy process in the era of 'top-level design': the contribution of 'political steering theory'. J Chin Polit Sci 24:199–224
Sha Q (2018) Malformed monopoly: how nationalization of China's tobacco industry was shanghaied by a 1950s cigarette conference. In: Kohrman M, Gan Q, Liu W, Proctor R (eds) Poisonous pandas: Chinese cigarette manufacturing in critical historical perspectives. Stanford University Press, Stanford, pp 56–72
Shan J (2015) E-cigarette controls considered for safety. China Daily, January. http://www.chinadaily.com.cn/china/2015-01/07/content_19255939.htm

Sheer V, Mao C (2018) Cigarette initiation among Chinese male teenagers and early smoking interaction. Health Commun 33(4):392–400
Shih V, Adolph C, Liu M (2012) Getting ahead in the communist party: explaining the advancement of central committee members in China. Am Polit Sci Rev 106(1):166–187
Singh G (2007) The emerging civil society in China. World Aff 11(4):108–117
Smith D, Zhao I, Wang L (2012) Tobacco smoking among doctors in mainland China: a study from Shandong province and review of the literature. Tob Induc Dis 10(1):14
Sohu News (2009) Seventeen Hope Primary Schools in Sichuan disaster area named for tobacco. 4th December 2009 (In Chinese). http://news.sohu.com/20091214/n268925079.shtml
State Council (2007) Approval of the Establishment of an Inter-Ministry Coordination and Steering Committee for Implementation of WHO FCTC (the Steering Committee) by the State Council. (In Chinese). http://www.gov.cn/zhengce/content/2008-03/28/content_1635.htm
State Council (2019) State Council measures to enhance people's fitness, health. Updated: July 15. http://english.www.gov.cn/policies/latestreleases/201907/15/content_WS5d2c6f8fc6d05cbd94d67a02.html
Su H, Deng Q (2019) Analysis on policy agenda of tobacco control in Haikou City from perspective of multi-source current model. Adv Soc Sci Edu Hum Res 294:270–274
Tan X, Wu Q, Shao H (2018) Global commitments and China's endeavours to promote health and achieve sustainable development goal. J Health Popul Nutr 37:8
Teets J (2018) The power of policy-networks in authoritarian regimes: changing environmental policy in China. Governance 31(1):125–141
Teets J, Hasmuth R (2020) The evolution of policy experimentation in China. J Asian Public Policy 13(1):49–59
TTH (2015) (Think Tank for Health) ThinkTank Research Center for Health Development – a new explorer in Health Research Arena. TTH, Beijing. http://www.healthtt.org.cn/Item/Show.asp?m=1&d=2684
Walgate R (2002) China sets up Centres for disease control and prevention. Bull World Health Organ 80(4):335
Wang H (2006) Tobacco control in China: the dilemma between economic development and health improvement. Salud Public Mexico 48(Suppl 1):s140–s147
Wang J, Li C, Jia C, Liu Y, Liu J, Yan X, Fang Y (2014) Smoking, smoking cessation and tobacco control in rural China: a qualitative study in Shandong Province. BMC Public Health 14:916
Wang L, Lu B, Wewers M, Foraker R, Xie M, Ferketich A (2017) Are retailers compliant with zoning regulations that ban tobacco sales near schools in Changsha, China. Tob Control 26(4):446–451
Wang S (2008) Changing models of China's policy agenda setting. Modern China 341(1):56–87
Wang W, He Z, Feng N, Cai Y (2019) Electronic cigarette use in China: awareness, prevalence and regulation. Tob Induc Dis 17:30
Wang X, Zhang X, Xu X, Gao Y (2018) Electronic cigarette use and smoking behavior among adolescents in China. Addict Behav 82:129–134
Wipfli H (2015) The global war on tobacco: mapping the World's first public health treaty. Johns Hopkins University Press, Baltimore
WHO (World Health Organization) (2000) Intergovernmental negotiation body on the WHO framework convention on tobacco control. World Health Organization, Geneva. http://apps.who.int/gb/fctc/E/E_it1.htm
WHO (World Health Organization) (2003) Framework convention on tobacco control. World Health Organization, Geneva. https://www.who.int/fctc/text_download/en/
WHO (World Health Organization) (2004) Tobacco industry corporate responsibility – an inherent contradiction. World Health Organization, Geneva. https://escholarship.org/content/qt6kf7q7v9/qt6kf7q7v9.pdf
WHO (World Health Organization) (2008) MPOWER: a policy package to reverse the tobacco epidemic. World Health Organisation, Geneva

WHO (World Health Organization) (2015) GATS Atlas. World Health Organization, Geneva. http://gatsatlas.org/
WHO (World Health Organization) (2017) The bill China cannot afford; health economic and social costs of China's Tobacco Epidemic. WHO Western Pacific/UNDP, Manila
WHO (World Health Organization) (2020) E-cigarette Ban and Regulation: Global Status as of February 2020. Global Center for Good Governance in Tobacco Control (GGTC), Pathumthani
Wu D (2019) Accountability relations and market reform in China's electric power sector. Glob Transit 1:171–180
Wu F (2013) Environmental activism in provincial China. J Environ Policy Plan 15(1):89–108
Wu J, Deng Y, Huang J, Morck R, Yeung B (2014) Incentives and outcomes in China's economic policy. Cap Soc 9(1):2
Wu Y, Li J, Pang J (2018) Manuals of obstruction. In: Kohrman M, Gan Q, Liu W, Proctor R (eds) Poisonous pandas: Chinese cigarette manufacturing in critical historical perspectives. Stanford University Press, Stanford, pp 254–275
Xiao L, Parascandola M, Wang M, Jiang Y (2019) Perception and current use of e-cigarettes among youth in China. Nicotine Tob Res 21(10):1401–1407
Xu X, Chen C, Abdullah A, Sharma M, Liu H, Zhao Y (2016) Knowledge about and sources of smoking-related knowledge and influencing factors among male secondary school students in Chongqing, China. Springer Plus 5:1879
Yang G (2014) Marketing 'less harmful, low-tar' cigarettes is a key strategy of the industry to counter tobacco control in China. Tob Control 23(2):167–172
Yang G (2018a) Action subject and surroundings of tobacco control and anti-tobacco control in China. In: Yang G (ed) Tobacco control in China. Springer, Singapore, pp 47–71
Yang G (2018b) International assistance and tobacco control in China. In: Yang G (ed) Tobacco control in China. Springer, Singapore, pp 73–88
Yang G (2018c) Introduction: China and the negotiation of WHO FCTC. In: Yang G (ed) Tobacco control in China. Springer, Singapore, pp 1–16
Yang G, Hu A (2011) Kongyan yu Zhongguo weilai: Zhongwai zhuanjia Zhongguo yancao shiyong yu yancao kongzhi lianhe pinggu baogao (Tobacco control and China's future: Chinese and foreign experts joint evaluation report on China's tobacco consumption and tobacco control). Jingji ribao chubanshe, Beijing
Yang G, Li Q, Wang C, Hsia J, Yang J, Zhao L, Zhang J, Xie L (2010a) Findings from 2010 global adult tobacco survey: implementation of MPOWER policy in China. Biomed Environ Sci 23(6):422–429
Yang L, Zhang M, Liu S, Jiang Y, Huang Z, Zhao Z, Deng Q, Li Y, Zhou M, Wang L et al (2019) Prevalence of smoking and knowledge about the hazards of smoking among 170,000 Chinese adults, 2013-2014. Nicotine Tob Res 21(12):1644–1651
Yang X, Yan J (2018) Top-level design, reform pressures, and local adaptations: an interpretation of the trajectory of reform since the 18th CPC Party Congress. J Chin Gov 3(10):25–48
Yang Y, Hammond D, Driezen P, Fong G, Jiang Y (2010b) Health knowledge and perception of risks among Chinese smokers and non-smokers: findings from the Wave 1 ITC China Survey. Tob Control 19(Suppl 2):i18–i23
Yang Y, Wang J, Wang C, Li Q, Yang G (2010c) Awareness of tobacco related hazards among adults in China. Biomed Environ Sci 23(6):437–444
Yao T, Jiang N, Grana R, Ling P, Glantz S (2016) A content analysis if electronic cigarette manufacturer websites in China. Tob Control 25(2):188–194
Yi-chong X (2018) China's giant state-owned enterprises as policy advocates: the case of the State Grid Corporation in China. China J 79(January):21–39
Zhang X, Lin S, Hu T (2016) The epidemic of cigarette gifting: a social barrier to tobacco control in China. In: Economics of tobacco control in China, World scientific series in global economics and public policy, vol 4. World Scientific, Singapore, pp 129–146
Zheng H, De Jong M, Koppenjan J (2010) Applying policy network theory to policy-making in China: the case or urban health insurance reform. Public Adm 88(2):398–417

Zheng P, Xiao L, Wang F, Yang G (2018) Comprehensive ban on tobacco advertising, promotion and sponsorship. In: Yang G (ed) Tobacco control in China. Springer, Singapore, pp 245–284
Zheng Q (2016) Control, discretion and bargaining; provincial leader rotation in China. Chin Polit Sci Rev 1:623–644
Zhou R, Cheng Y (2006) Countermeasures against the WHO's FCTC and approaches to deal with the FCTC's impacts on the tobacco industry of China, 1st edn. Economic Science Press, Beijing
Zhu X, Zhang Y (2016) Political mobility and dynamic diffusion of innovation; the spread of municipal pro-business administrative reform in China. J Public Adm Res Theory 26(3):351–551
Zhuang P, Owermohle S (2019) Online sales ban hits China's vaping industry – but don't expect it to disappear with a puff of smoke! South China Morning Post, 26 November. https://www.scmp.com/news/china/society/article/3039423/online-sales-ban-hits-chinas-vaping-industry-dont-expect-it

Chapter 8
Evaluating China's Record of Tobacco Control

Ross Barnett, Shuhan Jiang, and Sihui Peng

Abstract This chapter evaluates tobacco control policies in China in four key areas: the use of price increases to dampen demand; the introduction of smokefree legislation; controls on advertising and the effect of health system interventions. In all cases the dominant conclusion is that China has 'dragged its feet' in implementing effective tobacco control measures. Its interventions have been weak compared to other middle-income countries and countries in South and South East Asia. Nevertheless prospects are improved by the recent restructuring of the state bureaucracy. This has diminished the power of the CNTC and, through the *Healthy China* 2030 vision, has begun to emphasise national health rather than economic goals. The chapter concludes that current policy implementation should not focus on 'one size fits all' but must take account of geographic differences within China's vast landscape.

Keywords Advertising · Cigarette affordability · Health professionals · Smoking bans · Point of sale (POS) · Policy enforcement · Smoking cessation · Taxation · Tobacco control advocacy · World Health Organization

R. Barnett (✉)
School of Earth and Environment, University of Canterbury, Christchurch, New Zealand
e-mail: ross.barnett@canterbury.ac.nz

S. Jiang
School of Humanities and Management, Zhejiang Chinese Medical University, Hangzhou, China

S. Peng
School of Medicine, Jinan University, Guangzhou, China

R. Barnett et al. (eds.), *Smoking Environments in China*,
Global Perspectives on Health Geography,
https://doi.org/10.1007/978-3-030-76143-1_8

8.1 Introduction

This chapter builds on the institutional context of tobacco control discussed in the previous chapter by focusing upon the process of policy implementation and its outcomes. The main aim of the chapter is to evaluate the actual tobacco control policies introduced by China following the signing of the FCTC in 2005. The World Health Organization (WHO) FCTC and its guidelines provide the foundation for countries to implement tobacco control policies and monitor their outcomes. To assist in this process in 2008, the WHO introduced the MPOWER package. As outlined in the previous chapter, the acronym MPOWER stands for: M, Monitor tobacco use and prevention policies; P, Protect people from tobacco smoke; O, Offer help to quit tobacco use; W, Warn about the dangers of tobacco; E, Enforce bans on tobacco advertising, promotion and sponsorship; and R, Raise taxes on tobacco (WHO, 2008, 2013, 2014, 2019a). MPOWER identifies, and promotes, proven measures of tobacco control that correspond to one or more articles contained in the foundation FCTC document.

This chapter reviews the key interventions adopted in China to reduce smoking initiation, increase cessation and reduce exposure to second-hand smoke. Four key policy interventions are discussed: taxation and pricing, smoking bans, controls over advertising including cigarette packaging and health system initiatives emphasising education and cessation treatments. The first two interventions are important because they directly impact the smoker by changing the economic and social context of tobacco consumption by influencing the affordability of cigarettes and by placing restrictions on where smoking can take place. The second two interventions indirectly influence smokers by constraining the marketing activities of the tobacco industry and by forcing the industry to warn the public of the dangers of tobacco use. The final set of policies involve health system initiatives which, among other things, attempt to strengthen educational campaigns to encourage smoking cessation, but also to create positive role models and promote tobacco control advocacy within the health system.

8.2 Taxation and Pricing Interventions

An important strategy promoted by the WHO has been price increases via taxation. Tobacco taxation has long been a cornerstone of global tobacco control efforts and, together with smoking bans, is seen as one of the two most successful intervention priorities (Hoffman and Tan 2015). Price and tax increases have been most effectively used in high-income countries where tobacco taxes tend to be higher and account for a greater proportion of price than in low- and middle-income countries (Chaloupka and Pacula 2012). The 2013 *Global Action Plan for Non-communicable Diseases (NCDs)* further emphasised the importance of tobacco taxation as a key part of United Nations Sustainable Development Goals in reducing premature mortality from NCDs (Marquez et al. 2017). Between 2012 and 2014, the WHO's 2015 Report on the Global Tobacco Epidemic noted that 106 countries had raised their excise rates on cigarettes, with increases of more than 50% occurring in one fifth of those cases (WHO, 2015a). However, as excise taxes in many of these countries were very low to start with, there was scope for larger increases. China followed suit in 2015 but the increase was small.

8.2.1 The 2009 and 2015 Tax Increases

Between 2006 and 2015, China increased tobacco taxes twice, but only the most recent increase had a significant impact on consumer price levels. The first tax increase in May 2009 had no impact on consumption since the increase in the producer/wholesale price was not passed on to consumers. If it had been, the weighted tax rate would have increased by approximately 11.7% (Hu et al. 2010). In fact, Wang (2017) showed that the number of cigarettes smoked per day increased slightly following the tax adjustment. The main aim of the 2009 adjustments was not tobacco control but revenue generation designed to reduce the adverse impacts of the 2008 global financial crisis on central government revenues. Thus, economic, not social, goals dictated the structure of the tax increase (Xu et al. 2019).

China has a complicated tax structure (Shang et al. 2014; Zheng 2018) which involves five different tobacco taxes. However, only one of these, the excise tax, directly affects the retail price of cigarettes (Hu et al. 2016). The excise tax has two components, a producer price and a wholesale price. Within each component there first is a specific excise tax based on product quantity and then another ad valorem tax based on a percentage of the product price. The 2009 tax increase occurred only at the producer and wholesale levels, but central government, wishing to maintain high rates of consumption, forbade changes in retail prices. At the producer level, the ad valorem tax was raised particularly for higher priced cigarettes, while a new 5% ad valorem tax was introduced at the wholesale level (Xu et al. 2019).

By contrast the May 2015 tax increase represented a change in direction with tobacco control and health and social goals now becoming more important factors in shaping the policy change. Unlike 2009, the 2015 increase directly affected consumers. The 2015 increase resulted in an increase of ¥0.10 per pack at the wholesale level and increased the ad valorem tax by 6% from 5–11%. This time the CNTC was allowed to shift the increase to retailers, resulting in an average retail price increase of 10.3% (from 11.61 RMB per pack in 2014 to 12.81 RMB per pack in 2015) (Marquez and Zheng 2016). Nevertheless, while provincial branches of the CNTC recommended set retail prices, retailers had certain flexibility in order to retain profit margins of at least 10% (Goodchild and Zheng 2018).

Unlike 2009, it could be argued that the change was significant in the sense that it represented progress in living up to the commitments of Article 6 of the FCTC (Marquez and Zheng 2016) indicating that it was 'a step in the right direction'. However, the impact was modest. The Tobacco Atlas (Drope et al. 2018) estimated that, as a percentage of the average retail price, the Chinese excise tax of 36% was well below the WHO benchmark of 70% and much lower than almost all other countries in South and South East Asia (Table 8.1). Even when compared to the average tax rate for all lower- and upper middle-income countries in 2018 (41.8%), or just to other countries with larger populations (>10 million) in the upper middle-income group (47.2%), China still performs poorly (Fig. 8.1).

Given past CNTC interference in the policy process, China's low ranking is not unexpected. Hu et al. (2016) noted that the CNTC, in an attempt to undermine the scale of the tax increase, claimed that its implementation would cause a marked reduction in sales. While sales decreased in 2015, a trend also reflecting the impact of the new advertising law, they rebounded soon afterwards (Yu et al. 2019). This suggests that the CNTC achieved its goal of achieving a modest tax increase and limiting any significant adverse effects on smoking prevalence. The CNTC has also attempted to stress the regressive aspects of increased taxation claiming that any tax increase would lead to an unfair burden on lower-income smokers (Zheng and Huang 2015). However, such distributional arguments fail to acknowledge social inequalities in the health harms of smoking. Nor do they take account of subsidised lower tax rates on discount brands that keep them affordable for the poor.

Table 8.1 Excise taxes (ranked) as a % cigarette retail price for countries in South and South East Asia, 2018

Country	World Bank income classification	Excise tax % retail price[a]
Thailand	UMIC	66.9
South Korea	HIC	66.8
Singapore	HIC	60.6
Bangladesh	LMIC	56.0
Japan	HIC	55.7
Sri Lanka	UMIC	55.5
Philippines	LMIC	51.9
Indonesia	LMIC	48.7
Malaysia	UMIC	47.1
Pakistan	LMIC	41.8
China	**UMIC**	**36.0**
Myanmar	LMIC	32.5
India	LMIC	31.2
Vietnam	LMIC	29.8
Cambodia	LMIC	13.4
Lao PDR	LMIC	9.7

[a]Mean = 44.4% (SD = 17.3)

Credit: Compiled from country data in Drope et al. (2018)

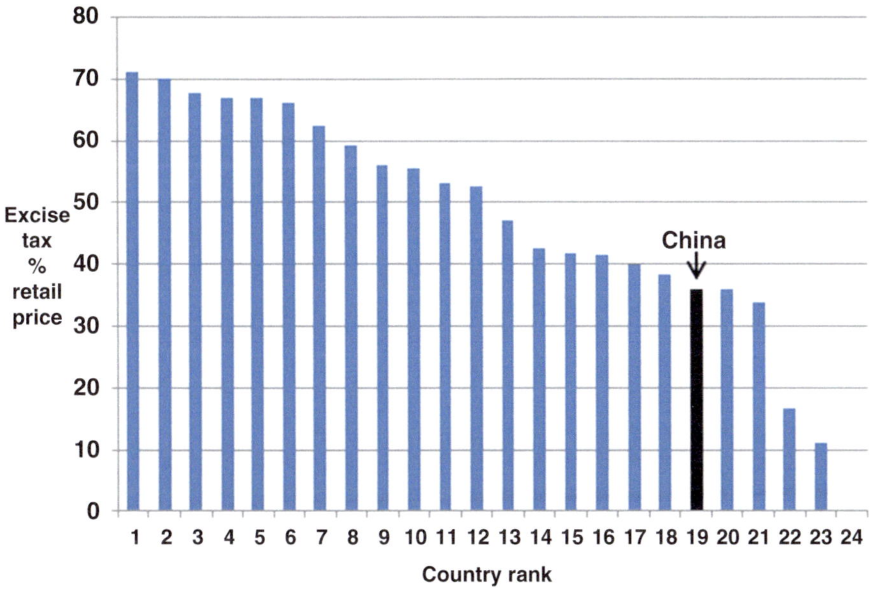

Fig. 8.1 Excise tax as a proportion of average cigarette retail price, larger upper middle-income countries with populations exceeding 10 million, 2018
Rankings: 1–5, Argentina, Cuba, Venezuela, Thailand, Jordan; 6–10, Turkey, Colombia, Ecuador, Georgia, Sri Lanka; 11–15, Mexico, Romania, Malaysia, Russian Federation, Kazakhstan; 16–20, South Africa, Brazil, Guatemala, China, Dominican Republic; 21–24, Peru, Algeria, Iran, Iraq
Credit: Authors. Compiled from country data in Drope et al. (2018). The mean tax rate for the 24 UMICs was 47.2% (SD =19.0) compared to 36.2% (SD = 18.9%) for 23 LMICs and 41.8% (SD = 19.6%) overall

8.2.2 *Impacts on Consumption*

The retail price increase was not uniform across different brands, with the biggest increases of 10% and 20% occurring among the lower price categories (categories 4 and 5), compared to smaller increases for mid-priced (category 3) (5%) and higher-priced brands (categories 1 and 2) (5 and 9%). The means that the price gaps between categories has been reduced, thus encouraging smokers of cheaper brands to switch to higher-quality products. This shift is not unexpected given the CNTC's premiumisation strategy discussed in Chap. 4. The differential price shift not only reflects the CNTC's goal of producing higher-quality products, which can compete internationally, but at the same time the modest price increase and the maintenance of retailer margins have meant that cheaper offerings have been maintained on the domestic market. This was necessary in order to maintain consumption among lower SES groups and to encourage initiation by new smokers (Goodchild and Zheng 2018). This was an important goal especially since, from 2014 onwards, the CNTC faced declining sales.

At first glance the May 2015 tax increase resulted in a large drop in consumption. Yu et al. (2019) reported that about one in five of male Chinese smokers reported reductions in smoking, while Xu et al. (2019) reported a decline of 16.54% in the 26-month period following the May 2015 tax increase (May 2015 to June 2017). The decline, however, was perhaps more apparent than real; in Yu's sample of 5782 male smokers, 77.9% reported no reduction in smoking, a proportion that was higher among daily (80.9%) than occasional (68.6%) smokers. If only no change and small decreases are combined, 96.6% daily smokers reported little or no change compared to 91.3% for occasional smokers and 95.3% for both groups. Overall the results tend to support the frequent comment that Chinese smokers have low price elasticity (Lance et al. 2004; Huang et al. 2015; Yang et al. 2016).

While raising cigarette prices is widely regarded as an effective measure of tobacco control, reductions in cigarette consumption will only occur when price increases reduce cigarette affordability, that is, where cigarette prices are weighted as a proportion of disposable household or personal income (Hu et al. 2019). Where affordability is high, price increases will have little effect on consumption. This was evident in the Yu et al. (2019) study where, although more daily smokers considered cigarettes to be expensive and very expensive (40.7%) than occasional smokers (34.8%), the former were much less likely to report any changes in consumption. On the other hand, while low-income smokers displayed greater price sensitivity, in reality this would have been minimal given the now well-established tendency of this group to trade down and switch to less expensive brands (Liu et al. 2015). Kostova et al. (2017) argue that brand switching has been one of the reasons why higher prices (estimated on the basis on variations in cigarette prices in different regions of China over time) have affected initiation rather than smoking cessation as in western countries. They concluded that the absence of a cessation effect was consistent with growing evidence of brand down-switching in China (White et al. 2014) which was greater among lower-income smokers who traditionally have tended to purchase less expensive cigarettes (Li et al. 2010, 2015; Yao et al. 2013; Zheng and Huang 2015).

Unfortunately Yu et al. (2019) did not consider interactions between income, smoking status, attitudes towards current cigarette prices, brand substitution and changes in consumption. The main finding of the study was that smokers with *higher* not lower household incomes per capita (¥80,000–99,999), who smoked more often, and who purchased more expensive brands, were much *less* likely to be affected by the price increase than those with low household incomes per capita. The results, therefore, suggest that the effects of the 2015 tax increases were socially selective, having less effect on higher-income groups. Those who were most affected tended to be lower-income smokers, occasional smokers and people who had intended to quit. These findings are similar to those reported by writers such as Nargis et al. (2019) and Hu et al. (2019) who showed, for samples of all (>18 years of age) and older (aged 40 and over) Chinese adults, that smokers of higher SES demonstrated greater affordability than those of lower SES. However, the very low price of discount cigarette brands in China, which are more popular among daily than occasional smokers (Hu et al. 2019), is likely to buffer low-income groups from significant reductions in consumption. Even the 20% average price increase for discount cigarettes in 2015 would have made little difference to cigarette affordability for lower-income smokers. As a number of writers (e.g. Kostova et al. 2017; Hu et al. 2019) have suggested, given the lack of sensitivity to changes in affordability among younger and low-income smokers, the predominant users of discount brands, it may be important to adopt minimum price regulations in order to further limit consumption by these more vulnerable groups.

Compared to other East Asia countries, cigarettes have become far more affordable in China nullifying the effects of any modest price increases. Zheng et al. (2017) showed that the average price of cigarettes in China was 1.85 times more affordable in 2016 than in 2001, while the cheapest category of cigarettes, typically consumed by low-income persons, became 2.09 times more affordable. Cigarette affordability in China also increased in comparison with other countries internationally. In just two decades, China moved from the fourth to the second quartile affordable country group as its cigarette affordability surged with the fastest growth rate in the world (Zheng et al. 2017, p. 27).

8.2.3 *Explaining Behavioural Responses to Increased Cigarette Prices*

The price elasticity of cigarettes (i.e. the purchasing response to increase in price) is not uniform, with the impact of price increases varying according the nature of the economy and other possible influences on purchasing behaviour. In a study of low- and middle-income countries in the Asia-Pacific region, Ho et al. (2018) reported that simulated price elasticity was greatest for the higher-income countries (Malaysia and China) of that region. Elasticity was, nevertheless, still relatively low and subject to other factors that reduced the impact of increased taxation, such as poor implementation of other tobacco control strategies or manipulation of retail pricing.

While smokers may change their tobacco consumption in response to price, there is a need to understand how specific variables might affect susceptibility to price rises and either amplify or minimise the expected decline in consumption. These influences fall into two broad groups: individual variables such as socio-demographic characteristics or patterns of consumption and environmental variables that have an impact on the decision to smoke. With personal factors, there are differential responses to price increases from smokers in China, with younger people more likely to be influenced by price (Ding 2003; Kostova et al. 2017). Those on higher incomes are less likely to be affected (Yu et al. 2019), although the evidence here is ambiguous (Yang et al. 2016), and those on lower incomes more likely to switch to cheaper brands (Huang et al. 2014). There is also evidence that females and minority non-Han minority groups are much less sensitive to price increases (Yang et al. 2016).

While many studies have explored Chinese smokers' behavioural responses to cigarette prices and individual-level influences, few studies have explored the effect of contextual factors and how they may influence this relationship. One recent exception is by Yang et al. (2016) who considered a price behavioural response and a strong price behavioural response to the question of whether smokers were consuming fewer cigarettes because current cigarette prices were too high. The study was undertaken in 21 Chinese cities in 2014 before the implementation of the 2015 price increase. Of the environmental factors considered, regional smoking restrictions in workplaces, cigarette production and media news coverage were all associated with price sensitivity (SPRF). The presence of workplace restrictions had a strong effect on increasing SPRF (OR = 3.30; $p < 0.001$) as did media news coverage (OR = 3.35; $p < 0.01$), while higher provincial cigarette production (pieces/100 Yuan GDP) decreased sensitivity. Smokers living in provinces with the highest cigarette production levels were 76% less likely (OR = 0.24; $p < 0.001$) to have had a SPRF response compared to smokers living in areas with less production.

Price sensitivity thus has an environmental context with reductions in consumption being reflective of workplace social norms and strong antitobacco messages. On the other hand, the pro-tobacco social norms of cigarette-producing regions and the presence of provincial differences in retail pricing and the availability of cigarettes are likely to have made consumers less sensitive to price. In a later study conducted in the year following the 2015 price increase, Yang et al. (2016) report similar findings. To assess price-induced smoking behavioural response (PSBR), current smokers were asked whether they had reduced their cigarette consumption due to the cigarette price increase. This study indicated that 6.2% of smokers reported a significant smoking reduction after the tax increase, slightly higher than the 5.2% who reported a price-related reduction in consumption prior to the 2015 tax increase. Besides the 6.2% who reported a significant decline in consumption (and this included 0.9% who reported giving up completely), an additional 19% reported a 'little reduction' in smoking which is an encouraging sign.

The study found that increased exposure to anti-smoking information was strongly associated with PSBR. This suggests that in China, anti-smoking information may influence people's beliefs, awareness and behaviours about smoking and tobacco control to make smokers more sensitive to cigarette price increases. In common with findings for many middle-income and poorer countries, the overall public awareness of the hazards of smoking in China is relatively low. This lack of awareness reduces 'readiness' to limit consumption in response to higher prices and encourages smokers to respond in other ways, such as changing to cheaper brands. This is highly likely in China where cigarettes are cheap and there is a wide range of prices available. The study also found that the home and workplace environmental restrictions were also related to the PSBR. The mechanism for this is likely to be the combined impact of these environmental restrictions and anti-smoking information on the denormalisation of smoking (Eriksen and Cerak 2008). Denormalised settings will create greater sensitivity to any 'trigger' to reduce consumption, including the 'trigger' of increased price (Im et al. 2015).

Interest in context effects has also recently been extended to studies of cigarette affordability. Building on the work of Zheng et al. (2017), Hu et al. (2019), using individual-level data from the China Health and Retirement Longitudinal Study (CHARLS), examined the effect of affordability on consumption for middle-aged and elderly smokers (45 years and over) living in four different regions in China. Particularly important were the regional and urban-rural variations uncovered. Regional differences in responses to cigarette affordability were most marked in the Northeast Region; smokers living there were less likely to reduce their daily cigarette consumption compared to those living in the Eastern Region when cigarette affordability decreased. This was seen as a response to economic recession in the Northeast, described by the authors as China's 'rust belt'. Years of low economic growth had led to higher levels of stress limiting the effect of decreased affordability on consumption. Similarly urban-rural differences were also apparent. The study found that cigarettes were less affordable for rural than urban smokers and that when cigarette affordability decreased, rural smokers reduced their consumption more than urban smokers. Despite higher rates of rural smoking, this trend most likely reflected high rates of rural poverty and economic marginalisation, especially among older members of the rural community.

8.2.4 Summary

In summary tax increases, while an important instrument of tobacco control, are in their infancy in China, and the smoking and public health gains from the 2015 tax increase have been modest. A one-off tax increase is unlikely to have a significant lasting impact on smoking prevalence and, as Goodchild and Zheng et al. (2018) have argued, did not go far enough to significantly alter smokers' expectations about future tax increases and the likely impact of these on their consumption. Consequently, the smoking cessation and health gains were minimal. In addition a recent World Bank Assessment of tobacco tax reform (Marquez et al. 2017) criticised the inefficiency of Chinese-style tiered excise tax systems because of their inability to stop smokers from switching to cheaper products, thus diluting the intended public health impact of raising tobacco taxes. Higher excise rates on cheaper cigarettes may partly limit such trends. However, writers such as Verguet et al. (2015) and Wang et al. (2015) have estimated that it would take a 50% increase in the tobacco price to produce significant changes in consumption and positive health impacts. Given China's past reluctance to impose higher taxes, such an outcome seems unlikely.

Research undertaken to date has increasingly moved away from studies of price elasticity to issues of cigarette affordability and behavioural studies which have increasingly linked smoker price behaviour to a range of contextual factors. This is an important step as it is important to highlight the links between price-induced changes in consumption with a range of contextual variables, including other tobacco control measures, and how these may modify consumption. For example, the tax increase of 2015 occurred alongside other important policy changes, such as the implementation of advertising and smoking bans which also influenced consumption. Thus, it is difficult to claim that price increases alone had a significant impact. Nevertheless, as Kostova et al. (2014) have noted, because China has both high affordability and wide price ranges, there are multiple opportunities to further significantly reduce consumption through tax adjustments.

8.3 Smokefree Policies

The development of smokefree policies has, along with substantial tax increases, been one of the most important components of tobacco control legislation. Smokefree legislation is a keystone of the FCTC and MPOWER policy packages particularly in countries, like China, where tobacco remains cheap and price mechanisms have done little to curb consumption. Regulating the places where people can smoke can be a powerful tool, not only to reduce exposure to second-hand smoke (SHS) but also to change social norms via promoting the unacceptability of smoking in public places. In addition, increased spatial and venue restrictions over the location of smoking can be effective in encouraging smokers to quit.

8.3.1 National Policy Developments

In China, although the health effects of SHS exposure have been increasingly documented, after ratification of the FCTC in 2005, it took some time to develop smokefree policies. These followed a predictable pattern with key areas such as healthcare and education first receiving attention, followed later by a focus on transportation, other public spaces and places of employment. It was not until 2009 that the first major national policy, aimed at banning smoking in healthcare facilities, was developed (Table 8.2). The Smokefree Healthcare Facilities Policy, issued by the former Ministry of Health, decreed that all healthcare facilities should be smokefree by 2010 (Gan 2018). Given that many doctors continued to smoke, the policy aimed not only to make hospitals and other health facilities smokefree but also to change the smoking habits of the medical profession so that doctors could be role models and advocates for smokefree living. This also eventually led to the development of tobacco control curricula for medical students (see Sect. 8.5.4). Despite such initiatives, while smoking in hospitals declined from 37.9% in 2010, it was still relatively high (24.4%) 8 years later (China CDC 2019).

The next major initiative was the Smokefree Schools Policy released by the Ministry of Education and the former MOH in 2010. Although the policy banned indoor smoking in all schools and outdoor smoking on the grounds of elementary and middle schools, the reliance on self-compliance meant that implementation was poor and SHS exposure remains high, 23.4% in 2018 (China CDC 2019). In 2011

Table 8.2 Timeline of national smokefree legislation and subnational initiatives, 2007–2014

National policies
2009 Smokefree Healthcare Facilities Policy
2011 Smokefree Schools Directive
2011 12th Five-Year Plan (2011–2015) National commitment to ban smoking in public places.
2011 Executive Order on Public Sanitation
2013 Joint Notice on Smoking Ban of Public Officials in Public Places
2014 Ministry of Education and National Health and Family Planning Commission notifications
2014 China State Council draft regulation. 'Smoking Control in Public Places'
Subnational urban initiatives
Associated with international events
2008–2010 Beijing Olympic Games, Qinghao co-hosting city, Shanghai World Expo, Guangzhou Asian Games
Provincial smokefree policies
2011 Harbin become the first major city to enact a comprehensive smokefree law
Philanthropic initiatives
2007 China CDC, Peking Union Medical College, John Hopkins University, 'Toward a Smoke-free China'
2008 Emory Global Health Institute-China Tobacco Control Partnership – 17 cities
2014 Pfizer grant to support new smokefree initiatives in five cities

the MOH issued an executive order on public sanitation requiring that some public places be smokefree, but again the provisions for enforcement were weak and lacked penalties.

More important was the 'Joint Notice on Smoking Ban of Public Officials in Public Places', released by the Communist Party of China Central Committee and the State Council, China's cabinet, in 2013. This focused upon restricting smoking by public officials in their offices. Launched as part of a general anti-corruption campaign by President Xi Jinping, it called on government officials to 'walk the talk' by not smoking in public places, including their offices, at business meetings and public government events (CATC 2018a). It also forbade the common practice of public officials purchasing cigarettes using government funds (Gan 2018). Although the China CDC also attempted to provide guidance to local governments to issue smokefree government office policies, enforcement was poor (Mehta et al. 2014; Kegler et al. 2014). Indeed it was not until 2018 that such policies were finally implemented in cities in some provinces, such as Wuhan (CATC 2018b).

Mindful of the difficulties and lack of progress in enforcing smokefree policies in educational institutions, in 2014 the Ministry of Education reissued a notice calling for a national ban in schools, colleges and universities (WHO 2015b). In 2013–2014 the China City Adult Tobacco Survey of 14 cities reported that exposure (indoor and outdoor) to SHS at primary, secondary and high schools, including technical schools, was above 20% in all cities, with the highest prevalence (45%) occurring in Nanchang (China CDC 2015, p. 16). Although prevalence was lower in universities, it still ranged from a high of 32.5% in Nanchang, the capital of Jiangxi Province, to 13.8% in Tangshan, an industrial city in Hebei Province close to Beijing. In 2014 the Ministry of Education directive was followed by a notification of the National Health and Family Planning Commission that called for the provision of smokefree healthcare and family planning facilities.

In late 2014 the State Council of China issued a draft regulation, 'Smoking Control in Public Places' calling for a complete ban on smoking in public places, workplaces and public transport without exceptions (China CDC 2015, p. 4). Lin et al. (2019a, b) report that, as of 2018, the law still had not been adopted. Consequently because of a lack of national leadership and interference from the tobacco industry, less than half of the smokefree laws in China comprehensively ban all indoor smoking.

While the implementation of 'national' policies has been a fitful experience, it is at the subnational level where more substantial gains have occurred with the implementation of smoking bans in particular cities. A detailed case study of the geography of these developments and their health and social impacts is provided in Chap. 9. For the moment it is worth noting that the urban innovations have occurred in particular types of places. These have been largely been provincial capitals, in cities aided by international philanthropic agencies, but also in places closely connected to international tourist flows and the global economy. It has been in these places where anti-smoking coalitions have had their greatest impact. However, even at the urban level, progress has been slow and geographically uneven. Lin et al. (2019a, b. p. 3) in a study of 125 of the main cities in China claimed that 21 cities had FCTC compliant smokefree laws. Yet of these 21 cities, the authors later noted that only 9 (Beijing, Shanghai, Xian, Qingdao, Shenzhen, Lanzhou, Nanning, Tangshan and Changchun), or just 7.2% of their total sample, had reached, or were very close to reaching, FCTC Article 8 requirements (full coverage of smokefree laws with no exceptions). Of the other 12 cities, only 4, Harbin, Tianjin, Hangzhou and Anshan, were close to reaching compliance, giving a total of only 10.4% of the 125 cities. Of the other eight cities, most had low levels of policy coverage, with the majority having venues with only partial smokefree laws or no laws at all. When the degree of policy coverage for the 7 types of venues (covering indoor public places and workplaces and public transportation), identified by Lin et al. (2019a, b), was measured on a 3-point scale (1 = full coverage, 0.5 = partial bans, 0 = no bans), 3 cities scored 5.0–5.5, but the remaining cities all scored 4 or less.

8.3.2 *Assessing Gains*

While some cities have made substantial gains in implementing smokefree legislation, these are not widespread. However, of the three main goals of smoking bans, we can judge their effects particularly in reducing SHS exposure (see Chap. 9), whereas smoking cessation and denormalisation are longer-term goals. We can also examine the coverage of smokefree laws and whether these apply to all areas or only partially apply. Of course, the problem with such data is its reliability for even in places with full smokefree laws they are not necessarily enforced. Ideally it would be good to examine the effects of smokefree laws at different geographic scales focusing on the impact of smoking bans in changing exposure and different behavioural outcomes across different types of venues: indoor public places, outdoor public places, workplaces and private spaces such as the home, in different types of places (e.g. urban vs rural). To date most Chinese research on the effects of smoking bans has been cross-sectional, but, with the exception of particular cities, has not considered temporal changes in outcomes other than at the national level.

Lin et al. (2019a, b) provide an analysis of the degree of coverage (full, partial, none) of smokefree laws in 21 cities for indoor public spaces, transport and indoor workplaces. Although only reporting data for 2015, their findings are similar to those of SHS exposure. It was estimated that places with full coverage included health facilities (85.7%), schools (81%) and indoor workplaces (76.2%) with lower totals for government buildings (66.7%), public transport (66.7%) and restaurants and bars (both 47.6%). The latter venues were also most likely to report no coverage (19.0 and 23.8%, respectively) and, unexpectedly, so were government buildings (19.0%). Although the urban sample is small and for 1 year only, it indicates that both smokefree coverage and SHS exposure data are needed to obtain reliable estimates of policy outcomes.

With respect to policy compliance, MPOWER, using a variety of different measures, has attempted to score the performance of different countries in terms of the compliance of their smokefree policies (WHO 2019b). Compliance is scored 0–10, where 10 represents the highest level of compliance in meeting MPOWER targets. Compliance was measured by recording the number of venues where indoor smoking is completely prohibited along with other data on fines and enforcement. The problems of making such comparisons are discussed in detail in the following chapter. Nevertheless, in 2019 China received a score of five overall indicating good progress in some areas, such as public transport, but highly variable progress in others. When compared to the other countries in South and South East Asia discussed previously, China ranked fifth, along with Thailand, South Korea, the Philippines, Myanmar and Laos, out of the 14 countries for which data was available. When compared to other larger upper middle-income countries, China ranked 13th, together with Algeria, Guatemala and Thailand, out of 21 countries. China's compliance score of five thus was lower than the average MPOWER score of 6.1 (SD = 2.0) for this UMIC group.

8.3.3 Summary

Currently China lacks a national smokefree law that bans smoking in all indoor public places, and less than half the laws comprehensively ban all indoor smoking. The reluctance of central government to impose a national ban, and the absence of province-wide comprehensive bans, has meant that policy development has been limited to a number of key cities. These cities have been either provincial capitals or adjacent smaller cities reflecting the possible effects of policy transfer in particular regions. In addition cities having international connections, whose economies depend upon tourism or places which have hosted international events, have also been important places where stronger smokefree policies have been implemented. It is tempting to conclude that policy development up to the present has been motivated by China's desire to present a civilised face to the world, a world that has become increasingly intolerant of the harms caused by second-hand smoke. But the hard fact is that most of China is not covered by smokefree legislation with rural areas and smaller towns being the most obvious examples. Because these places are also poorer, then the absence of such legislation will only compound existing health inequalities. Because smokefree legislation is very visible in its spatial impacts, it is a key policy designed to foster denormalisation and smoking cessation. Widening the scope of the legislation would seem to be a very high priority.

8.4 Banning Tobacco Advertising, Promotion and Sponsorship

A third major area of tobacco control, the focus of Article 13 of the WHO FCTC, is the banning of tobacco advertising, promotion and sponsorship (TAPS). Advertising bans have an important negative effect on consumption (Yang et al. 2012, 2019), with one estimate for developing countries being a 6.7% decline following the imposition of comprehensive bans (Blecher 2008). Research also indicates that only comprehensive bans have an impact on consumption since the imposition of limited bans causes only a substitution away from those media that have been banned to media still allowed (Saffer and Chaloupka 2000). Only recently (2015) has China adopted a 'comprehensive' ban, but this appears to be in name only as not all media sources have been included under its umbrella, and is open to subversion.

8.4.1 Early Policy Changes

Although some initial attempts were made as early as 1987 (Zheng et al. 2018), it was not until the enactment of the *Advertising Law of the People's Republic of China* in 1995 that the first serious move was made to ban advertising in different types of media (broadcasting, films, TV, newspapers and periodicals) and in different types of public places, such as waiting rooms, cinemas or sports facilities. At the same time, it was stipulated that sponsorship advertising activities held in the name of tobacco enterprises should be subject to review. Further changes occurred in 1997 with advertising bans applying to vehicles and waiting rooms associated with public transportation and again in 2003 in a policy designed to produce tobacco-advertising-free cities (Lv et al. 2011). In 2009–2010 the Ministries of Education and Health forbade the placement of tobacco advertising on school campuses, while the following year, a notification from the State Administration of Radio, Film and Television prohibits advertisements in film and television (Xiao et al. 2015).

However, like so much Chinese legislation, attempts to control advertising occurred in name only, and little change occurred. This reflected key deficiencies such as the enactment of partial rather than comprehensive bans, poor operational definition and lack of enforcement and the absence of constraints being placed on sponsorship and indirect promotional activities. In addition, after the FCTC came into force in 2006, the STMA argued that no further legislation was necessary since the 1995 law already met the requirements of the Convention (Zheng et al. 2018). Consequently no further action was taken to regulate TAPS for another 10 years after the FCTC came into force.

The ineffectiveness of these early moves to regulate tobacco advertising and sponsorship is evident in the results of many studies that were conducted after China signed the FCTC in 2006. Many of these were surveys designed to measure advertising exposure often based on respondent recall during the past 30 days. For example, the 2010 GATS China Survey recorded that almost a fifth of adults had noticed tobacco advertisements and sponsorships during the last 30 days (China CDC 2011; Yang et al. 2010a), while other estimates, often based on surveys in particular cities, were much higher (41.3%–51%) both for adults (Yang et al. 2012; Zheng et al. 2013) and for middle school students (48.5–88.9%) (Xiao et al. 2011; Xiao et al. 2015). The main conclusions reached from these early studies, conducted before the 2015 change in the *Advertising Law*, were that:

(i) Problems inherent in early legislation meant that the level of exposure to tobacco advertising, not surprisingly, was high.
(ii) Awareness of tobacco advertising was higher in China than in other countries (Li et al. 2009).
(iii) Advertising was more prevalent in urban areas and in particular cities depending upon tobacco company activities and the strength of local tobacco control efforts (Yang et al. 2010b).
(iv) Within cities, such as Shanghai, outdoor tobacco advertisements were often highly visible and clustered along major roads and on buildings on key streets (Li 2008; Zheng et al. 2014), with one commentator to indicating that 'we were "quite impressed" by the ingenuity of the local tobacco companies in advertising their products' (Li and Yong 2009, p. 63).
(v) Advertising was highly concentrated in certain types of media, particularly films, TV and billboards (Fig. 8.2), and affected mainly young (15–24) urban men (China CDC 2011).
(vi) Adolescents were also affected leading to more favourable attitudes to tobacco companies and to tobacco use (Xiao et al. 2015).

In addition to direct advertising, indirect advertising, using alternative methods such as sponsorship of events or product tie-ins to public facilities, such as

Fig. 8.2 Patriotic billboard in Haikou, Hainan, in 2008 depicting the Zhonghua brand
The billboard says 'I love Zhonghua'
Credit: Alamy Image ID **B5CXJG Stock photo**

museums or private vehicles such as taxis, was very common (O'Connor and Casper 2011). Typical of such research was the use of tobacco museums as platforms for tobacco promotion (Wang et al. 2016) or the sponsorship of schools, sports events or social and cultural activities (Zheng et al. 2018).

As in the case of early moves to impose smoking bans, these early studies similarly provided a variety of evidence for the ineffectiveness of the early moves to limit advertising and provided much-needed evidence for groups advocating stronger controls. This was aided by MPOWER data showing that China, of the 17 countries reported in the 2010 GATS Survey, had the lowest scores for the implementation of bans on direct advertising (Yang et al. 2010a). Some improvement had occurred by the time of the 2014 WHO Global Progress Report, with China scoring 2 out of a maximum compliance score of 13 (WHO FCTC 2014). However, China's international position remained poor, with a ranking of 32nd out of 44 upper- and middle-income countries (UMICs) included in the Report and 13/19 of UMICs with populations exceeding 10 million. Among the 12 countries for which data was available in South and South East Asia, China ranked 8th. Similarly Borzekowski and Cohen (2013), in a study of the familiarity of young children (5–6-year-olds) with cigarette brands in six LMICs, found that China ranked first in terms of awareness, indicating the effectiveness of advertising in promoting tobacco use as a normal way of life. Given that higher media exposure was most strongly associated with smoking intentions in China, this indicated the necessity for stronger controls on advertising to limit initiation.

8.4.2 Policy Changes Since 2015

In keeping with other improvements in tobacco control, recent policy changes have witnessed increased regulation of advertising and sponsorship with the enactment of an amended *Advertising Law* in 2015 and the passing of the *Charity Law* in 2016 (Zheng et al. 2018). The former banned advertising of tobacco products in the mass media, in public places, on public transport and outdoors and prohibited any form of tobacco advertising to minors, while the *Charity Law* prohibits organisations and individuals from seeking to promote tobacco products through charitable donations (Xiao et al. 2015; Guo and Quan 2020).

Why did these changes occur at this particular time? Zheng et al. (2018, p. 269) argue that the law came about because of the influence of public health advocacy campaigns which 'finally overwhelmed the voice of the tobacco industry'. Readers are referred to the Zheng et al.'s account of these various campaigns, such as those by the Chinese Association on Tobacco Control's *Dirty Ashtray Award* given to films and TV programmes with excessive smoking footage. However, in a largely descriptive account of TAPS in China, the authors fail to situate recent changes within the wider policy context of what was unfolding in China at the time. For example, the amended *Advertising Law* in 2015 was not confined to tobacco. The updated legislation was part of a bigger picture, one in which wider marketing and consumer protection issues were increasingly emphasised. The amended law reflected the influence of public health advocacy groups in the sense that it was more comprehensive in scope than much other tobacco control legislation and also more prescriptive, thus, in theory, making compliance much more straightforward. The intention that advertising must 'not be detrimental to the physical and mental health of the people of China' (Chatterton and Jebbitt 2015) was explicitly included in the legislation and is consistent with the later Healthy China 2030 Guidelines issued in 2016. It is also apparent at this time that China's tobacco control policies were by far the weakest among all BRICS countries (Brazil, Russia, India, (China), South Africa) (Tobacco Free Kids 2014). A poor global image, therefore, and a need to 'save face' also may have been a contributing factor in the enactment of stronger controls over advertising and other policy changes that occurred in 2015. Because direct tobacco advertising was so visible, any restrictions had an immediate positive impact.

Yet to claim the tobacco industry had been 'finally overwhelmed' by public health advocates seems a simplification. For example, the influence of the tobacco industry was apparent in the enactment of the *Charity Law* which, because it only applied to tobacco products and not to tobacco companies, stopped short of a full sponsorship ban (Xiao et al. 2015). Consequently, because of this loophole, tobacco companies could continue using their company names for sponsorship and, therefore, were able to continue promoting their brands and company images under the guise of corporate social responsibility (see Chaps. 4 and 7). Similarly, because the amended Act did not address point-of-sale advertising, nor ban disguised forms of advertising on the Internet, the 2015 legislation, although an improvement on that existing in 1995, could hardly be defined as comprehensive. Internet platforms pose new challenges to tobacco control, with many advertisements circumventing the 2015 law against online tobacco advertisements by masquerading as shopping experiences or lifestyle posts (see Chap. 4).

Of course, the question arises over the effectiveness of the recent legislation in curtailing tobacco advertising. If one walks along East Nanjing Road in Shanghai or the main thoroughfare of any other large Chinese city, it soon becomes apparent that the large billboards of yesteryear are gone. But due to the paucity of recent studies since the amended law in 2015, it is difficult to estimate how effective compliance has been. Yang et al. (2019) in a six-city study conducted in 2016 showed that two thirds (67.8%) of male smokers reported never or seldom being exposed to advertising with smaller proportions indicating 'sometimes' (18.7%) or often/almost always (13.5%). However, the sources and types of advertising were not considered. By contrast, Wei et al. (2017) compared the prevalence of tobacco advertisements in tobacco sales places and other venues in Shanghai before and after the implementation of the 2015 law. They found that while the proportion of tobacco sales places having tobacco advertisements was significantly lower in 2016, the proportion of tobacco promotion behaviours was higher. The decline was also least, and not statistically significant, in tobacco stores which comprised a quarter of the 101 point-of-sale (POS) outlets sampled in 2016 (Fig. 8.3).

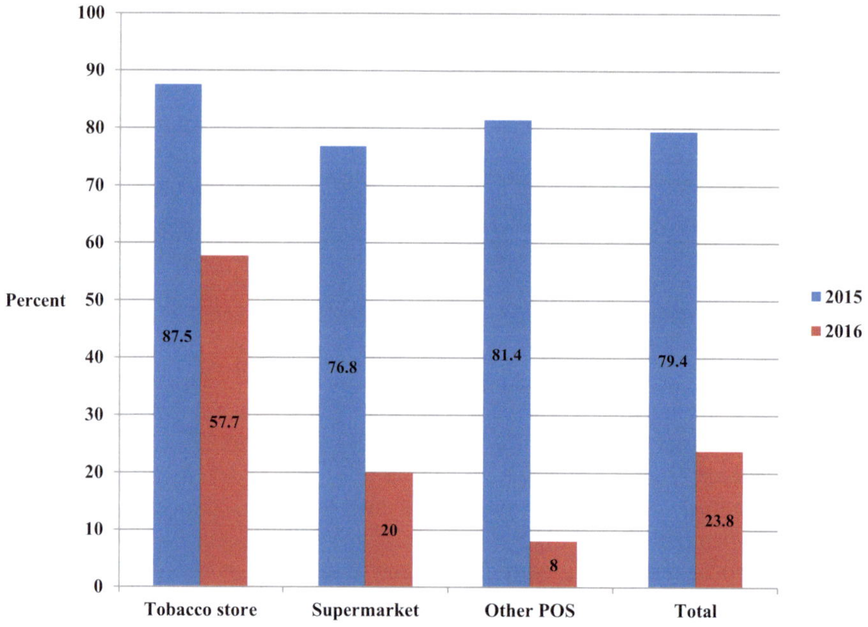

Fig. 8.3 Survey of tobacco advertisements at different types of tobacco sales places
Credit: Authors. Based on data in Wei et al. (2017), p. 139
Other POS includes convenience and food stores

A clearer indication of the effects of the amended *Advertising Law* in 2015 is evident in two recent surveys. The first is a city and county survey conducted in 2019 by the Chinese Association on Tobacco Control (CATC 2019a). This indicated that 64.9% retail stores selling cigarettes still openly advertised tobacco products, increasing from 45.7% in 2016 (Table 8.3). Promotional behaviour, such as vouchers, deals, discounts or free gifts, was also more apparent in 2019 than before. However, a positive trend was the decline in the proportion of stores not displaying signs indicating that sales were prohibited to minors. Of the 8 cities studied, the highest levels of POS advertising occurred in Guangzhou and Zhengzhou where advertisements and other promotional materials were present at all of the 641 sales points investigated, while Chengdu (97.5%) and Hefei (96.3%) had similar profiles. By contrast, sales outlets in Shenzhen (3.4%), Laiwu City in Shandong Province (3.3%) and, to a lesser extent, Shanghai (51.1%) were less likely to have POS outlets displaying tobacco advertising. When different types of sales outlets were considered, specialist tobacco stores were more likely to display advertisements and engage in promotional behaviour than convenience stores and supermarkets.

When viewed from the point of view of a longer timeframe, the China CDC GATS household survey for 2010 and 2018 (China CDC 2019) indicates that exposure to advertising, on the whole, has declined. For cigarette advertisements the proportion of respondents who had noticed any cigarette advertisements, any advertising promotions or sponsorship during the last 30 days dropped from 15.0 to 10.7% ($p < 0.05$) between 2010 and 2018. When promotions and sponsorship were included, there was less change (19.6–18.1%) with no significant difference between the 2 years. This suggests that while direct advertising has diminished, disguised forms of advertising and sponsorship remain.

Table 8.3 Point-of-sale advertising and promotional behaviour, 2016 and 2019

Year	Survey points	% stores with tobacco ads	% stores with promotional behaviour	% stores failing to display signs indicating sales prohibited to minors
All outlets				
2019	641	64.9	25.0	58.5
2016	1236	45.7	4.0	70.0
Convenience stores				
2019	347	62.5	20.7	65.4
2016	622	37.3	3.2	78.1
Supermarkets				
2019	179	62.6	20.1	50.3
2016	306	35.0	2.9	62.4
Tobacco stores				
2019	115	75.7	45.2	50.4
2016	308	77.9	6.8	61.0

Credit: Data derived from Chinese Association on Tobacco Control (CATC 2019a) http://www.catcprc.org.cn/index.aspx?menuid=4&type=articleinfo&lanmuid=122&infoid=11858&language=cn

Nevertheless other improvements are evident in films and TV shows. Data from the Chinese Association on Tobacco Control shows that when compared to 2007, the average number and length of smoking scenes in films in 2018 dropped by 47 and 37%, respectively, while an 82% decrease occurred in both the average number and length of smoking scenes in a TV series (Xinhuanet 2020). Despite this, in the recent movie, *Dying to Survive*, winner of CATC's 'Dirty Ashtray' award in 2019, tobacco-related scenes comprised 14.5% of the total length of the film.

8.4.3 Tobacco Packaging

Article 11 of the FCTC stipulates that tobacco packaging must carry health warnings about the dangers of cigarette use. The minimum requirements are that such warnings should cover 30% of the front and back of the package, be large and clearly visible and ideally include warning pictures. While Chinese tobacco companies have paid much attention to increasing the visual appeal of their products (see Chap. 4), much less attention has been devoted to health warnings on packaging, which have failed to reach the standards set down by Article 11. The current warning, introduced in 2008, that 'smoking is harmful to your health' replaced the early health warning which was located on the side of the pack. The current warnings are text-only warnings on 30% of the bottom of the front and back of the pack.

Compared to other health warnings internationally, Fong et al. (2010) found that Chinese text warnings were rated as less effective than visual warnings and far less effective than even text warnings from other countries (Qin et al. 2011). The current warnings are small, positioned at the bottom of the pack and fall far below international best practice. Other studies support the efficacy of graphic pictorial warnings, the dominant conclusion being that these are more effective in influencing awareness of health risks than text-only warnings (Elton-Marshall et al. 2015; Mannocci et al. 2019). This appears to be most important among low-income smokers (van Mourik et al. 2020) and is independent of cultural context (Hammond et al. 2019). There is also increasing evidence that pictorial warnings, which increase awareness and knowledge of health harms (Chiosi et al. 2016), are important in affecting the intention to quit smoking (Droulers et al. 2017; Wu et al. 2015) and encouraging quit behaviour (Chow et al. 2017), especially among less educated smokers (Ngoc Bich et al. 2019). Indeed it has been concluded that the present text-only health warnings on cigarette packets are among the least effective media presentations of anti-smoking messages for influencing quit rates among Chinese smokers (Li et al. 2014).

As in other areas of tobacco control, China's record in implementing effective packaging ranks below most other middle-income countries. In 2018 75% of the larger MICs, with populations over ten million, had picture warnings (Canadian Cancer Society 2018). When comparisons are made with countries in South and South East Asia, China fares even worse as 14/16 countries in this group (87.5%) had picture warnings. Coverage of the front and back of cigarette packets was often very high with India and Thailand both recording an average coverage of 85%, Sri Lanka 80% and Laos and Myanmar 75%. The Territory of Hong Kong stands in contrast to the rest of China by having both pictorial warnings and a high average coverage of 85%. China and Japan are the two exceptions in this part of Asia.

Clearly there is much room for improvement. Li et al. (2017), for example, found that non-smokers were strongly supportive of strengthening warning labels by including more information and pictures, a call increasingly being voiced by some key role models in the health sector (CATC 2019b). Warnings should also be increased in size to meet the FCTC recommendation that they be more prominent and cover at least 50% package. Given that GATS survey data shows that Chinese smokers generally have a poor knowledge of the health harms of smoking (Xu et al. 2013; Cheng et al. 2015), large pictorial warning labels on cigarette packaging are one means to raise levels of health awareness. They may also increase the likelihood of reducing consumption and increasing cessation.

8.4.4 Summary

The amended *Advertising Law* in 2015 appears to have had an initial positive effect on reducing exposure to tobacco advertising, but the response to the law varies by region and differs between different types of retail outlets. Surveys indicate that the increased focus on POS advertising, as other avenues have increasingly been restricted, is against both the spirit and intention of the amended 2015 law. In the words of one CATC representative, 'these results are alarming and point to poor implementation of the *Advertising Law* in China' (CATC 2019c) and flout the Law on the Protection of Minors. In Zheng et al.'s (2018, p. 284) words, 'China still has a long way to go' regarding the implementation of a comprehensive TAPS ban.

Such evidence suggests that China does not fit the pattern of low- and middle-income countries which have been leaders in adopting strong TAPS bans (WHO 2019a, p. 103). China's 2015 reform is far from comprehensive with loopholes in the legislation being quickly exploited by the STMA/CNTC which have also been persistent in blocking efforts to introduce stronger warning labels (Wu and Yang 2018; Guo and Quan 2020), claiming that the use of picture-based health warnings is incompatible with Chinese cultural traditions (Burki 2016). In the light of such trends, key priorities for future research should continue to monitor industry resistance to improved health warnings, to evaluate the effects especially of disguised advertising (Yang 2019) and sponsorship on smoking prevalence, especially among youth, and to further explore the types of social contexts in which such activities are most prevalent.

8.5 Health System Interventions

An important goal of MPOWER is building national capacity to a carry out effective tobacco control. The health system forms an important part of this strategy particularly through the establishment of smoking cessation clinics and of the importance of health professionals as non-smoking role models who provide smoking cessation advice to their patients (Zhang et al. 2019). In western countries, while the use of such services has been uneven, reflecting either the absence of services (Twigg et al. 2009) or the presence of financial barriers which limit access (Honjo et al. 2006; Roddy et al. 2006), the provision of cessation advice by health providers, who also serve as important role models, has been an important development.

8.5.1 Smoking Cessation Services; Patterns of Provision

China presents a similar picture. Despite the requirement that all healthcare facilities were required to provide cessation services and establish quitlines in 2006 (Xiao et al. 2015), soon after China signed the FCTC, and improved funding for such services in 2014 (Guo and Quan 2020), the situation remains geographically and institutionally uneven. Using data from a national survey of smoking cessation activities in all 366 provincial hospitals and primary care centres, conducted by the National Health and Family Planning Commission, Lin et al. (2019a, b) show that less than half (43%) provided smoking cessation medications. Geographically, 5 of the 31 regions (Henan, Sichuan, Anhui, Guangdong and Jiangsu), all important tier one cigarette producers (see Chap. 3, Table 3.3), had much more limited coverage than the rest of the country. Institutionally, unlike western countries, almost three quarters (73%) of the clinics were based in general hospitals, predominantly in respiratory departments. Quality of care varied between tertiary hospitals, most of which had enough treatment options, compared to secondary hospitals and lower-level health institutions which were characterised by a limited availability of medication and a lack of basic inspection equipment (Lin et al. 2019a, b, p. 7). While recognising these differences, it was concluded that specialist cessation training was performed poorly in hospitals at all levels with poor knowledge resulting, in some cases, in misinformation in the advice doctors gave to patients (Zhang et al. 2012).

Furthermore, hospital-based treatments, in contrast to many other countries, are not yet integrated into primary care, resulting in poor follow-up and monitoring of treatment outcomes. Data on the number of patients attending the clinics and treatment outcomes are not currently collected by all clinics. Despite a rapid rise in the number of clinics established after 2014, the number of smoking cessation medications provided has declined, partly due to cost. Although part of the initial clinic registration fee may be reimbursed, smokers attending the clinics, with few exceptions (CATC 2018c), have to pay the full cost of their medications, which is often high. The same is true of the use of national quitlines where the costs of prescribed medications are incurred by patients.

Quitlines have the potential to play a greater role in smoking cessation in China, but the number of smokers using central and local government helplines is small (Yang and Yang 2012). Since the establishment of the first telephone quitline in Hong Kong in 2000 (Abdullah et al. 2004), a number of quitlines were established, but their results proved disappointing. For example, in Hefei City where a quitting helpline was established in 2015, or in Jiangxi Province, where helplines were established earlier on World Smokefree Day in 2009, few people made contact (Yang et al. 2019, p. 2). These trends indicate the fact that most Chinese smokers wishing to quit do not use official sources, but do so without assistance. While unassisted quit attempts and rates of success have been shown to be most prevalent in situations where smokers have been exposed to antitobacco information and environmental smoking restrictions (Yang et al. 2019), the lack of use of official services raises questions about how factors such as cultural norms, the effect of cost barriers and the healthcare treatment environment affect the use of such services.

8.5.2 Healthcare Professionals: Adequate Role Models?

Because many doctors continue to smoke in China, it could be argued that they are poor role models for patients wishing to quit. Estimates of smoking among medical professionals vary, but the 2015 GATS Survey indicated that 43% of male health professionals were current smokers (China CDC 2015). This estimate is similar to other studies such as Jiang et al. (2007) for six cities (41%) and Smith et al. (2012) for Shandong Province (46.7%), but less than the 55% recorded for a large teaching hospital in Beijing in 2006 (Ceraso et al. 2009), and the 61.3% recorded a decade earlier (Lindorff and Hill 2010). Generally older doctors tend to smoke more. Across specialities prevalence has been shown to be high among surgeons, and there is some evidence that doctors working in township hospitals located in lesser urban areas have higher smoking rates than those working in city-level hospitals (Abdullah et al. 2013).

In many respects these patterns echo the smoking behaviour of health professionals in western countries at an earlier point in time. For example, in 1955 around half of all doctors (mainly male) in the United States were smokers (Smith 2008, p. 4), and even as late as 1972, 35% and 15%, respectively, of male and female doctors in New Zealand continued to smoke (Christmas and Hay 1976). It is hard to estimate temporal trends in smoking prevalence among health professionals. Smith et al. (2012), in their review of various studies in China published between 1987 and 2011, report that the overall rate of smoking among Chinese male doctors has remained relatively high. They also concluded that quit rates tend to be low, reflecting broader cultural processes such as cigarette gifting and accepting cigarettes as gifts from patients. High smoking prevalence among medical professionals is not unique to China but is typical of most upper middle-income countries. In a meta-analysis of studies published between 2000 and 2016, Nilan et al. (2019) reported a pooled prevalence rate of 35% for male doctors practising in UMICs, prevalence rates which were similar to, or higher than, the general population.

For China, however, current high rates are problematic especially when a high proportion of doctors (15–43%) smoke in front of their patients (Abdullah et al. 2013), and those who smoke tend to have less knowledge of smoking-related health risks (excluding lung cancer), including harms of second-hand smoke (Liao et al. 2016). Doctors who smoke are less likely to provide cessation advice or recommend quitting to their patients (Abdullah et al. 2014; Ceraso et al. 2009) even when they themselves are parents (Liao et al. 2016). In the Ceraso et al. (2009) survey (conducted in 2006), it was also found that, despite appointments at teaching hospitals, doctors received little support in their efforts to quit. Nor did the hospitals provide training in how to intervene with smoking patients. However, since 2007 the *Clinical Smoking Cessation Guideline* has provided guidance for the certified training of Chinese doctors in the area of smoking cessation (Xiao et al. 2015). While this may have improved the provision and quality of professional cessation advice, it has not stopped the decline in the number of patients seeking that advice (Lin et al. 2019a, b).

The problem of doctors continuing to smoke and setting a poor example to patients wishing to quit goes beyond the health system itself. The general absence of strong role models in health or in other sectors such as education and government affects general public support for tobacco control policies. Yang et al. (2014b) found that (positive) expectations of role models were strong predictors of support for FCTC tobacco control policies among urban residents. Support was lower where expectations were not met, thus suggesting that stronger emphasis should be placed in developing health service role models to promote greater acceptance of tobacco control measures.

8.5.3 Current Smoking Behaviour Among Medical Students

While a number of studies have focused on smoking prevalence and cessation practices among Chinese doctors (Abdullah et al. 2013), less attention has been devoted to smoking behaviour among medical students. Evidence suggests that smoking prevalence increases after the freshman year and that medical students smoke at comparable rates to non-medical students (Niu et al. 2018). Therefore it is important to boost tobacco control efforts among new entrants, both clinical and non-clinical, to the health sector, in order to strengthen the delivery of more effective smoking cessation services.

Until recently national data on tobacco use among medical students was missing. To address this gap, Niu et al. (2018) conducted a systematic review and meta-analysis of published studies up to 2018 on current smoking among Chinese medical students. The following comments draw heavily on this analysis. Perhaps their most positive finding was that overall (pooled) smoking prevalence dropped between 1988/1999 and 2010/2014 from 17.2% to 7.6%. The drop was greatest for male students (−13.4% from 33.1 to 19.7%), while female rates remained constant, but low (1.7–1.0%). Male prevalence rates were highest in western provinces with significant regional differences occurring between this region and central and eastern provinces (west 32.9%, central 25.9%, east 19.8%). No significant regional differences occurred for females. Despite these positive trends, smoking prevalence for all students increased with the year of medical school training from 8.9% for first year students to 22.1% among fifth year students, rates that were *higher* than the national figure of 18.6% recorded for persons aged 15–24 in 2018 (China CDC 2019). Such trends indicate that a significant proportion of students will become smokers after entering college and that more targeted programmes are needed to prevent this outcome.

Exploring factors that may increase susceptibility to smoking initiation among Chinese medical students is an essential first step in planning preventative actions. In a study of 50 universities, Peng et al. (2019) found that susceptibility to smoking on the part of medical students at different course levels was 23.0%, similar to the rate recorded by Niu et al. (2018) for fifth year students. Using a four-point ordinal scale, susceptibility was estimated by assessing the importance of social influences, including those at work, on the likelihood of smoking initiation in the future. Students highly susceptible were those most exposed to second-hand smoke in domestic and public places, to cigarette advertising and to promotional activities on campus. By contrast, positive attitudes towards health professionals' role in tobacco control were commonly associated with less susceptibility. These findings underscore the importance of strengthening medical education relating to the role of physicians in tobacco control and including this content early in medical programmes before students become smokers.

The attitude of medical and other health professionals towards providing smoking cessation services is an important follow-up to this research. Yang et al. (2015), in a 2013 survey of medical students and students in other health programmes in 50 universities, reported that although the majority of (third) year students were taught about the health effects of smoking, few (13%) reported having received any formal training in smoking cessation approaches. Tobacco advertising on campus lessened support for smoking cessation services, and many (35.9%) students believed industry propaganda that light cigarettes were less harmful to health. Increased support, on the other hand, was associated with increased knowledge of the health harms of smoking, receiving professional advice and intention to quit. Taken together these results indicate that current training in Masters of Public Health (MPH) programmes is insufficient to provide an informed health workforce capable of delivering smoking cessation programmes. However, since the time of Yang et al.'s survey, considerable progress has been made in developing an enhanced advocacy capability in China.

8.5.4 *Tobacco Control Advocacy*

Tobacco control advocacy is a key WHO strategy designed to build up the capacity of medical and other health professionals to develop teaching and research programmes aimed in improving the overall delivery of smoking cessation services (WHO 2004). Such programmes are relatively common in high-income countries and have played an important role in reducing smoking. However, in LMICs, less is known about the effectiveness of training materials and teaching methodologies in improving smoking cessation. Recently research into different modalities of training medical students in smoking cessation methods has concluded that none of the programmes to date 'provide any conclusive evidence on their effectiveness' (Sreeramareddy et al. 2018, p. 9). This global review of such programmes reports that lack of time, poor long-term retention of counselling skills and lack of practice are the main barriers to the implementation of cessation training. It could also be said that such programmes need institutional support which may not always be forthcoming either in hospitals (Chan et al. 2017) or in educational institutions that are not smokefree. Indeed in an early attempt to increase tobacco control capacity in three Chinese provinces in 2004, few school or hospital personnel viewed smoking control as an important issue or relevant to their work (Stillman et al. 2006).

With such barriers in mind, more recently a number of programmes have developed in China to build capacity for tobacco control among medical and other healthcare professionals (HCPs) within Chinese universities. The first of these began in Hong Kong in 2006 involving the University of Hong Kong, and three major hospitals in Guangzhou and Beijing (Chan et al. 2017), with the establishment of training programmes for health professionals and three model cessation clinics. In a follow-up review in 2014, at the local level, it was reported that HCPs had ventured beyond the clinics and into the community to publicise smoking cessation in the three initial sites and in some other cities, and had helped in the drafting of smokefree legislation in Guangzhou. At the national and international levels, HCPs had also, in conjunction with the American Cancer Society, produced self-help books on smoking cessation, evaluated different techniques in smoking cessation and extended their research capability. Particularly important was the evaluation, and confirmation of the greater effectiveness of face-to-face counselling services with telephone follow-up on quit rates, compared those with no follow-up (Wu et al. 2016). These successes, however, were negated by the clinics attracting few clients, resulting in the closure of two of the three clinics soon afterwards (Chan et al. 2017).

8.5.4.1 Local Innovation: The FCTC Tobacco Control Advocacy Programme

Training programmes in medical universities provide a unique opportunity to build FCTC tobacco control advocacy capacity among new cadres of healthcare professionals so that they can act as role models for non-smoking norms, provide smoking cessation services and advocate for tobacco control policies at local and national levels. Until recently, however, the absence of tobacco control curricula in medical schools meant that students were ill-equipped to undertake such tasks. This deficit in curriculum and training was reinforced by the absence of funding for research into tobacco control and cessation.

To address the lack of capacity, the FCTC tobacco control advocacy (FCTC-TCA) programme, which focused on boosting the tobacco control education and skills of public health (PH) students, was developed by Wang and associates at the Center for Tobacco Control Research at the Zhejiang University School of Medicine in Hangzhou (Yang et al. 2009, 2011). Funded initially by the International Union Against Tuberculosis and Lung Disease and later by the Bloomberg Foundation Global Bridges Project, an 18-month pilot non-clinical training programme was implemented in seven universities in 2008 to enhance the tobacco control capacity of the PH workforce, many of whom work for the China Centers for Disease Control and Prevention as well as in government and universities. This was an important development since the education of PH students in China had been criticised for its outdated curriculum, an excessive focus on teaching biomedical science and insufficient or low-quality practical training. These deficiencies had left PH students, one of the least favoured medical specialities in China, with a poor awareness of disease prevention and population health topics (Hou et al. 2018).

The basis of Yang's approach was an 'advocacy behavioural capacity' (ABC) model involving three components: attitudes and awareness of tobacco control, interest and motivation and practice. The latter was particularly important since it focused not only on students' own smoking behaviour but also advocacy for change in smoking habits and reductions in SHS exposure in their own family and wider community. The pilot study showed that, compared to a control group, tobacco control advocacy training improved PH students' ABC, leading to the recommendation that it be applied in a larger number of universities, eventually 98 in all (Peng et al. 2019) (Fig. 8.4). In conjunction with this growth, the emphasis of the programme increasingly shifted from public health to a broader range of medical students. By the time of the final report of the Global Bridges Project in 2016, participants included 25,163 medical students in 54 universities, including public health, nursing, clinical medical and some other health science students. Clinical students comprised almost half (45.8%) of those who completed the final course evaluation (Yang and Team Members 2016). In addition 434 medical teachers and physicians in 10 affiliated hospitals took part in this stage of the programme.

8.5.4.2 Programme Implementation and Results

There were a number of programme components. For example, a Chinese cultural module in tobacco control advocacy was developed and combined into a training programme. This aimed to improve students' attitudes, awareness and motivation towards tobacco control as well as to highlight the unique social constructs and norms characteristic of Chinese culture. In addition, students were taught practical

Fig. 8.4 Geographic distribution of 98 universities implementing the tobacco control advocacy programme across China
Credit: Authors

skills by being required to draft smokefree plans to apply to university campuses. To ensure the sustainability of the programme, training occurred among faculty members at various institutions, and course content was integrated into the national public health curriculum.

As the first major advocacy training programme in China, the programme had to overcome some substantial barriers to implementation. Not all partner universities and hospitals were equally responsive to the goals of the programme. Among universities, tobacco control material was added to existing courses, and the success of this depended upon the co-operation of academic staff. Similarly among teaching hospitals, implementation depended upon the support of the hospital director and staff which was not always forthcoming (Yang and Team Members 2016). In hospitals many physicians claimed that they did not have enough time to implement cessation programmes with patients.

Despite these initial difficulties, the programme achieved the establishment of a national training infrastructure, culturally sensitive programmes and guidelines, with a critical mass of 165,000 medical students and 475 trained faculty members. FCTC modules are now incorporated into textbooks and core content in physician examinations. Thirty-six universities officially announced, and enforced, smokefree campuses which are gradually shaping the social norm. Programme evaluation shows that student knowledge and skills and behavioural capacity have all improved.

8.5.5 *Summary*

In summary the development of tobacco control advocacy in a large number of universities and affiliated hospitals has significantly improved the quality of medical education, provided a growing core of tobacco control advocates and improved the ability of the health sector to increase rates of smoking cessation. It has strengthened the research capacity of such institutions by international networking and collaboration and has resulted in a greater capacity of public health and medical practitioners to provide greater input into the policy process and to more effectively deal with patients and their need to quit smoking. While randomised controlled trials initially indicated that the involvement of public health students in the programme did not result in any significant reductions in smoking (Wu et al. 2018), this was less true of hospital-based clinic programmes which appear to have had more impact (Yang and Team Members 2016). Given the recent decline in the smoking prevalence of Chinese doctors (although still high rates for males), there is a need for a more focused attempt on extending the programme to clinical students and primary care doctors. This will be a challenge since the medical care system (hospitals and primary healthcare) and the public health system operate in different 'silos' and are often poorly integrated (Wang et al. 2019).

It is also clear that the development of smoking cessation clinics has not proved particularly effective. This is partly due to cultural factors, such as the tendency for quitting to be the result of self-reliance rather than due to professional assistance (Yang et al. 2019) and the fact that most clinics are located at tertiary or secondary hospitals, rather than in primary care settings, which have the potential to reach more smokers. Despite attempts to improve the attractiveness of general practice, primary care remains unattractive to most Chinese medical undergraduates (Wang et al. 2018). Similarly the presence of significant cost barriers for cessation medications has limited the effectiveness of quitlines. In comparison, many western settings and middle-income countries, such as India, have increasingly subsidised such services (WHO 2017a).

8.6 Conclusions

This chapter has provided an assessment of China's recent record in tobacco control with respect to four key areas: the use of price increases to dampen demand, the introduction of smokefree legislation, controls on advertising including point-of-sale and tobacco packaging and the effect of health system interventions. In all cases the dominant conclusion is that China has 'dragged its feet' in implementing effective tobacco control measures in the spirit of the FCTC with its interventions in all policy areas being weak in comparison with other middle-income countries and other countries in South and South East Asia. This is not surprising considering the discussion of the policy process in Chap. 7. Nevertheless prospects are improved by the recent restructuring of the state bureaucracy which has diminished the power of the CNTC in decision-making and through the Healthy China 2030 vision that emphasised health rather than economic goals (Tan et al. 2017). Healthy China 2030 stipulates one of the specific health goals to be reached by 2030 that the smoking rate of Chinese adults be reduced from its 2018 rate of 26.6% to 20% (China CDC 2019; Tan et al. 2019). This is a modest target but one, given the scope of current policies, that will be difficult to achieve. To consider how China may improve on its performance in reducing tobacco consumption, we conclude by focusing on two main issues: (1) the scope and effectiveness of current policy and (ii) policy implementation and place.

8.6.1 The Scope and Effectiveness of Current Policy

Currently Chinese smokers have very low quit rates when compared to other middle-income and even some low-income countries. This outcome reflects decades of interference by the tobacco industry and, until recently, the predominance of economic over health objectives in the development of tobacco control policies. While it is too soon to see the effects of the recent (2018) shift in responsibility for the FCTC from the Ministry of Industry and Information Technology, where the tobacco industry resides, to the Health Commission, this should be a positive move towards the development of more comprehensive and effective tobacco policies which are necessary to reduce a smoking prevalence that has changed little since the WHO FCTC came into force in 2005.

International experience indicates that a comprehensive suite of tobacco control policies is most effective in reducing prevalence. Chow et al. (2017), using a variant of the Joossens index tobacco control scale (Joossens and Raw 2006), showed that more extensive policy implementation was associated with higher quit ratios, the ratio of former to ever smokers (the latter including current and former smokers) among the 17 countries examined. Of the 17 high-, upper middle-, lower middle- and low-income countries, China ranked 15/17 for males and 16/17 for females with its quit rate eclipsed even by some LICs. The findings support the need to enact and enforce more comprehensive tobacco control strategies to increase quitting.

International research suggests that taxing tobacco products and smokefree legislation are the two most effective measures in reducing smoking prevalence (Hoffman and Tan 2015; Chung-Hall et al. 2019; Levy et al. 2018a, b). However, there is a lack of understanding of differential policy impacts, that is, how strong policy implementation in one area creates expectations and changes in behaviour that, in turn, affect the effectiveness of other tobacco control policies. As Li et al. (2014) found in the case of exposure to tobacco advertising, the effect of combined exposure to anti-smoking messages on quit attempts was greater than the sum of the parts. To date there is not enough understanding of these key policy interactions, other than the view that multiple policies need to be present and enforced to achieve the greatest overall impact.

Nevertheless some attempts have been made to assess the policy changes needed if China is to reach its goal of reducing overall consumption to 20% by 2030. Goodchild and Zheng et al. (2018) present a number of different policy scenarios in which 'business as usual' would see no change in prevalence from the 26.6% by recorded in 2018. When non-price measures, such as smokefree laws, are included, the rate of smoking falls to 22.0% (20.9%–23.1%), while under a third scenario, involving a 50% increase in price, the rate falls to 19.7% (18.2%–21.3%). Like other such predictions (e.g. Verguet et al. 2017), the effect sizes are open to debate especially as they do not consider interactions between different policy effects, nor the differential geographic and social variations of any reduction in prevalence. Thus, as the WHO (2017b, p. 26) concluded with respect to China, while the modelling commissioned for the report showed that tobacco tax and price increases would make the greatest contribution to reducing smoking prevalence, their effect would be weakened in the absence of weaker non-tax interventions. Tobacco control policies are most effective when they reinforce each other, because consistent measures across a variety of policy domains are most likely to change attitudes and produce behaviour change.

8.6.2 Policy Implementation and Place

A second consideration, especially in a country as large and diverse as China, is that the implementation and impact of many different tobacco control policies has been far from uniform. Smokefree policies, for example, are largely absent from the countryside and tobacco-growing and manufacturing regions. National and political barriers to policy implementation and the absence of a consistent national smokefree law affecting all public places have resulted in a mixture of regional and local policies which, as they stand, will not have the desired impact of reducing smoking prevalence. Similarly within cities, point-of-sale advertising as a factor in promoting consumption varies between cities and neighbourhoods, and again the lack of an effective national law controlling POS advertising and cigarette packaging means that the effects of the latter on consumption will remain.

Similarly the reliance of healthcare interventions based on hospitals rather than community primary healthcare approaches means that their effects on smoking cessation will be marginal. As in western countries, these will need to target particularly smokers from lower socio-economic backgrounds who experience greater barriers to quitting than their counterparts from higher SES backgrounds (Yong et al. 2013). Similarly tobacco control policies will need to pay greater attention to the personal and environmental triggers that prompt smoking cessation (Im et al. 2015), be these stress and the health concerns on the part of different groups of smokers (Fu et al. 2017) or environmental triggers such as price increases or smoking bans, which also influenced cessation (Yang et al. 2014a).

Outcomes such as these suggest a need for greater targeting of tobacco control policies in the future – to rural regions, poorer cities and tobacco-growing regions.

The barriers, outlined in the previous chapter, are enormous, but a more geographically sensitive approach to tobacco control will help produce more effective gains. Depending upon the context, the target populations are likely to differ across China's regions. Policies that work well in wealthy eastern cities and provinces may well be inappropriate for other cultural and economic contexts such as the southern tobacco-growing regions, the western steppes or the rust-belt cities of the north.

References

Abdullah AS, Lam TH, Chan SS, Hedley AJ (2004) Which smokers use the smoking cessation Quitline in Hong Kong, and how effective is the Quitline? Tob Control 13(4):415–421

Abdullah AS, Qiming F, Pun V, Stillman FA, Samet JM (2013) A review of smoking and smoking cessation practices among physicians in China: 1987–2010. Tob Control 22(1):9–14

Abdullah AS, Stillman FA, Yang L, Luo H, Zhang Z, Samet JM (2014) Tobacco use and smoking cessation practices among physicians in developing countries: a literature review (1987–2010). Int J Environ Res Public Health 11(1):429–455

Blecher E (2008) The impact of tobacco advertising bans on consumption in developing countries. J Health Econ 27(4):930–942

Borzekowski DLG, Cohen JE (2013) International reach of tobacco marketing among young children. Pediatrics 132(4):e825–e831

Burki TK (2016) Graphic warnings on cigarette packaging in China. Lancet Respir Med 4(5):350

Canadian Cancer Society (2018) Cigarette package health warnings: international status report, 6th edn. Canadian Cancer Society, Toronto

Ceraso M, McElroy JA, Kuang X, Vila PM, Du X, Lu L, Ren H, Qian N, Jorenby DE, Fiore MC (2009) Smoking, barriers to quitting, and smoking-related knowledge, attitudes, and patient practices among male physicians in China. Prev Chronic Dis 6(1):1–6

Chaloupka FJ, Pacula F (2012) Tobacco taxes as a tobacco control strategy. Tob Control 21(2):172–180

Chan SS, Cheung YT, Wong DC, Jiang CQ, He Y, Yang L, Jiang B, Wu L, Tan SY, Cheung KK, Lam T (2017) Promoting smoking cessation in China: a foot-in-the-door approach to tobacco control advocacy. Glob Health Promot 26(3):41–49

Chatterton E, Jebbitt, I (2015) China enacts its first major changes to its advertising laws in twenty years. Intellectual Property and Technology News, Asia Pacific Edition, November, pp 12–14. www.dlapiper.com/ipt_news/

Cheng HG, McBride O, Phillips MR (2015) Relationship between knowledge about the harms of smoking and smoking status in the 2010 global adult tobacco survey. Tob Control 24(1):54–61

CATC (Chinese Association on Tobacco Control) (2018a) Law ought to ban smoking in all public places. Domestic News, 1 December. (CATC, Beijing) http://www.catcprc.org.cn/index.aspx?menuid=25&type=articleinfo&lanmuid=186&infoid=10882&language=en

CATC (Chinese Association on Tobacco Control) (2018b) Wuhan bans public smoking inside party, government offices. Domestic News, 3 July. (CATC, Beijing) http://www.catcprc.org.cn/index.aspx?menuid=25&type=articleinfo&lanmuid=186&infoid=10394&language=en

CATC (Chinese Association on Tobacco Control) (2018c) Beijing smokers offered free clinical help to kick the habit. Domestic News, 18 October. (CATC, Beijing) http://www.catcprc.org.cn/index.aspx?menuid=25&type=articleinfo&lanmuid=186&infoid=10728&language=en

CATC (Chinese Association on Tobacco Control) (2019a) Tobacco sales, tobacco advertising, promotion and other phenomena. 3 September. (CATC, Beijing). http://www.catcprc.org.cn/index.aspx?menuid=4&type=articleinfo&lanmuid=122&infoid=11858&language=cn

CATC (Chinese Association on Tobacco Control) (2019b) NPC Deputy wants graphic health warnings on cigarette packs. CATC Domestic News 13 March. (CATC, Beijing) http://www.catcprc.org.cn/index.aspx?menuid=25&type=articleinfo&lanmuid=186&infoid=11168&language=en
CATC (Chinese Association on Tobacco Control) (2019c) Tobacco association says industry ignoring sales rules. CATC Domestic News 5 September. (CATC, Beijing) http://www.catcprc.org.cn/index.aspx?menuid=25&type=articleinfo&lanmuid=186&infoid=11861&language=en
China CDC (Centers for Disease Control and Prevention) (2011) Global adult tobacco survey (GATS) China 2010 Country Report 2011. China San Xia Press, Beijing, p 11
China CDC (Centers for Disease Control and Prevention) (2015). China GATS Global Adult Tobacco Survey. China CDC, Beijing
China CDC (Centers for Disease Control and Prevention) (2019). China GATS Global Adult Tobacco Survey, Comparison Fact Sheet, China 2010 & 2018. China CDC, Beijing
Chiosi JJ, Andes L, Asma L, Asma S, Palipudi K, McAfee T, Regional GATS, Country Authors, On behalf of the GATS Collaborative Group (2016) Warning about the harms of tobacco use in 22 countries: findings from a cross-sectional household survey. Tob Control 25(4):393–401
Chow CK, Corsi DJ, Gilmore AB, Kruger A, Igumbor E, Chifamba U, Yang W, Wei L, Iqbal R, Mony P et al (2017) Tobacco control environment: cross-sectional survey of policy implementation, social unacceptability, knowledge of tobacco health harms and relationship to quit ratio in 17 low-income, middle-income and high-income countries. BMJ Open 7(3):e013817
Christmas BW, Hay DR (1976) The smoking habits of New Zealand doctors: a review after ten years. N Z Med J 83(565):391–394
Chung-Hall J, Craig L, Gravely S, Sansone N, Fong GT (2019) Impact of the WHO FCTC over the first decade: a global evidence review prepared for the Impact Assessment Expert Group. Tob Control 28(Suppl 2):s119–s128
Ding A (2003) Youth are more sensitive to price changes in cigarettes than adults. Yale J Biol Med 76(3):115–124
Drope J, Schluger N, Cahn Z, Drope J, Hamill S, Islami F, Liber A, Nargis N, Stoklosa M (2018) The tobacco atlas, 6th edn. American Cancer Society and Vital Strategies, Atlanta, Atlanta. https://tobaccoatlas.org/countries/
Droulers O, Gallopei-Morvan K, Lacoste-Badie S, Lajante M (2017) The influence of threatening visual warnings on tobacco packaging: measuring the impact of threat level, image size and type of pack through psychophysiological and self-report methods. PLoS One 12(9):e0184415
Elton-Marshall T, Xu S, Meng G, Quah AC, Sansone GC, Feng G, Jiang Y, Driezen P, Omar M, Awang R, Fong GT (2015) The lower effectiveness of text-only health warnings in China compared to pictorial health warnings in Malaysia. Tobacco Control 24(Suppl 4):iv6–iv13
Eriksen MP, Cerak RL (2008) The diffusion and impact of clean indoor air laws. Annu Rev Public Health 29(April):171–185
Fong GT, Hammond D, Jiang Y, Li Q, Quah ACK, Driezen P, Yan M, for the ITC China Project Team (2010) Perceptions of tobacco health warnings in China compared with picture and text-only health warnings from other countries: an experimental study. Tob Control 19(Suppl 2):i69–i77
Fu H, Feng D, Tang S, He Z, Xiang Y, Wu T, Wang R, Shao T, Liu C, Shao P, Feng Z (2017) Prevalence of tobacco smoking and determinants of success in quitting smoking among patients with chronic diseases: a cross-sectional study in rural Western China. Int J Environ Res Public Health 14(2):167
Gan Q (2018) Protecting China from secondhand smoke. In: Yang G (ed) Tobacco control in China. Springer, Singapore, pp 167–182
Goodchild M, Zheng R (2018) Early assessment of China's 2015 tobacco tax increase. Bulletin of the World Health Organisation 96(7):506–512
Guo H, Quan G (2020) Tobacco control in China and the road to healthy China 2030. International Journal of Tuberculosis & Lung Disease 24(3):271–277

Hammond D, Reid JL, Driezen P, Thrasher JF, Gupta PC, Nargis N, Li Q, Yuan J, Boudreau C, Fong GT, Cummings KM, Borland T (2019) Are the same health warnings effective across different countries? An experimental study in seven countries. Nicotine Tob Res 21(7):887–896

Ho L-M, Schafferer C, Lee J-M, Yeh C-Y, Hsieh C-J (2018) Raising cigarette excise tax to reduce consumption in low-and middle-income countries of the Asia-Pacific region: a simulation of the anticipated health and taxation revenue impacts. BMC Public Health 18(1):1187

Hoffman SJ, Tan C (2015) Overview of systematic reviews on the health-related effects of government tobacco control policies. BMC Public Health 15(1):744

Honjo K, Tsutsumi A, Kawachi I, Kawakami N (2006) What accounts for the relationship between social class and smoking cessation? Results of a path analysis. Soc Sci Med 62(2):317–328

Hou J, Wang Z, Liu X, Luo Y, Sabharwal S, Wang N, Meng Q (2018) Public health education at China's higher education institutions: a time series analysis from 1998 to 2012. BMC Public Health 18(1):679

Hu T-W, Mao Z, Shi J, Chen W (2010) The role of taxation and its potential economic impact in China. Tob Control 19(1):58–64

Hu T-W, Zhang X, Zheng R (2016) China has raised the tax on cigarettes: what's next? Tob Control 25(6):609–611

Hu X, Wang Y, Huang J, Zheng R (2019) Cigarette affordability and cigarette consumption among adult and elderly Chinese smokers: evidence from a longitudinal study. Int J Environ Res Public Health 16(23):4832

Huang J, Zheng R, Chaloupka FJ, Fong GT, Li Q, Jiang Y (2014) Chinese smokers' cigarette purchase behaviours, cigarette prices and consumption: findings from the ITC China Survey. Tob Control 23(Supp 1):i67–i72

Huang J, Zheng R, Chaloupka FJ, Fong GT, Jiang Y (2015) Differential responsiveness to cigarette price by education and income among adult urban Chinese smokers: findings from the ITC China Survey. Tob Control 24(Suppl 3):iii76–iii82

Im PK, McNeill A, Thompson MW, Fong GT, Xu S, Quah AC, Jiang Y, Shahab L (2015) Individual and interpersonal triggers to quit smoking in China: a cross-sectional analysis. Tob Control 24(Suppl 4):iv40–iv47

Jiang Y, Ong MK, Tong EK, Yang Y, Nan Y, Gan Q, Hu YW (2007) Chinese physicians and their smoking knowledge and practices. Am J Prev Med 33(1):15–22

Joossens L, Raw M (2006) The tobacco control scale: a new scale to measure country activity. Tob Control 15(3):247–253

Kegler MC, Hua X, Solomon M, Wu Y, Zheng PP, Eriksen M (2014) Factors associated with support for smoke-free policies among government workers in six Chinese cities: a cross-sectional study. Public Health 14:Article No. 1130

Kostova D, Chaloupka FJ, Ross H, Cherukupalli R, Andes L, Asma S, on behalf of the GATS Collaborative Group (2014) A cross-country study of cigarette prices and affordability: evidence from the Global Adult Tobacco Survey. Tob Control 23(1):e3

Kostova D, Husian MJ, Chaloupka FJ (2017) Effect of cigarette prices on smoking initiation and cessation in China: a duration analysis. Tob Control 26(5):569–574

Lance PM, Akin JS, Dow WH, Loh CP (2004) Is cigarette smoking in poorer nations highly sensitive to price? Evidence from Russia and China. J Health Econ 23(1):173–189

Levy DT, Yuan Z, Luo Y, Mays D (2018a) Seven years of progress in tobacco control: an evaluation of the effect of nations in meeting the highest level MPOWER measures between 2007 and 2014. Tob Control 27(1):50–57

Levy DT, Tam J, Kuo C, Fong GT, Chaloupka F (2018b) The impact of implementing tobacco control policies: the 2017 tobacco control policy scorecard. J Pub Health Manag Pract 24(5):448–457

Li J, White JS, Hu TW, Fong GT, Yuan J (2015) The heterogeneous effects of cigarette prices on brand choice in China: implications for tobacco control policy. Tob Control 24(Suppl 3):iii25–iii32

Li L, Borland R, Yong H-H, Fong GT, Jiang Y, Li Q, Hammond D, Quah ACK (2014) Reported exposures to anti-smoking messages and their impact on Chinese smoker's subsequent quit attempts. Int J Behav Med 21(4):667–676

Li L, Yong H-H (2009) Tobacco advertising on the street in Kunming, China. Tob Control 18:63

Li L, Yong H-H, Borland R, Fong GT, Thompson ME, Jiang Y, Yang Y, Sirirassamee B, Hastings G, Harris F (2009) Reported awareness of tobacco advertising and promotion in China compared to Thailand, Australia and the USA. Tob Control 18(3):222–227

Li M (2008) Survey on outdoors tobacco advertisement status in Shanghai. China J Health Educ 9:728–729. (in Chinese)

Li Q, Hyland A, Fong GT, Jiang Y, Elton-Marshall T (2010) Use of less expensive cigarettes in six cities in China: findings from the International Tobacco Control (ITC) Survey. Tob Control 19(Suppl 2):i63–i68

Li Z, Elton-Marshall T, Fong GT, Quah ACK, Feng G, Jiang Y, Hitchman SC (2017) Noticing cigarette health warnings and support for new health warnings among non-smokers in China: findings from the International Tobacco Control project (ITC) China survey. BMC Public Health 17(1):476

Liao J, Winickoff JP, Nong G, Huang K, Yang L, Zhang Z, Abdullah AS (2016) Are Chinese pediatricians missing the opportunity to help parents quit smoking? BMC Pedriatr 16(1):135

Lin H, Xiao Z, Liu Z, Shi Q, Hajek P, Wang C (2019a) National survey of smoking cessation provision in China. Tob Induc Dis 17(April):25

Lin H, Chang C, Liu Z, Zheng Y (2019b) Subnational smoke-free laws in China. Tob Induc Dis 17(November):78

Lindorff KJ, Hill DJ (2010) Physicians' role in advancing tobacco control in China. Thoracic Cancer 1(1):17–22

Liu H, Rizzo JA, Sun Q, Wu F (2015) How do smokers respond to cigarette taxes? Evidence from, China's cigarette industry. Health Econ 24(10):1314–1330

Lv J, Su M, Hong Z, Zhang T, Huang X, Wang B, Li L (2011) Implementation of the WHO framework convention on tobacco control in mainland China. Tob Control 20(4):309–314

Mannocci A, Mapatrini D, Troiano G, Milazzo F, Langiano E (2019) The impact of pictorial health warnings on tobacco products in smokers behaviours and knowledge: the first quasi-experimental field trial after the implementation of the tobacco law in Italy. Annali 1st Super Sanità 55(2):186–194

Marquez PV, Zheng R (2016) China's 2015 tobacco tax adjustment: a step in the right direction. World Bank, Washington, DC. https://blogs.worldbank.org/health/china-s-2015-tobacco-tax-adjustment-step-right-direction

Marquez PV, Moreno-Dodson B, World Bank Group (2017) Tobacco tax reform at the crossroads of health and development. A multisectoral perspective. World Bank, Washington, DC, 222p. https://openknowledge.worldbank.org/handle/10986/28494

Mehta N, Zhang C, Hua X, Redmon P, Eriksen M, Koplan J, Ali M (2014) Tobacco smoking among government employees in six cities in China. Heart Asia 6(1):179–183

Nargis N, Zheng R, Xu S, Fong G, Feng G, Jiang Y, Wang Y, Hu X (2019) Cigarette affordability in China, 2006-2015: findings from International Tobacco Control China Surveys. Int J Environ Res Public Health 16(7):1205

Ngoc Bich N, Thu Ngan T, Bao Giang K, Thi Hai P, Thi Thu Huyen D, Ngoc Kue L, Tuan Lam N, Van Minh H, , Thi Quynh Nga P, The Quan N, Loan VH (2019) Salience and impact of health warning label on cigarette packs in Vietnam: findings from the Global Adult Tobacco Survey 2015. Behav Med 45(1): 30–39

Nilan K, Mc Keever TM, McNeill A, Raw M, Murray RL (2019) Prevalence of tobacco use in healthcare workers: a systematic review and meta-analysis. PLoS One 14(7):e0220168

Niu L, Liu Y, Luo D, Xiao S (2018) Current smoking behaviour among medical students in China: a systematic review and meta-analysis. Asia Pac J Pub Health 30(7):610–623

O'Connor A, Casper MF (2011) The role of indirect advertising in establishing social legitimacy: an analysis of tobacco sponsorship in China. In: Alozie EC (ed) Advertising in developing and

emerging countries: the economic, political and social context. Routledge, Abington/Oxford, pp 116–127
Peng S, Yu L, Yang T, Wu D, Bottorff JL, Barnett R, Jiang S (2019) Susceptibility to smoking and determinants among medical students: a representative nationwide study in China. Tob Induc Dis 17:36
Qin Y, Wu M, Pan X, Xiang Q, Huang J, Gu Z, Shi Z (2011) Reactions of Chinese adults to warning labels on cigarette packages: a survey in Jiangsu Province. BMC Pub Health 11:Article No. 133
Roddy E, Antoniak M, Britton J, Molyneux A, Lewis S (2006) Barriers and motivators to gaining access to smoking cessation services amongst deprived smokers – a qualitative study. BMC Health Serv Res 6:Article No.147
Saffer H, Chaloupka H (2000) The effect of tobacco advertising bans on tobacco consumption. J Health Econ 19(6):117–137
Shang C, Chaloupka FJ, Zahra N, Fong GT (2014) The distribution of cigarette prices under different tax structures: findings from the International Control Policy Evaluation (ITC) Project. Tob Control 23(Suppl 1):i23–i29
Smith DR (2008) The historical decline of tobacco smoking among United States physicians: 1949–1984. Tob Induc Dis 4(1):9
Smith DR, Zhao I, Wang L (2012) Tobacco smoking among doctors in mainland China: a study from Shandong Province and review of the literature. Tob Induc Dis 10(September):14
Sreeramareddy CT, Ramakrishnareddy N, Rahman M, Mir IA (2018) Prevalence of tobacco use and perceptions of student health professionals about cessation training: results from Global Health professions students survey. BMJ Open 8(5):e017477
Stillman F, Yang G, Figueiredo V, Hernandez-Avia M, Samet J (2006) Building capacity for tobacco control research and policy. Tob Control 15(Suppl 1):i18–i23
Tan X, Liu X, Shao H (2017) Healthy China 2030: a vision for health care. Value in Health Regional Issues 12(C):112–114
Tan X, Zhang Y, Shao H (2019) Healthy China 2030, a breakthrough in improving health. Glob Health Promot 26(4):96–99
Tobacco Free Kids (2014) China falling behind in tobacco control among BRICS. Tobacco Free Kids, Washington, DC. https://www.tobaccofreekids.org/assets/global/pdfs/en/BRICS_brief_en.pdf
Twigg L, Moon G, Szatkowski L, Igguldenn P (2009) Smoking cessation in England: intentionality, anticipated ease of quitting and advice provision. Soc Sci Med 68(4):610–619
van Mourik D-JA, Nagelhout GE, Vries H, van den Putte B, Cummings M, Borland R, Fong GT, Willemsen MC (2020) Quasi-experimentally examining the impact of introducing tobacco pictorial health warnings: findings from the International Tobacco Control (ITC) 4C and Netherlands surveys in the Netherlands, Australia, Canada, United Kingdom, and the United States. Drug Alcohol Depend 207(1):107818
Verguet S, Gauvreau CL, Mishra S, MacLennan M, Murphy SM, Brouwer ED, Nugent RA, Zhao K, Jha P, Jamison DT (2015) The consequences of tobacco tax on household health and finances in rich and poor smokers in China: an extended cost-effectiveness analysis. Lancet Glob Health 3(4):e206–e216
Verguet S, Tarr G, Gauvreau CL, Mishra S, Jha P, Liu L, Xiao Y, Qiu Y, Zhao K (2017) Distributional benefits of tobacco tax and smoke-free workplaces in China: a modelling study. Glob Health 7(2):020701
Wang F, Sun S, Yai X, Fu H (2016) The Museum as a platform for tobacco promotion in China. Tob Control 25(1):118–121
Wang L, Wang Z, Ma Q, Fang G, Yang J (2019) The development and reform of public health in China from 1949 to 2019. Glob Health 15:Article No. 45
Wang Q (2017) The relation between cigarette taxes and older adult smoking in Zhejiang and Gansu: what happened following the 2009 Chinese tax adjustments? Environ Health Prev Med 22:Article No. 21

Wang R, Zhu L, Yan W, Zeng G, Engelgau M (2015) The influence of estimated retail tobacco sale price increase on smokers' smoking habit in Jiangxi Province, China: a cross-sectional study. Tob Induc Dis 13(1):25

Wang S, Fu X, Liu Z, Wang B, Tang Y, Feng H, Wang J (2018) General practitioner education reform in China: Most undergraduate medical students do not choose general practitioner as a career under the 5+3 model. Health Prof Educ 4(2):127–132

Wei X, Li M, Chen D, Cai Y, Gao J (2017) Effects of strict control of tobacco advertisement after implementation of the Advertising Law. J Shanghai Jiaotong Univ (Med Sci) 37(2):137–140. (in Chinese)

White JS, Li J, Hu T-W, Fong GT, Jiang Y (2014) The effect of cigarette prices on brand switching in China: a longitudinal analysis of data from the ITC China survey. Tob Control 24(Suppl 1):i54–i60

WHO (World Health Organisation) (2004) Capacity building and initiatives. World Health Organisation, Geneva. https://www.who.int/tobacco/control/capacity_building/en/

WHO (World Health Organisation) (2008) WHO Report on the Global TOBACCO Epidemic 2008, The MPOWER Package. https://www.who.int/tobacco/mpower/2008/en/

WHO (World Health Organisation) (2013) MPOWER in action. Defeating the global tobacco epidemic. World Health Organisation, Geneva. https://www.who.int/tobacco/mpower/publications/brochure_2013/en/

WHO (World Health Organisation) (2015a) WHO report on the Global Tobacco Epidemic, 2015: raising taxes on Tobacco. World Health Organisation, Geneva, 103pp. https://www.who.int/tobacco/global_report/2015/report/en/

WHO (World Health Organisation) (2015b) World Health Organisation Western Pacific region and University of Waterloo, ITC Project. Smoke-free Polices in China. Evidence of Effectiveness and Implications for Action. World Health Organisation, Western Pacific Region, Manila. https://iris.wpro.who.int/handle/10665.1/12047

WHO (World Health Organisation) (2017a) WHO Report on the Global Tobacco Epidemic, 2017. World Health Organisation, Geneva, 103pp

WHO (World Health Organisation) (2017b) The Bill China Cannot Afford: Health, Economic and Social Costs of China's Tobacco Epidemic. WHO Office for the Western Pacific, Manila. https://www.who.int/tobacco/global_report/2017/en/

WHO (World Health Organisation) (2019a) WHO report on the global tobacco epidemic, 2019. World Health Organisation, Geneva. https://www.who.int/tobacco/global_report/en/

WHO (World Health Organisation) (2019b) WHO Report on the Global Tobacco Epidemic, 2019 Country Profile – China. World Health Organisation, Geneva. https://www.who.int/tobacco/surveillance/policy/country_profile/chn.pdf?ua=1

WHO (World Health Organisation) FCTC (2014) 2014 Global Progress Report on implementation of the WHO Framework Convention on Tobacco Control. World Health Organisation, Geneva, pp 81–82. https://www.who.int/fctc/reporting/summary_analysis/en/

Wu D, Yang T, Cottrell RR, Zhou H, Yang XY, Zhang Y (2015) The effects of tobacco-related health-warning images on intention to quit smoking among urban Chinese smokers. Health Educ J 74(3):287–298

Wu D, Yang T, Stillman FA, Guo W, Zhou H, Feng X (2018) Assessment of tobacco control advocacy behavioural capacity among public health faculty and students: a cluster intervention study. Public Health 165:95–105

Wu L, He Y, Jiang B, Zuo F, Liu Q, Zhang L, Zhou C, Liu M, Chen H, Cheng KK, Chan SS, Lam TH (2016) Effectiveness of additional follow-up telephone counselling in a smoking cessation clinic in Beijing and predictors of quitting among Chinese male smokers. BMC Pub Health 16:Article No. 63

Wu Y, Yang G (2018) Health warnings of cigarette package in China. In: Yang G (ed) Tobacco control in China. Springer, Singapore, pp 221, 323p–244

Xiao D, Bai C-X, Chen Z-M, Wang C (2015) Implementation of the World Health Organization framework convention on tobacco control in China: an arduous and long-term task. Cancer 121(Suppl 17):3061–3068

Xiao L, Jiang Y, Zhang YB (2011) Tobacco advertisement exposure of adolescents among three cities in China. Chinese J Prev Control Chronic Dis 19(2):131–133. (in Chinese) 8 Evaluating China's Record of Tobacco Control 284

Xinhuanet (2020) "Winners" of Dirty Ashtray Award announced in China. Xinhuanet 24 June, 2015. http://www.xinhuanet.com/english/2019-06/24/c_138168553.htm

Xu X, Zang X, Hu TW, Miller LS, Xu M (2019) Effects of global and domestic control policies on cigarette consumption per capita: an evaluation using monthly data in China. BMJ Open 9(6):e025092

Xu Y, Xu S, Wu Q, Guo Y (2013) Tobacco knowledge among adults in Zhejiang Province, China. PLOS One 8(3):e59172

Yang G-H, Li Q, Wang C-X, Hsia J, Yang Y, Xiao L, Yang J, Zhao L-H, Xie L (2010a) Findings from 2010 global adult tobacco survey: implementation of MPOWER policy in China. Biomed Environ Sci 23(6):422–429

Yang T, Abdullah AS, Rockett IRH, Li M, Zhou Y, Ma J, Ji H, Zheng J, Zhang Y, Wang L (2011) Assessment of tobacco control advocacy behavioural capacity among students at schools of public health in China. Tob Control 20(1):20–25

Yang T, Mao A, Feng X, Jiang S, Wu D, Bottoroff JL, Sarbit G, Wang X (2014a) Smoking cessation in an urban population in China. Am J Health Behav 38(6):933–941

Yang T, Peng S, Yu L, Jiang S, Stroub W, Cottrell I, Rockett I (2016) Chinese smokers' behavioral response toward cigarette price: individual and regional correlates. Tob Induc Dis 14(1):13

Yang T, Rockett IRH, Li M, Xu X, Gu Y (2012) Tobacco advertising, environmental smoking bans, and smoking in Chinese urban areas. Drug Alcohol Depend 124(1–2):121–127

Yang, T and Team Members (2016) Building tobacco dependence treatment capacity in medical universities and affiliated hospitals in China. Center for Tobacco Control, Zhejiang University, Global Bridges Award No. 13498319), Hangzhou

Yang T, Yang G (2012) Smoking cessation strategies in tobacco control in China. J Tuberc Lung Health 1(1):45–47

Yang T, Yang X, Lv Q, Zhao Q, Ke X (2009) Special communication: China's first historic efforts to develop a tobacco control advocacy workforce via schools of public health. Tob Control 18(5):422–424

Yang T, Yu L, Bottoroff JL, Wu D, Jiang S, Peng S, Young KJ (2015) Global Health Professions Student Survey (GHPSS) in tobacco control in China. Am J Health Behav 39(5):732–741

Yang T, Zhu Z, Barnett R, Zhang W, Peng S, Jiang S (2019) Tobacco advertising, anti-tobacco information exposure, environmental smoking restrictions, and unassisted smoking cessation among Chinese male smokers: a population-based study. Am J Mens Health 13(3):1557988319856152

Yang W (2019) Sneaky tobacco ads show smoking as chic lifestyle. China Daily 17 April. http://www.chinadaily.com.cn/a/201904/17/WS5cb67f7ea3104842260b6a1e.html

Yang XY, Anderson JG, Yang T (2014b) Impact of role models and policy exposure on support for tobacco control policies in Hangzhou, China. Am J Health Behav 38(2):275–283

Yang Y, Li L, Yong H-H, Borland R, Wu X, Li Q, Wu C, Foong K (2010b) Regional differences in awareness of tobacco advertising and promotion in China: findings from the ITC China survey. Tob Control 19(2):117–124

Yao T, Huang J, Sung H-Y, Ong MK, Mao Z, Jiang Y, Fong GT, Max W (2013) Who purchases cigarettes from cheaper sources in China? Findings from the ITC China survey. Tob Control 23(Suppl 1):i97–i101

Yong H, Siapush M, Borland R, Li L, O'Connor RJ, Yang J, Fong GT, Jiang Y (2013) Urban Chinese smokers from lower socioeconomic backgrounds face more barriers to quitting: results from the International Tobacco Control-China Survey. Nicotine Tob Res 15(6):1044–1051

Yu L, Cohen JE, Hoe C, Yang T, Wu D (2019) Male smoking reduction behaviour in response to China's 2015 cigarette tax increase. Tob Control 29(4). https://doi.org/10.1136/tobaccocontrol-2019-055053
Zhang J, Chan S, Fon DY, Malone RE, Lam T (2012) The social context of smoking cessation in China: an exploratory interview study. Tob Control 21(1):57–58
Zhang M, Liu S, Yang L, Jiang Y, Huang Z, Zhao Z, Deng Q, Li Y, Zhou M, Wang L, Chen Z, Wang L (2019) Prevalence of smoking and knowledge about the hazards of smoking among 170000 Chinese adults, 2013–2014. Nicotine Tob Res 21(12):1644–1651
Zheng P, Ge X, Qian H, Wang F, Fu H, Berg CJ, Kegler MC (2014) 'Zhonghua' tobacco advertisement in Shanghai: a descriptive study. Tob Control 23(5):389–394
Zheng P, Qian H, Wang F, Sun S, Nehl EJ, Wong FY (2013) Awareness of tobacco advertising, perceived harms of smoking, and beliefs about tobacco control among a sample of Shanghainese in China. Health Educ Res 28(5):879–887
Zheng P, Xiao L, Wang F, Yang G (2018) Comprehensive ban on tobacco advertising, promotion and sponsorship. In: Yang G (ed) Tobacco control in China. Springer, Singapore, pp 245, 323p–284
Zheng R (2018) Tobacco tax and tobacco economy in China. In: Yang G (ed) Tobacco control in China. Springer, Singapore, pp 285, 323p–309
Zheng R, Huang J (2015) Would raising tobacco tax in China unfairly burden the poor? Lancet Glob Health 3(4):e181–e182
Zheng R, Wang Y, Hu X, Marquez PV (2017) Cigarette affordability in China 2001–2016. World Bank Group, Washington, DC

Chapter 9
Case Study: Smoking Bans and Secondhand Smoke

Ross Barnett, Sihui Peng, and Shuhan Jiang

Abstract Comprehensive tobacco control policies must take account of not only smoking prevalence but also problems associated with exposure to secondhand smoke (SHS). This chapter provides a detailed case study of the implementation of smokefree policies in China since 2005. It first outlines patterns of exposure to SHS and why comprehensive smokefree legislation is necessary. The chapter then provides a detailed analysis of the geography of smoking bans and problems of implementation. The chapter uses a modified version of the conceptual model developed by Hyland et al. (Tob Control 21(2):154–161). This is a useful framework which evaluates smokefree policies in terms of the locations where they have been developed, their degree of comprehensiveness and level of enforcement and their wider health and social effects.

Keywords Denormalisation · Health inequalities · Philanthropic agencies · Policy enforcement · Secondhand smoke (SHS) exposure · Smokefree policies · Stigmatisation · Sub-national initiatives

R. Barnett (✉)
School of Earth and Environment, University of Canterbury, Christchurch, New Zealand
e-mail: ross.barnett@canterbury.ac.nz

S. Peng
School of Medicine, Jinan University, Guangzhou, China

S. Jiang
School of Humanities and Management, Zhejiang Chinese Medical University, Hangzhou, China

R. Barnett et al. (eds.), *Smoking Environments in China*, Global Perspectives on Health Geography,
https://doi.org/10.1007/978-3-030-76143-1_9

9.1 Introduction

Comprehensive tobacco control policies must take account of not only smoking prevalence but also problems associated with exposure to secondhand smoke (SHS). As we indicated in the previous chapter, smokefree legislation not only aims to encourage a denormalisation of smoking but is also an important policy instrument to reduce both SHS exposure and smoking prevalence itself. This chapter provides a detailed case study of the implementation of smokefree policies in China since 2005. We begin by first outlining patterns of exposure to SHS in China and why comprehensive smokefree legislation is necessary. To a western observer, this should be self-evident, but in China acceptance of such policies is far from widespread. We then provide a detailed analysis of the geography of smoking bans in China and problems in the implementation of smokefree policies.

In order to accomplish this task, we use a modified version of the conceptual model developed by Hyland et al. (2012) (Fig. 9.2). This is a useful framework because it suggests that smokefree policies can be evaluated in a number of different ways, in terms of (1) the locations where they have been developed; (2) their degree of comprehensiveness and level of enforcement; and (3) their wider health and social effects. With respect to the first step, it is evident that, although China has signed the FCTC, there is still no nationwide ban on smoking in public places and nor have any provinces adopted provincial-wide smoking bans. Instead implementation of smokefree policies has largely occurred at the city scale and for particular public spaces, such as airports or hospitals. Second, the uneven development of SHS legislation in China is matched by low levels of compliance with the legislation and marked variations in enforcement. Thus it is necessary to consider why this has occurred. In the final step in the model, we consider the health and social implications of smokefree policies and to evaluate the research that has been undertaken in these areas. We focus on the effects of smokefree policy on reductions in SHS exposure, smoking cessation, and health inequalities. In terms of social implications, we pay particular attention to two issues: the extent to which public smokefree policies have encouraged denormalisation and influenced smoking in the private space of the home.

9.2 Individual and Environmental Correlates of SHS Exposure

Exposure to secondhand smoke has, until very recently, been quite ubiquitous in China (Liu et al. 2011; Zheng et al. 2009) and, as in western countries in the 1980s and 1990s, has attracted considerable attention. It is important to outline the different correlates of SHS exposure as a prelude to examining the implementation of smokefree policies. Most studies of SHS in China have focused on individual predictors of exposure and its health effects. Less attention has been directed at environmental determinants of exposure. At the individual level, age, gender, marital status, educational level, knowledge of health harms and the smoking habits of spouses have been the dominant predictors of exposure.

Increasing age has generally been seen to be a protective factor. Zeng et al. (2016) in a meta-analysis of Chinese studies (1997–2015) showed that people aged 15–59 years were more likely to be exposed to SHS than people aged 60 years and over. Similarly Gao et al. (2019) showed that in Jilin Province, age was a protective factor in both urban and rural areas with 'elderly' people (60+) reporting only 60% exposure levels of 'youth' (18–44) in urban areas and 62% in rural areas. However, the protective effects of age may be context dependent, as between 2001 and 2010 SHS exposure among older people (60+) increased in Beijing. The main source of SHS changed from the spouse to family offsprings living in the household, a finding of importance especially for cities with housing shortages. For instance, in Jilin Province, Li et al. (2015a, b) showed that separated/divorced and widowed women were, respectively, almost three quarters and half as likely to be subject to SHS than married women, thus suggesting that the absence of family was a protective factor. High levels of child exposure to SHS remain a continuing concern. Internationally, China ranks third (with 67%), after Indonesia (79%) and Vietnam (76%), in levels of child exposure (Mbulo et al. 2016).

Gender differences are also important with women often being reported as having higher levels of exposure especially in the home (Li et al. 2015a, b). Cai et al. (2014), for example, in a study of rural areas showed that even though women's smoking prevalence was minor compared to males, their exposure to SHS was higher. Yao et al. (2012) in a study of six rural areas of China showed that females were over three times more likely to be exposed to SHS than males, a trend also evident among non-smoking (rural) pregnant women (Yang et al. 2010). However, Zeng et al. (2016) in their 1997–2015 meta-analysis found no significant gender differences in either urban or rural areas as did Cai et al. (2013) in Tonghai County, one of the poorest parts of rural Yunnan. An earlier meta-analysis of studies conducted between 2000 and 2009 (He et al. 2011), however, showed that females were more likely to be exposed to passive smoke. This no doubt reflects women's difficulty in avoiding SHS exposure in the more patriarchal social environment of the time. Recent studies suggest that gender differences are particularly important in rural areas where females are more likely to be exposed to SHS than males, the reverse being true in the urban context (Chen et al. 2018; Gao et al. 2019). In rural areas, with traditional social values, it is noted that women seldom force smoking restrictions upon their husbands, even if they have knowledge of the adverse effects of SHS (Abdullah et al. 2012). By contrast, in urban areas SHS exposure may be higher among males than females reflecting greater female emancipation and economic independence and the ability to distance themselves from male social environments where smoking is the norm.

As in western countries, most LMICs have a socio-economic gradient in SHS exposure (Hajizadeh and Nandi 2016). At the individual level, lower educational attainment and income are associated with higher SHS exposure (Yao et al. 2012) and poorer knowledge of the harmful effects of tobacco (Jin et al. 2014). Yang et al. (2015b), for example, using 2011 data for 21 cities, showed that SHS exposure was lowest among individuals with greater education and income. Li et al. (2015a) report similar findings for women non-smokers in Jilin Province. Yao et al. (2012) in a study of 6 rural areas found significant income differences in exposure for adults, but especially for children, while Chen et al. (2018) using data from 122 cities showed that people of lower socio-economic status were more likely to be exposed to passive smoking. Xu et al. (2016), although not examining links to socio-economic status, showed that the cultural practice of cigarette gifting and sharing was also associated with higher levels of SHS exposure. Yang et al. (2015b) found that higher-income people engage in more social activities, thus inhibiting home smoking restrictions. These differences are indicative of the different dimensions of socio-economic status and how education and income may have different consequences for exposure. Social differences in SHS exposure, however, are likely to be affected by the implementation of smokefree policies and the extent to which they are enforced in different types of homes and workplaces.

Compared to individual variations in SHS exposure, less attention has been paid to the influence of environmental predictors of exposure. Other than the influence of smokefree policies on exposure, most attention has been devoted to urban-rural and regional factors, especially the former. It could be argued that, because of higher smoking prevalence and an almost complete absence of smokefree policies in smaller towns and rural areas (e.g. see Stillman et al. 2007), SHS exposure should be highest in such places. A meta-analysis of studies carried out up to 2009 suggests that this appears to have been the case (He et al. 2011). However, more recent studies, such as Gao et al. (2019) of Jilin Province in northern China, indicate that rural-urban differences in SHS exposure are similar (60.2% urban and 61.8% rural). Zeng et al. (2016) in their meta-analysis concluded that urban-rural differences were not significant; that the pooled prevalence rate for 21 urban studies of SHS (48.6%) was only slightly higher than the pooled rate for 20 studies for rural areas (43.5%); and that this gap remained over the time period studied (2002–2013). They suggest that several factors may have contributed to this divergence; the greater scope of their later meta-analysis, the higher level of importance attached to social interactions in urban settings and a greater health consciousness in urban residents may have led to more self-reports of passive smoking. However, no mention is made of tobacco industry marketing and increased consumption, the practice of cigarette gifting, nor of the possible effects of the implementation of smokefree policies on urban-rural differences in exposure. The geographically variable nature of such policies will have affected urban-rural differences in exposure, complicating any such comparisons. Also important has been rural-urban migration which, for some groups, is associated with both high rates of SHS exposure. Zheng et al. (2018) found higher rates of exposure among migrants who worked in agricultural occupations or in business/services and low rates among those who were widowed or divorced. Duration of stay also appeared to be important with longer-term migrants having slightly less exposure.

Among larger cities Yang et al. (2015b) have shown that city-level contextual factors, particularly urban GDP per capita, tobacco advertising and anti-smoking media news coverage, had an impact on SHS exposure above and beyond the influence of individual characteristics. Poorer cities, independent of individual SES, had higher rates of exposure. This may reflect the greater tolerance of smoking and fewer resources available for anti-smoking programmes, compared to richer cities with stronger anti-smoking norms, which are more likely to have enforced smokefree policies. Not surprisingly respondents living in cities with higher levels of tobacco advertising were almost three times more likely to be exposed to SHS than those reporting no advertising. Also important was the relationship between anti-smoking media coverage at the city level and individual SHS exposure. Where more negative media coverage occurred at the city level, individual non-smokers were less exposed to tobacco advertising and consequently to less SHS.

There have been a few studies of the geography of SHS prevalence in China which is surprising given the urban-rural differences in socio-economic development that have long characterised the country. However, two studies are of note. The first is by Jin et al. (2014) who noted the lack of attention to the variability in SHS exposure as well as in smoking and SHS-related knowledge. Using data from the 2010 GATS China survey, they found that older less educated people had less knowledge but did not examine how this was affected by place of residence. However, when viewing the proportion of people who observed smoking in various public places by urban-rural and regional location, they found that the prevalence of viewing smoking in healthcare facilities and on public transport was higher in rural than urban areas (33.0 vs 67.0% and 27.0 vs 42.3%) and distinct regional variations occurred as well in the public places where smoking was observed. Significant differences occurred across six regions (North, Northeast, Northwest, East, Mid-south and Southwest): for government buildings, healthcare facilities and public transportation. In all cases SHS exposure (as measured by the prevalence of observed smoking) was highest in the southwest region reaching a maximum of 75.7% in government buildings. Exposure in schools and restaurants was also highest in that region although regional differences, as a whole, were not significant. Astell-Burt et al. (2018), using 2010 data from the Chinese Non-communicable Disease Surveillance Survey, also considered regional variations in SHS exposure but at the provincial level. Provincial variations in passive smoking were generally higher in southern provinces, some with a strong presence of the tobacco industry, and Tibet (Fig. 9.1). Smaller gender differences in SHS exposure were observed compared to

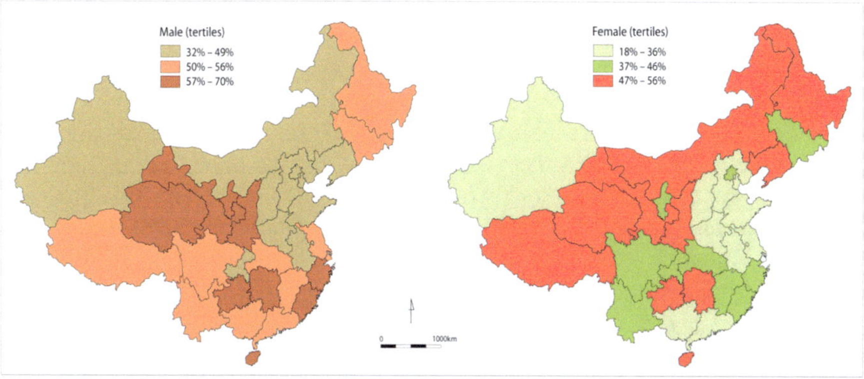

Fig. 9.1 Exposure to passive smoking for at least 15 min a week, 2010. (Credit: Modified from Astell-Burt et al. (2018), p. 758. Reproduced courtesy of Oxford University Press. (Rightslink Licence No. 4878561188341))

smoking, with the age-adjusted prevalence for women being lower than for men (39.2% vs 51.2%). However, generally in provinces where male smoking rates were highest, so was the level of SHS exposure among women.

The lack of attention to the effect of the tobacco industry on SHS exposure is surprising given its prominence in many communities. However, one recent study by Hoe et al. (2021) showed that province-level per capita cigarette production and city level tobacco retail outlet density were both associated with higher levels of SHS exposure. The results suggest that both factors affect attitudes and behavioural responses to SHS exposure as a result of higher rates of tobacco advertising and its influence on community social norms.

9.2.1 The Need for Smokefree Policies

Given the high levels of SHS exposure across China, the health consequences of SHS and of the implementation of smokefree policies (Xiao et al. 2020) have received increased attention. Although the health effects and economic costs of SHS have been well documented in western countries, it has taken some time for a body of research to be developed in the Chinese context (Xia et al. 2020; He et al. 2020). The health risks of SHS exposure have been noted among a wide variety of groups with persons working in at-risk occupations, women, adolescents and young children receiving particular attention. Liu et al. (2014a), for example, have emphasised the significant number of lung cancer and heart disease deaths among servers and patrons which could be prevented by more effective smoking bans. In Tianjin, Xiao et al. (2020) estimated that the implementation of smokefree legislation in 2012 saved an estimated 10,000 deaths due to acute myocardial infarction within 3.5 years of the implementation of the law. Others, such as Cai et al. (2014), have emphasised the disproportionate burdens faced by women. In their study of the economic burdens of smoking in rural SW China, they found that while 73% males smoked, only 38% were affected by SHS exposure. On the other hand, among women, while only 1.4% smoked, 43.4% were exposed to SHS.

Exposure to SHS has also been related to the prevalence of depression and mental disorder. Yang et al. (2015a), using data from 21 cities, showed that the prevalence of mental disorder was more than twice as high among respondents who were ‘always’ exposed to SHS in their workplaces and in the public places they frequented. It was also higher among women although the interactions between gender, exposure and mental disorder were not examined. Guo et al. (2020) showed that women in long-term relationships with smoking husbands had significantly more depressive symptoms as did women before and during pregnancy (Song et al. 2019). Furthermore a higher risk of hypertension was associated with the amount and cumulative exposure of husbands smoking (Yang et al. 2017). Passive smoking at home has also been shown to increase the depressive symptoms of older people with the effects being stronger for non-smokers than smokers (Lu and Wu 2019).

In a similar vein, Wang et al. (2019), in a study of children aged 6–13 years in Shanghai primary schools, showed that SHS exposure was an independent risk factor for psychological problems experienced by this group. Among adolescents in Guangzhou, Chen et al. (2018) showed that students had an SHS exposure rate of 69% and that there were significantly increased risks of having respiratory symptoms corresponding to SHS exposure in public places (OR = 1.60), homes (OR = 1.53) and indoor (OR = 1.43) and outdoor (OR = 1.37) campuses. Other evidence also suggests that childhood SHS exposure can have long-lasting effects. Lee et al. (2018), using data from the Chinese Longitudinal Healthy Longevity Study, showed that older adults who reported SHS exposure in childhood had lower odds (OR = 0.69) of reporting better health status, compared to those not exposed. In general, higher SHS exposure at home has been associated with lower levels of family harmony, happiness, health and family well-being (Wang et al. 2016). Bearing in mind the western experience of the health effects of SHS, these findings are not surprising, but given the high rates of SHS in China, they point to the need for comprehensive national smokefree policies.

Protection of the population from exposure to secondhand smoke is a core component of tobacco control and, as a strategy, has featured prominently in the FCTC efforts to promote more effective tobacco control efforts in LMICs. As we outlined in the previous chapter, while significant moves have been made to promote national legislation, this has varied in its effectiveness in limiting exposure with few substantial advances since 2014. By contrast, it has been at the local rather than the national level where the most significant advances have occurred. Many cities have adopted sub-national smokefree laws and, while varying in the level of enforcement and effectiveness, have nevertheless taken the initiative in promoting smoking bans in particular regions. To accomplish this review, we use a modified version of Hyland et al.'s framework to explore the implementation of smoking bans and their impact (Fig. 9.2). The remainder of this chapter first explores the geography of policy

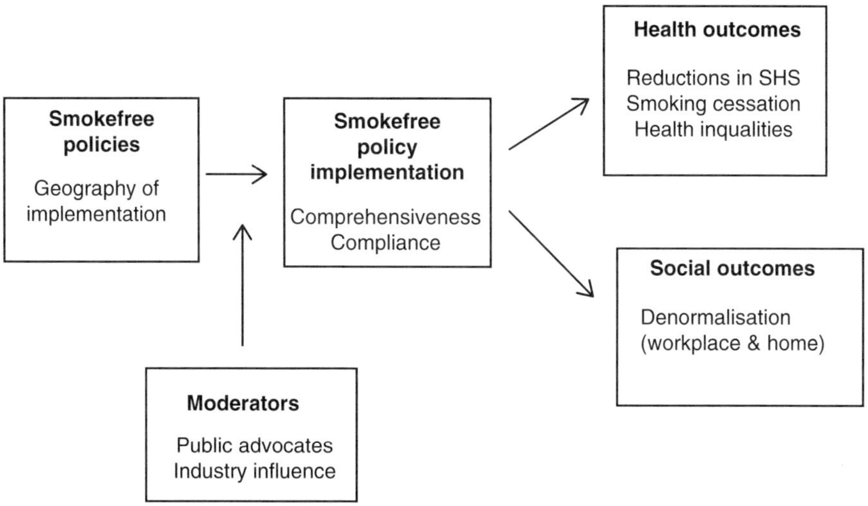

Fig. 9.2 Conceptual framework for the evaluation of smokefree policy. (Credit: Authors. Adapted from Hyland et al. (2012), p. 158)

implementation (Sect. 9.3) and problems of comprehensiveness and compliance (Sect. 9.4). This is followed by a discussion of health and social outcomes (Sects. 9.5 and 9.6).

9.3 Geographies of Implementation: Sub-national Smokefree Initiatives

Beginning with Harbin, the capital of Heilongjiang Province in 2011, it is at the sub-national level where most progress has occurred. The reasons for this are complex, but local initiation of smoking bans has arisen from a variety of different sources. First, public health and other smokefree advocates have been able to achieve local success stories, whereas this was not possible at the national level because of greater interference from the tobacco industry (Gan 2018). Thus policy development at grass-roots level had a much greater chance of success when its effects were more geographically restricted and not nationwide in scope. The 'bottom-up' development of such initiatives represents a new stage in the development of tobacco control in China and, given recent events, seems to be more successful in enforcing smokefree regulations and changing social norms through targeted health campaigns.

A second major force for the early development of local smokefree initiatives was the holding of major international events in some cities and the opportunity for China to present a positive image to the world and highlight its commitment to the FCTC. Smokefree laws were passed in Beijing in 2008 in anticipation of the Beijing Olympics and again 2 years later in Shanghai and Guangzhou before the World Expo and Asian Games. The Guangzhou law was important because not only did it ban smoking in government offices for the first time but also enforced stronger penalties for violations (Gan 2018). The Shanghai law (the Shanghai Public Places Smoking Control Regulations) was also significant because it was generally well enforced with visitors being satisfied with the smokefree environment (Li et al. 2013).

A third major force stimulating the development of local smokefree policies has, since 1986 (Redmon et al. 2013), been the impact of philanthropic agencies such as the American Bloomberg and Gates Foundations. In 2008 these corporations provided funding to the Emory Global Health Institute in Atlanta to support the development of tobacco control capacity in China. As a result of such funding, the Emory Global Health Institute-China Tobacco Control Partnership (GHI-CTCP) was established in 2008. The major focus of the partnership was the Tobacco-Free Cities (TFC) programme which aimed to change social norms of tobacco use through a 'bottom-up' approach in Chinese cities (Redmon et al. 2014). Cities selected for the programme were ones that seemed most receptive to tobacco control and likely to be considered 'innovators' or 'early adopters' as described by diffusion of innovation theory (Rogers 1962). The programme also aimed to build tobacco control research capacity and has played an important role in assisting Chinese universities and research institutions conduct tobacco control research and build up local expertise.

Fig. 9.3 Emory Global Health Institute-China Tobacco Control Partnership cities. (Credit: Authors. Data sourced from Redmon et al. (2014) and Emory University (2018))

Initially the programme focused on developing and implementing tobacco control initiatives in 17 large- and medium-sized cities, with an additional 5 included at a later date (Fig. 9.3). The grantee cities were allowed to decide where to focus their efforts based on each city's unique context. A variety of city organisations, covering hospitals and community clinics, educational institutions, major employers and religious sites were selected to participate because of their local responsibilities in enforcing health policies, monitoring and promoting health and educating the public on health-related issues (Redmon et al. 2014). For instance, in Anshan an exemplary smokefree workplace programme occurred in the Angang Steel units reaching more than 105,000 workers to provide a widespread smokefree environment in the Company. A high degree of compliance within production units was achieved through workplace educational programmes and inspections by the Company's Environmental Protection Department. By encouraging employee involvement and competition between production units, it was possible to improve employee knowledge of the harms of smoking and SHS, both leading to increased attempts at cessation (Emory University 2013).

Given the philanthropic origins of the programme and the limitations of its scope, the initial goal was to focus on specific policies in a selected group of cities. Within each city the aim was not to develop comprehensive smokefree policies but rather to work closely with local stakeholders to ensure a high chance of success in particular areas. Seven of the cities initially included a legislated smokefree public policy component in their plans including some larger cities (e.g. Shanghai and Hangzhou) covered by the programme. In this way the demonstration projects could adapt to local circumstances and serve as role models for the rest of the community. As Redmon et al. (2014, p 10066) explained:

The 17 TFC grantees were provided training opportunities, technical assistance, and grant funds to develop, implement, and evaluate strategies aimed at changing the social norm of tobacco use in their cities. The TFC grantees were required to select tobacco control strategies based on their unique situation and to draft an evaluation plan to assess the impact of their program. The TFC program model was designed to acknowledge the expertise and experience of the grantee as well as the unique situation of the city rather than utilizing a 'one-size-fits-all' model of tobacco control.

Each city conducted a systematic situational analysis to understand the extent of tobacco use problems and existing tobacco control efforts and to identify potential interventions and partners. The grantees selected the programme goals based on their perceived ability to achieve the greatest initial impact and the potential for social norm change, and each grantee had to demonstrate the impacts of their programme through monitoring and evaluation (Mehta et al. 2014). More recently, through additional funding from the New York-based Pfizer Corporation, the programme has expanded in size to include an additional five cities (Chengdu, Chongqing, Wuhan, Xiamen and Xi'an) (Emory University 2018).

9.3.1 Smokefree Initiatives in Major Cities

The launch of the Emory GHI-China TCP partnerships in 2009 predated the first major comprehensive smokefree policy in Harbin in 2011 (implemented in 2012) and has played an important role in helping many cities move towards this goal. However, it is relatively difficult to determine the smokefree status of various cities in China. Until recently, most cities did not have comprehensive smoking bans, and the extent to which partial bans have been enforced has been difficult to quantify. For instance, Hu (2008, p. 39) noted that by the end of 2006, 154 cities had passed bans on smoking in public areas; however, all of these were partial bans that relied upon public education to achieve enforcement. Penalties, if they existed, were small. For example, Li et al. (2010) reporting on Wave 1 of the International Tobacco Control (ITC) survey conducted in six cities in 2006, before the introduction of more stringent national legislation, show the variability of policy implementation within the cities. Most of the city ordinances, if they existed, were weakly enforced partial bans (Table 9.1). It was only after China's ratification of the WHO FCTC in 2005 that cities began to develop more comprehensive smokefree laws (Gan 2018). The problem, however, still remains of determining the extent of enforcement that has occurred. As Hu (2008, p. 39) noted, there has been no official report or

Table 9.1 Smoking policies of different venues in six cities, 2006

Venues	Beijing	Shenyang	Shanghai	Changsha	Guangzhou	Yinchuan
Hospitals	Partial ban	Partial ban	Partial ban	Partial ban	Partial ban	Partial ban
Public transport	***Total ban***	***Total ban***	***Total ban***	Partial ban	Partial ban	***Total ban***
Schools	Partial ban	Partial ban	Partial ban	Partial ban	Partial ban	Partial ban
Workplaces	No rules	No rules	No rules	No rules	Partial ban	No rules
Restaurants/bars	No rules	No rules	No rules	No rules	Partial ban	No rules
Bars	No rules	No rules	No rules	No rules	No rules	No rules
Conference rooms	***Total ban***	Partial ban	***Total ban***	***Total ban***	Partial ban	***Total ban***

Credit: Adapted from Li et al. (2010), Table 1 p.i42

published research on the implementation and enforcement of these regulations, and, at the time of writing, this is still largely true.

At the local level, laws and regulations can originate from two sources: the People's Congress Council (PCC) and the 'People's Government' (PG). The former can refer to congresses at the county, city and township level. At the city level, the PCCs are elected by lower-level congresses, while county and township bodies are elected directly by local voters. Compared to their national counterpart, local urban legislatures have a great deal of autonomy, and local laws are seldom referred centrally for approval from the Party (Cho 2006). Strictly speaking local smokefree regulations should only be passed by the legislative procedure of the PCC to have legal effect and to be classified as smokefree 'laws'. Urban governments, here referred to as People's Governments, can pass administrative regulations, but they are not as strong and as binding as PCC laws. Hence, whether a smokefree city policy has its origins with PCCs or PGs is a good indicator of the likely extent of enforcement.

In order to provide an updated list of cities which have enacted smokefree laws, beyond that provided by Gan (2018) and Lin et al. (2019), who each differ in their estimates, a comprehensive literature and document search was undertaken of cities which had implemented smokefree laws or regulations since China ratified the FCTC in 2005. We were interested in the legislative body responsible for the regulations and whether the city had been part of the Emory GHI-CTCP initiative. Other than determining the source of the legislation, we did not attempt to assess the extent to which the laws/regulations were compliant with FCTC guidelines. Unlike Lin et al. (2019), we chose not to do this because of variations in the extent of enforcement and a lack of confidence in the quality of the data. However, searches were made of each city website to determine the nature of penalties for lack of compliance.

Since 2005 we estimated that 30 cities have enacted, or are currently enacting, smokefree laws or regulations, many of them, such as Shenzhen, revising earlier laws (Table 9.2 and Fig. 9.4). The number of cities is greater than the 21 places listed by Lin et al. (2019) because we also included Hong Kong as well as a number of other cities which are in the process of enacting smokefree legislation. At the time of writing, one megacity (Chongqing) and two provincial capitals (Chengdu

Table 9.2 Main smokefree cities in China by date and type of legislation

City	Province	Date of implementation[a]	Legislature[b]	Provincial capital	Part of Emory GHI-CTCP[c]	City population (million) (2019)[d]
Qinhuangdao	Hebei	2019	PG			0.76
Zhangjiakou	Hebei	2019	**PCC**			0.69
Hangzhou	Zhejiang	2010 (revised 2019	**PCC**	Yes	Yes	6.24
Shenzhen	Guangdong	1998 (revised 2019)	**PCC**			10.36
Chongqing	Municipality	1998 (under revision 2019)	PG	Yes	Yes	7.46
Chengdu	Sichuan	1997 (under revision 2019)	**PCC**	Yes	Yes	7.42
Dalian	Liaoning	2019 (drafting legislation)	PG	Yes		4.08
Xi'an	Shaanxi	2018	PG	Yes	Yes	6.5
Shanghai	Municipality	2010 (revised 2017)	**PCC**	Yes	Yes	22.32
Fuzhou	Fujian	2015	**PCC**	Yes		1.18
Xining	Qinghai	2015	**PCC**	Yes		0.77
Beijing	Municipality	2015	**PCC**	Yes		11.72
Lanzhou	Gansu	2014	**PCC**	Yes		2.63
Nanning	Guangxi	2014	PG	Yes	Yes	0.80
Tangshan	Hebei	2014	PG		Yes	3.37
Changchun	Jilin	2014	PG	Yes	Yes	4.19
Qingdao	Shandong	Temporary ban 2008 2013	**PCC**		Yes	3.72
Kelamayi	Xinjiang	2013	PG		Yes	0.26
Shaoxing	Zhejiang	2013	PG			0.42
Anshan	Liaoning	2013	PG		Yes	1.20
Tianjin	Municipality	2012	**PCC**	Yes		11.09
Harbin	Heilongjiang	1996 (revised 2012	**PCC**	Yes		5.88
Guangzhou	Guangdong	2010 (revised 2012)	**PCC**	Yes		11.07
Shijiazhuang	Hebei	1997 (revised 2010)	**PCC**	Yes		2.83
Jinan	Shandong	1996 (revised 2001, 2010)	**PCC**	Yes		4.34
Nanjing	Jiangsu	2010	PG	Yes	Yes	7.17
Hohhot	Inner Mongolia	2008	PG	Yes		0.77
Yinchuan	Ningxia	2009	**PCC**	Yes	Yes	0.48

(continued)

Table 9.2 (continued)

City	Province	Date of implementation[a]	Legislature[b]	Provincial capital	Part of Emory GHI-CTCP[c]	City population (million) (2019)[d]
Hong Kong	Special Administrative Region	1982 (last revision 2017)	n/a	Yes		7.44
Wuhan	Hubei	1995 (revised 1998, 2005)	PG	Yes	Yes	9.79

[a]Date the legislation was implemented not promulgated
[b]People's Congress Council or People's Government. Local laws are enacted by the PCC, while administrative regulations are enacted by PGs (urban governments)
[c]Emory Global Health Institute-China Tobacco Control Partnership cities
[d]World Population Review data for China http://worldpopulationreview.com/countries/china-population/

Fig. 9.4 Cities which have implemented smokefree legislation since 2005. (Credit: Authors)

and Dalian) are currently revising former laws/regulations or drafting new legislation. Most of the cities (22/30) that passed, or are currently drafting, more comprehensive smokefree laws have been capital cities, and almost half (14/30) had been part of the Emory GHI-CTCP. Together these 30 cities have a combined population of 157 million, approximately 10% of China's total population in 2019.

There are also distinct regional clusters of cities that have developed more comprehensive smokefree policies. With the exception of recent developments, particularly in large cities such as Xi'an, Chongqing and Chengdu, most of the early developments were located in more affluent eastern provinces with clusters centred on global cities such as Beijing, Guangzhou and Shanghai. Few cities are located in the west apart from Kelamayi, an affluent oil-rich city with a professional workforce. In Central China there is a small cluster of five provincial capitals that have been recent 'adopters'. Xi'an, an international cultural city and home of the Terracotta Warriors, has received much attention because of its growing prominence in global tourism and its efforts also to promote itself as a healthy city. In the tobacco-growing provinces of the south and southwest, smokefree cities are largely absent.

The urban distribution is interesting since the cities range in size from large urban agglomerations such as Shanghai to some smaller cities such as Zhangjiakou in Hubei Province. Most smokefree initiatives in smaller cities have occurred because of their status as provincial capitals, being part of the Emory GHI-CTCP or being close to other much larger innovators, such as Shaoxing City adjacent to Hangzhou, or Shenzhen City close to Hong Kong. This appears to be the case for other smaller cities, such as Qinhuangdao, adjacent to the Beijing-Tianjin axis. This suggests that policy transfer is important with the adoption of smokefree policies diffusing from larger to smaller places in a hierarchical diffusion process (Yang et al. 2019). Most of the 'smaller' cities (here defined as <2 million) which have adopted smokefree legislation have either been provincial capitals or been part of the Emory GHI-CTCP smokefree initiative, and only three, Zhangjiakou City one of the host cities for the 2022 Winter Olympics, Shaoxing adjacent to Hangzhou and Hohhot in Inner Mongolia, fall outside these categories (Table 9.2). Most of the recent policy enactments have all occurred in larger cities, and less commonly in some tier 2 cities such as Lanzhou in west China's Gansu Province where revision of the city's 2014 legislation is underway (Yuan 2019). However, in the absence of national smokefree policies, it will take some time before smaller cities will become part of the adoption cycle.

While there is some evidence of the beginnings of a diffusion of smokefree policies from provincial capitals to adjacent cities, 12 provincial capitals still lack smokefree legislation, most notably in the remoter provinces of Xinjiang and Tibet in western China, in Shanxi and Liaoning provinces in northern China and in a belt of provinces in central and southern China where the tobacco industry is strong. In Liaoning Province it has been two smaller cities (Anshan, Dalian) that have enacted smokefree legislation rather than the provincial capital, Shenyang, where legislation failed to pass. Similarly in Nanchang, capital of Jiangxi Province, a low awareness level among both the public officials about the harms of SHS and the inadequacy of the public health campaign resulted in a political environment unfavourable to tobacco control (Gan 2018). As is the case nationally, at present no provinces have comprehensive province-wide smokefree legislation. Guangdong Province, however, is drafting a regulation that would require indoor public places and workplaces to be 100% smokefree, with the regulations, as currently proposed, applying only to urban areas (Galea 2018).

In summary smokefree legislation has been restricted to certain types of cities; provincial capitals and other cities in their hinterlands, cities which have hosted international events and cities linked to global tourist flows. However, as was pointed out in the previous chapter, other than the study by Lin et al. (2019), there have been a few systematic attempts to quantify the importance of different city contextual factors in influencing the emergence of smokefree legislation.

9.4 Smokefree Policy Enforcement: Comprehensiveness and Compliance

Smokefree policies have not only varied across different types of cities but so too has their level of enforcement. This section first examines some of the factors which have influenced variations in enforcement and compliance with smokefree legislation. Second, using data from the China City Adult Tobacco Survey (China CDC 2015), we illustrate variations in the strength of smokefree legislation and its relationship to SHS exposure.

9.4.1 Factors Influencing Variations in Enforcement

The presence of smokefree laws and regulations, however, does not guarantee policy enforcement, which varies substantially across cities. For example, evidence from Beijing where a 100% smoking ban, the *Beijing Smoking Control Ordinance*, was passed in 2015, suggests that a particular set of factors was important in the development of the strongest smokefree law in China. Beijing is home to the greatest concentration of public health advocates in China, and, being a global city, many of its residents are familiar with smokefree norms in other countries. There was a high level of political commitment, and the local PCC placed pressure on the Beijing government to ensure compliance with the ordinance. In many other cities, a variety of different agencies have been responsible for enforcement, but this was not the case in Beijing where sole responsibility was assigned to a health inspection team (Gan 2018). However, effective implementation of the ordinance also called for public input, and, to encourage this, the Beijing Tobacco Control Association, with the support of a local technology company, developed the Beijing Tobacco Control Map, a digital system that involves information collection, data visualisation and mapping (Zhang et al. 2019). A key component of this is a Complaint Map, which records reported violations in the city in real time. The map is used by tobacco control staff to address complaints and to ensure compliance. Other cities, such as Lanzhou, are developing similar applications (Yuan 2019). Nevertheless even in

Fig. 9.5 Designated outdoor smoking area on Wangfujing Street, Beijing. (Credit: China Association of Tobacco Control, Domestic News, 12 January, 2019 (Permission obtained))

Beijing outdoor smoking areas, which are still allowed by the 2015 legislation, provided they are not located in areas where people congregate and have proper signage warning people of the health risks of smoking, have still been constructed in the city (Wang and Ning 2019). One such area, located outside the Beijing Department Store in one of the city's iconic shopping areas in 2019 (Fig. 9.5), drew ire from tobacco control advocates for the health risks it posed and was later dismantled since it ran counter to the city's regulations on tobacco control (CATC 2019).

The recent experience of cities like Beijing and Shanghai raises the broader question of what factors have influenced the implementation of urban smokefree policies. Unfortunately research on this question is sparse with most Chinese studies continuing to focus on individual-level factors determining SHS exposure, rather than on broader structural socio-economic and political factors which have affected policy implementation, a focus more typical of western research (Barnett et al. 2017). At the city level, Gan (2018) highlights the importance of the political commitment of local city governments to support the development of more comprehensive legislation as was the case in Beijing where the local legislature, the People's Congress, placed pressure on the Beijing City government to enforce the legislation. Similarly, in Shenzhen the local PCC monitored the performance of city and district enforcement agencies to ensure compliance. But these efforts did not happen by chance. One could argue that international pressure on Beijing to seriously implement the FCTC regulations may have been a factor, while in Shenzhen the close relationship with Hong Kong may also have played a part. But in other first-tier cities, such as Nanchang, which failed to pass smokefree legislation in 2010, the role of the local legislature which vigorously opposed the legislation also deserves attention (China Daily 2010).

There has been little research into the local role of the tobacco industry in influencing the enactment of smoking bans. This is surprising especially given the past influence of the industry on national policymaking via its influence within the STMA. Nevertheless the example of Hangzhou provides some insights into the impact of the industry in limiting the development of comprehensive smokefree policies. In Zhejiang Province, of which Hangzhou is the capital, the Zhejiang Industrial Company paid the highest taxes among all enterprises, far more than the global leading e-commerce giant, Alibaba (Jiang et al. 2019). Because of the power of the CNTC, it was no accident that the draft comprehensive smokefree policy was defeated in 2018. Instead China Tobacco issued a national agreement called *Building Civilised Smoking Environment to Promote Beauty in China.* This meant that provincial tobacco administrations were required to invest in building smoking rooms in a variety of indoor and outdoor public places and to try to work with local governments to implement this national instruction (CATC 2018a). For example, in Hohhot, capital of China's Inner Mongolia Autonomous Region, it was reported that the city aimed to establish more than 100 such rooms in places such as train stations, airports, malls, parks and hospitals. By the time of Lin et al.'s (2019) analysis, smokefree laws in almost two thirds of the 21 cities studied did not include bans of indoor smoking rooms, thus failing to reach compliance with FCTC guidelines which stipulate that only a complete ban on indoor smoking can guarantee protection from SHS. In addition, many rooms also allow cigarette advertising and contain tobacco vending machines. As such, they provide an example of how the CNTC is seeking to maintain consumption and subvert smokefree laws.

Other research has focused on issues related to bureaucratic non-compliance, both at the central and local government levels, when implementing tobacco control policies (Qian 2017). Unfortunately there are few examples of Chinese empirical work on this issue other than the frequent comment that smokefree policies have been ineffective because they have not been enforced. Zhou et al.'s (2016) recent review of factors associated with compliance with smoking bans largely focuses on individual and interpersonal factors, but says little about the key organisational factors which have influenced the implementation and effectiveness of bans. One interesting study is that by Au et al. (2016) in Shantou, a medium-sized city in Guangdong Province. The authors concluded that 'our study clearly indicates that the fundamental causes of the smoking problem are due to the lack of effort from government regulatory agencies (as) most smokers would support a ban on smoking in public places' (Au et al. 2016, p. 52). As part of their research on attitudes towards smoking bans, the authors contacted local government agencies only to be informed that the departments responsible for smoking control either had no operating budgets or no priority assignments. The Shantou case reflects the more general situation where cities without specialised enforcement authorities are unlikely to take controlling SHS exposure seriously if this involves extra responsibilities beyond their normal duties (Gan 2018). Effective bureaucratic enforcement requires a more co-ordinated and committed central and local government approach, involving heavier penalties to ensure compliance, along with media and other educational programmes to inform smokers of the harm they are causing themselves and the people they interact with. Bureaucracies, on their own, cannot accomplish this. There are a variety of theoretical perspectives, such as the Institutional Analysis and Development (IAD) Framework which have the potential to guide the understanding and measurement of smokefree policy implementation (e.g. see Fallin et al. 2014), and these may be helpful for further studies.

At the individual level, it is worth exploring why individuals have not complied with smokefree regulations. Ma et al. (2010), in an early study of residents of six counties in Sichuan Province, found that even when respondents reported that their workplaces had a smokefree policy (and only 28.5% did), 41.1% indicated that they were non-compliant. Similarly the low level of compliance in restaurants will reflect the lack of support for bans from smokers or managers who may have a low level of knowledge of the adverse effects of smoking (Jiang et al. 2010). They may also reflect cultural norms of politeness, such as the reluctance of non-smokers to ask smokers not to smoke in places where smoking is restricted, especially when social status differences occurred (Kaufman et al. 2015). Some studies have indicated that incorporation of the word ‘please’ on no smoking signs has been helpful in getting smokers to obey the law (Yang et al. 2016b).

Low levels of compliance have also been typical of smokefree public places where many people continue to smoke. Those who did so had predictable social characteristics: having lower education and income, being daily smokers, persons who were more highly stressed and who had started smoking at a younger age and residents of ‘smaller’ (<5 million) cities (Yang et al. 2016b). While these are interesting findings, they need to be related to the wider question of the denormalisation of smoking in China. Compliance with smokefree regulations at the individual level will only occur when the unacceptability of smoking becomes the dominant social norm (Leis 2015). Perhaps this is best expressed in the Xiamen Special Economic Zone’s new smoking regulations which came into effect in October 2017. The law, China’s first law on social civilisation, listed ‘smoking in public places’ as one of the nine uncivilised behaviours (Xiamen People’s Government Smoking Regulations, October 1, 2017).

9.4.2 Strength of Smokefree Policies and SHS Exposure

In order to examine the issue of smokefree policy in more detail, data from the China City Adult Tobacco Survey (CCATS) (China CDC 2015) was used to explore links between the strength of smokefree policies and public exposure to SHS in 14 cities (Table 9.3). The survey, conducted in 2013–2014, supported by the Bloomberg and Gates Foundations, was conducted jointly by the Chinese and US Centers for Disease Control and Prevention. One of the study goals was to provide greater understanding of tobacco control measures and to provide rigorous data to evaluate tobacco control efforts and to support the passage of tobacco control regulations. Male smoking prevalence exceeded 40% most cities with only Harbin, Hangzhou, Qingdao and Shenzhen having lower rates. In each city SHS exposure (defined by

% adults aged 15 and above who were exposed to smoke in indoor public places in the last 30 days) was measured for different kinds of public places: workplaces, healthcare facilities, public transport, schools (covering indoor and outdoor spaces in primary, secondary and high schools), universities, restaurants and bars/nightclubs.

Overall, the mean rates of exposure were highest for bars and nightclubs (87.3%), restaurants (67.8%), followed by schools (33.8%) and government buildings (27.2%), with universities (22.4%) healthcare facilities (19.2%) and public transport (9.6%) having the lowest levels of exposure (Table 9.3). Levels of SHS exposure, however, varied markedly across the 14 cities. Mindful that the study was conducted before the implementation of the 2015 regulation, Beijing recorded the lowest rate of exposure for public transport (3.1%), but not for other public places. Kelamayi scored highest for government buildings (9.3%) and healthcare facilities (7.3%), Shenzhen highest for universities (14.5%) and restaurants (42.4%) and Qingdao for schools (21.7%) and bars/nightclubs (69.1%). Cities with no smokefree legislation generally scored worst, particularly Nanchang in Jiangxi Province, for government buildings (44%), healthcare facilities (43.9%), public transport (19.0%), schools (45%) and universities (32.5%) and Shenyang for bars and nightclubs (99.2%). Tianjin, with only partial smokefree protection, at the time for this indoor public place, surprisingly scored worst for restaurants (80.8%).

The relationship between the strength of smokefree policies and SHS exposure outcomes was examined by classifying cities according to the strength of their smokefree policies. For each city, public places where smoking was prohibited were allocated a score of two, where smoking was partially prohibited a score of one, and zero where there was no smokefree policy. Generally cities with stronger regulations recorded less SHS exposure with significant negative correlations occurring for transport (−0.73, $p < 0.01$), schools (−0.62, $p < 0.01$), universities (−0.52, $p < 0.05$), healthcare facilities (−0.51, $p < 0.05$) and bars (−0.51, $p < 0.05$). Correlations for government offices and restaurants, although negative, were not significant.

9.5 Health Outcomes

Due to the rapidly changing policy situation in China, an increasing amount of research is being undertaken on the outcomes of policies designed to reduce exposure to SHS. Policy outcomes can be viewed in terms of the direct effects of smokefree legislation on changes in the levels of SHS exposure, whether smoking bans have affected levels of smoking cessation and their implications for health inequalities. In Sect. 9.6 we also consider social outcomes paying particular attention to the impacts of smoking bans in promoting denormalisation and the extent to which wider smokefree policies have impacted smoking at home.

Table 9.3 Percentage of adults exposed to SHS in indoor public places in the last 30 days

	Government buildings	Healthcare facilities	Public transport	Schools	Universities	Restaurants	Bars/ nightclubs	Male smoking prevalence (%)	Smokefree regulation (2013–2014)
Anshan	21.7	20.0	6.3	41.7	19.6	74.0	95.7	44	PG
Beijing	19.9	13.4	3.1	33.4	26.0	65.3	89.1	41.5	**PCC**
Changchun	32.1	22.5	11.5	33.3	17.9	72.7	78.9	43.2	PG
Harbin	38.2	28.9	10.1	30.4	24.5	65.0	74.7	35.1	**PCC**
Hangzhou	20.6	16.3	5.5	30.7	20.5	58.1	92.5	38.6	**PCC**
Kelamayi	9.3	7.3	6.2	23.3	23.5	53.6	90.2	41.6	PG
Lanzhou	40.0	10.3	8.1	39.4	24.9	69.9	89.1	41.2	**PCC**
Luoyang	31.6	15.8	11.1	38.7	23.9	78.2	90.2	44.1	None
Nanchang	44.0	43.9	19	45	32.5	79.9	95.3	42.2	None
Qingdao	14.7	14.1	11.9	21.7	26.6	59.0	69.1	32.7	**PCC**
Shenyang	29.6	23.9	18.1	43.9	27.2	78.1	99.2	44.8	None
Shenzhen	23.6	18.3	7.3	27.9	14.5	42.4	76.9	33.9	**PCC**
Tangshan	21.2	16.1	6.7	23.9	13.8	71.8	86.0	40.6	PG
Tianjin	34.6	18.4	9.4	40.3	18.0	80.8	95.5	41.1	**PCC**
Mean	*27.2*	*19.2*	*9.6*	*33.8*	*22.4*	*67.8*	*87.3*	*40.3*	

Source: Table derived from data in Liang X (ed), Report of China City Adult Tobacco Survey 2013–14. 2015 CDC Foundation, Atlanta, Georgia, USA, Appendices, pp. 44 and 51

9.5.1 *Smokefree Policies and Changes in Exposure*

Change in the level of SHS exposure has been the main goal of smokefree policies. While there has been an increasing tendency for cities to adopt smokefree policies, their effectiveness will depend upon the degree to which policies are comprehensive, in terms of both the spaces they cover and their degree of enforcement. To date, with few exceptions, compliance and enforcement have been poor, and consequently only modest improvements have occurred in terms of air quality and lowering SHS exposure. Most early studies showed that partial bans had little impact on SHS exposure in particular cities (Fong et al. 2015). The following studies are illustrative of the ineffectiveness of earlier partial bans in reducing SHS exposure in selected cities:

Beijing (Liu et al. 2014b, c)

Data: Field observations of patron smoking behaviour and measurements of $PM_{2.5}$ fine particulate matter in 79 restaurants and bars in 2010 before and after the introduction of the 2008 smokefree regulations.

The 2008 governmental smoking restrictions, requiring large restaurants to designate no less than 50% of their dining area as non-smoking (without defining 'large' or specifying how the designated smoking sections and non-smoking sections should be separated), were poorly implemented and SHS exposure, as defined by $PM_{2.5}$ levels, in restaurants remained high. In 2008 and 2010 smoking was observed in 40% of the nominal non-smoking venues/sections. Cui et al. (2009), using household survey data, reported that only minimal changes (only −1.5%) in smoking prevalence occurred after the 2008 Government Order especially for those with less education. Beijing's legislation was revised in 2015.

Hangzhou and Jiaxing (Yang et al. 2013)

Data: Cross-sectional surveys in two adjacent cities: Hangzhou (with a smokefree law) and Jiaxing (no law).

Household surveys in 2009 and 2011 showed that the 2010 regulations implemented in Hangzhou resulted in a significant decline in SHS exposure in government buildings, buses/taxis and restaurants, but no change in healthcare facilities and schools. Overall the decline in SHS exposure in Hangzhou was modest (−4.9%; 34.7–29.8) and similar to Jiaxing (−1.9%; 31.6–29.7). The Hangzhou SFR at the time lacked specific enforcement guidelines or strong penalties. Hangzhou's legislation was revised in 2019.

Guangzhou (Ye et al. 2014)

Data: Household survey of self-reported SHS exposure 1 year before and after the 2010 legislation.

SHS exposure dropped significantly in places with full smoking bans (from 58.0 to 50.3%) but hardly changed in places with partial bans (89.5–87.4%), where there were designated smoking rooms. Partial bans were ineffective in reducing SHS. Guangzhou's legislation was revised in 2012.

Qingdao (Zhao et al. 2018)

Data: Population-based surveys before and after the 2008 Summer Olympics when a temporary smoking ban was implemented.

From 2006 to 2009, among non-smokers SHS exposure showed a modest decline of −5.4% from 62.2% to 56.8% with the decline being most evident among female non-smokers (−17%). Qingdao formally adopted a smokefree policy in 2013.

By contrast, taking Beijing and Shanghai as examples, decline in SHS exposure has been more evident since the introduction of more comprehensive bans.

Beijing (Xiao et al. 2017)

Data: Participant observation at 176 restaurants 5 months before and 1 month after the new 2015 Smoking Control Regulation took effect.

The occurrence of smoking dropped significantly from 40.3% to 14.8%. This was most evident in open dining areas (32.1%–5.1%) but still remained relatively high in men's restrooms (23.8–18.8%). Although the number of no smoking signs increased after the law came into effect, very few (38.5%) of these included the symptom hotline number, and only 5.6% mentioned the penalty for violations. No intervention by restaurant staff was observed when smoking occurred.

Shanghai (Wei et al. 2019)

Data: Surveys were undertaken in urban and suburban districts 3 months before and 12 months after the implementation of the new smokefree law in March 2017.

Measurements of evidence of smoking (via participant observation), density of lit cigarettes and air quality ($PM_{2.5}$ levels) were undertaken in eight types of venues (office buildings, hotels, restaurants, bars, karaoke venues, gaming rooms, farmer's markets, buses and railway stations). All decreased after the legislation (e.g. any evidence of smoking dropped from 66.2% to 49.7% 1 year after the law, but improvements were less in bars (83.3–63.2%), office buildings (90.0–70.0%) and farmer's markets (94.4–88.9%) where prevalence remained high. Reduction in smoking was affected by management performance in ensuring compliance.

The experience of recent trends in these cities is promising. The implementation of stricter smokefree laws has resulted in significant reductions in SES exposure (China CDC 2019) and parallel improvements in air quality. Despite this, management training to ensure that smokers comply with the legislation is still an issue, as is the displacement of smokers to other spaces, such as doorways, where the legislation does not apply. This has been a familiar theme in western countries where indoor smoking restrictions have raised concerns about increased exposure to secondhand smoke outdoors and how concentrations of outdoor smokers, especially around bars, may affect the patrons of such establishments as well as their indoor air quality (Brennan et al. 2010; Sureda et al. 2012, 2013). Exposure in indoor public spaces has not just been a concern of entertainment venues but has also affected lobbies and elevators in residential communities. In Shanghai the property management company which runs the Huijingyuan residential community was fined in 2018 after complaints from residents of the complex and ordered to fix the problem (CATC 2018b).

9.5.2 Smoking Bans and Smoking Cessation

While reducing SHS exposure is an important objective of smoking bans, so too is the aim of providing a supportive environment for people who wish to quit or not start smoking. By raising general awareness of the dangers of SHS, smokefree policies help create a supportive environment for health professionals or family members wishing to encourage smoking cessation. Yang et al. (2013), for instance, showed that the 2010 smokefree legislation in Hangzhou was associated with the giving of quitting advice to proximal family members.

Numerous studies of the effects of smokefree legislation in individual cities, such as Shanghai, Guangzhou and Qingdao, have reported that smokefree laws may substantially reduce smoking prevalence (Yang et al. 2012) and increase smoking cessation (Fu et al. 2017; Qian et al. 2018; Ye et al. 2015; Zhao et al. 2018; Zhang et al. 2018). Yang et al. (2011) in a six-city study showed that smoking cessation was associated with both public place and workplace smoking restrictions especially when measured at the city level (ORs = 1.82 and 2.33). The ITC study of seven cities in China showed that cities with stronger smokefree legislation were more likely to record larger decreases in smoking between 2007 and 2012. Similarly, in a study of 17 cities in Anhui Province, Luo et al. (2015) showed that the presence of smoking bans in public places increased the intention to quit (OR = 1.31) despite the fact that smokefree policies were largely partial and poorly enforced. However, Wan et al. (2013) in a seven-city study showed that their Strength of Tobacco Control (SOTC) index, while associated with a variety of mid-term outcomes, such as the presence of smokefree signage, was unrelated to long-tern indicators such as male smoking prevalence and freedom from SHS.

It is difficult to assess the independent effect of smoking bans on smoking prevalence because of the variety of individual and environmental factors which affect levels of consumption and cessation rates on the part of different groups. For instance, Zhao et al. (2018) showed that in Qingdao, the city's smoking ban affected light or occasional smokers the most and not heavy smokers who were least likely to quit. With few exceptions (e.g. Wei et al. 2019), there has been little research on patterns of displacement to other city spaces, such as parks or smoking rooms, in the face of environmental smoking restrictions. As indicated elsewhere (Barnett et al. 2017), there is a need to understand how smoking bans are producing new social geographies of smoking and the impact of such changes on different groups, and their smoking norms.

9.5.3 Smoking Bans and Health Inequalities

Many of the studies which have attempted to assess the effects of smoking bans on smoking prevalence were conducted prior to the implementation of more comprehensive smoking bans and often did not examine the effects of legislation on different population groups. This is an important issue because smokefree policies have the potential to widen health inequalities. For instance, Qian et al. (2018) found that the 2010 smoking ban in Shanghai had little effect on younger (15–29) people whose smoking rates, in contrast to some older population cohorts, remained relatively unchanged between 2010 and 2015. A concern with increased inequality in the burden of smoking is important, particularly given that tobacco use is a strong determinant of overall levels of health and patterns of health inequality. Most attention has focused on SES differences in health outcomes, but stage in the lifecycle is also important as early social inequalities are likely to be accentuated in later years. There is some indication that SES differences in smoking status may become more marked following smokefree legislation, but to date results are equivocal. In China Luo et al. (2015), in their 17-city study, showed that smoking bans had a greater impact on intention to quit among more highly educated residents and those who perceived that smoking bans had had a positive effect on their lives. By contrast, in the USA, Hernandez et al. (2019) showed that smoking bans were most effective at changing behaviours among the most educationally disadvantaged whether the latter was measured for individuals or their parents.

Support for comprehensive smokefree policies has been shown to be higher among people who have a better knowledge of the health harms of SHS and among people who work in smokefree worksites or who have frequented smokefree places of entertainment (Li et al. 2010). Since people of lower socio-economic status tend to have poorer knowledge of the harmful effects of tobacco (Jin et al. 2014), then the influence of smoking bans on prevalence is likely to be less for this group, thus increasing health inequalities in smoking cessation.

9.6 Social Outcomes

The social outcomes of smoking bans are also important. In this last section, we focus on two themes in particular: smoking bans and denormalisation and the impact of smokefree policies on smoking in the home. Important here are the extent of home smoking restrictions (HSRs), their association with different individual and household characteristics and the extent to which public smoking bans have had positive or negative impacts on SHS exposure.

9.6.1 Smokefree Policies and Denormalisation

In western countries smokefree policies, apart from reducing exposure to SHS and encouraging smoking cessation, have also aimed to denormalise smoking by restricting the spaces in which it can be practised. When comprehensive in scope, and well-enforced, smoking bans send clear messages about the need to separate less socially acceptable forms of behaviour. However, while such policies have often resulted in the almost complete abolition of SHS in public venues in western countries, they have been criticised because they have contributed to the stigmatisation of smokers and exacerbating social inequalities in health by disproportionately affecting disadvantaged groups and marginalising them even further (Frohlich et al. 2012). However, while there has been debate over the extent to which stigmatisation has occurred (e.g. Nagelhout et al. 2012), this is less true with respect to the impacts of smoking bans in encouraging denormalisation. The research findings suggest that smokefree policies, especially when used in conjunction with other tobacco control efforts, have played an important role in affecting community norms about the social acceptability of smoking.

In China the fragmented nature of smokefree legislation and the absence of strong enforcement suggest that recent policy changes have been too restricted to encourage denormalisation, at least on a national or provincial scale. However, social processes of denormalisation may well be underway in those cities which have enacted stronger smokefree laws, and it may just be a matter of time before officials and smokers become more accepting of the need to comply with smokefree laws.

9.6.2 Smokefree Policies and Smoking in the Home

Smokefree policies are especially important in the home as they decrease exposure to SHS for vulnerable groups such as pregnant women and children. Besides the obvious benefits of protection, smokefree homes may also install and reinforce antismoking norms among youth, and a number of studies have shown that living in a smokefree home reduces levels of adolescent smoking (Mills et al. 2009; Emory et al. 2010). Thus the absence of smoking in private domestic spaces is a good marker of denormalisation particularly when it has been triggered by the implementation of wider public smokefree policies.

It has been sometimes been argued that smoking bans have resulted in a displacement of smoking from the public sphere into the home. However, in western countries there is little evidence of this. Monson and Arsenault (2017), in a review of mainly western studies, dispel the notion of public bans displacing smoking into the home. They suggest that through their influence on social norms, legislative bans on smoking in public places are likely to encourage citizens to establish home smoking restrictions (HSRs) and to enforce them in an anti-smoking community environment (Mons et al. 2013). Such research findings may not be relevant to China, where weaker enforcement of public smokefree policies and a male-dominated smoking culture are more the norm. Such differences are also likely to be more evident in rural areas where the autonomy of women is weakest. It is also important to take account of gender inequalities within the domestic sphere which may limit the ability, especially of lower-income women or women living in extended family arrangements, to manage SHS exposure both for themselves and their families (Yao, Lee & Mao, 2009; Mao 2013).

In order to examine links between smokefree policies in public spaces and home smoking restrictions, we examine three issues: (1) the extent of home smoking restrictions; (2) what individual and household characteristics have been associated with the presence of HSRs; and (3) what has been the impact of public smokefree legislation on smoking at home.

9.6.2.1 Extent of Home Smoking Restrictions

The presence of home smoking bans is not widely documented, but a limited number of mainly urban studies indicate that HSRs ranged from a low of 6.3% in counties in Sichuan, Jiangxi and Henan Provinces in 2004 (Wang et al. 2009) and 14.3% in Nanning City in Guangxi Province in 2013–2014 (Huang et al. 2016) to a high of 35.3% in Shanghai in 2012–2013 (Zheng, Kegler et al. 2014). In fact the Shanghai figure was 71.6% when households with partial bans were included. Yang et al. (2015c) report a rate of 22% in 2011 for 21 cities, but unfortunately individual city variations are not documented. An earlier study conducted in Shanghai in 2006 (Ji et al. 2009) reported 26% respondents having complete home smoking bans, thus suggesting that rates have increased in that city. Xu et al. (2016) provide more detailed estimates for five cities in Zhejiang Province in 2012 where 25% respondents indicated that no one smoked in their home, but for the 75% who had no HSR, 34% experienced smoking on a daily basis. One of the problems of using partial bans as a measure is that, as Wang et al. (2009) found, there may be little difference in SHS exposure between households with partial bans and those who allow smoking in all indoor places. Even households who report having 'complete' smokefree home policies may still be exposed to SHS at home particularly when they give priority to the demands of older relatives and visitors (Zheng et al. 2014a).

9.6.2.2 Individual and Household Characteristics Associated with HSRs

What types of households have been more likely to implement home smoking bans? Most of the studies undertaken, not surprisingly, all had an urban focus where, as in the case of public smokefree areas, most private smokefree households are located. Typical of such studies is that by Zheng and colleagues who, in 2012–2013, investigated correlates of smokefree home policies in Shanghai (Zheng et al. 2014a, b). Households with complete bans tended to be occupied by females rather than males. They also tended to be higher income (Wei et al. 2014), having no other smokers present in the home, having children present and having fewer friends/relatives who permit home smoking. Concern about the health impacts of SHS was the most important reason for having a smokefree home (Berg et al. 2015). In this regard Mao and Robinson (2015), for example, in a study of home smoking management in rural China, showed that women used healthy narratives, supported by the (former) one-child policy, about the importance of raising a healthy child. Women were able to use their elevated status as bearers and carers of only children to subvert the pre-eminence of men in domestic environments.

Huang et al. (2016) reported similar findings in a 2013–2014 study which compared parents of hospitalised children in Nanning City who had smoking bans to those who had none. Home smoking bans were associated with feelings of confidence in being able to prevent others from smoking in the home. However, those who perceived difficulty in adopting a no smoking ban would not do so. This was more likely among women of lower socio-economic status who are less independent of their husbands and extended families. Huang et al. (2016) noted the influence of Chinese cultural norms where Chinese women are often treated as secondary citizens based on stereotypical roles in the household as mother, daughter-in-law, wife or daughter, rather than on their needs and preferences as distinct individuals. Such norms often dictate that women have less control, especially in rural areas, over setting this type of policy in the home (Mao 2013).

In western countries it has been noted that household smoking bans are less common among lower-income households because of their higher smoking prevalence but also because, in contrast to more affluent and better educated groups, they are less aware of the health hazards of SHS exposure (Wang et al. 2009). In China, however, the effect of income on the presence of HSRs appears to be inconsistent. In contrast to the findings of Zheng and Kegler et al. (2014), Yang et al. (2015b) found that HSRs were *less* common in higher-income households and no association existed between education and the presence of HSRs. The authors postulated that the negative association between the presence of HSRs and household income was due to the fact that high-income people engage in more social activities in the home, thus inhibiting the development of HSRs. It is also possible that the result reflects the way HSR was measured. In the Yang study, the responses were recorded as a simple dichotomy between households who had 1 = none or partial restrictions and 2 = complete restrictions, but the relative proportions of households in each group were not disclosed other than 22% households with complete bans. It may be that many higher-income households have partial rather than complete bans, thus affecting the result. The analysis also reported that HSRs were more common among males than females which is not consistent with the findings of most Chinese research.

Few studies have specifically examined the effect of community characteristics on the presence of home smoking restrictions. Not only will these vary between cities, but also within cities distinct socio-economic variations are likely to occur particularly where there are more women but also in more developed cities with higher levels of female participation (Yang et al. 2016a). It is in these places where distinct community anti-smoking norms will begin to develop and affect the level of smoke-free households. One of the few studies to examine this, was conducted by Ji et al. (2009) in Changqiao, an upmarket residential area in southwest Shanghai. The authors asked two questions to assess community influence on household smoking bans: 'How many families in this community are smokefree – none, few, some, most or all?' 'In this community, how likely is someone to be reprimanded for smoking at home – very likely, somewhat likely or not likely at all?'. The first question was designed to ascertain the perceived density of smokefree homes and the second to measure the perceived likelihood of being reprimanded by the community for smoking at home. The authors found that home smoking bans were positively associated with the perceived density of smokefree homes in the community and with the perceived likelihood of community reprimand for smoking in the home. Unfortunately there was no data on the socio-economic makeup of the community and how this may have affected the presence and transmission of anti-smoking norms (Christakis and Fowler 2008). At the international level, Nazar et al. (2014) found a strong link between the presence of smokefree workplaces and the proportion of respondents who lived in smokefree homes. However, of the 16 countries examined, China ranked last in terms of the proportion of respondents in smokefree workplaces (21%) who reported that they lived in smokefree homes.

There have been a few studies on intra-urban variations in the presence of home smoking bans. One of the few is by Zeng et al. (2014a) who examined smokefree homes in two areas of Shanghai. Residents who lived in the higher-density inner city area of Pudong were more likely to have complete bans than their suburban counterparts residing in Fengxian in south Shanghai (39.6% vs 30.8%). Unfortunately the wider community differences were not explored, but were likely to have reflected variations in socio-economic status of the two areas or differences in the nature of the housing stock which may have affected the accessibility of smokers to outdoor spaces where they could smoke. Only recently has one study, of six cities, confirmed that contextual social deprivation results in higher levels of SHS exposure of urban males independent of individual socio-economic factors (Yang et al. 2021). While such research is in its infancy, such results suggest that community social norms deserve greater attention in terms of the role they play in influencing individuals' tolerance of SHS in different types of urban communities.

9.6.2.3 Impact of Smokefree Legislation on Smoking in the Home

An important question is the impact of public smoking bans on smoking in the home. It has been argued that when smokefree legislation is strictly enforced, then increased home smoking is more likely than where enforcement is weak. However, the record in most western countries suggests that there has been little impact and that smoking bans appear to have increased the adoption of smokefree homes, particularly where children are involved (Mons et al. 2013).

In China there have been a few studies of links between public place or workplace smokefree policies and smoking in the home. The evidence is mixed with some studies suggesting no impact. Ye et al. (2015) concluded that the implementation of smokefree legislation in Guangzhou in 2010 did not lead to more smoking in homes which remained stable (91.0% vs 89.4%) as did levels of SHS exposure (39.6% in 2009 vs 40.0% in 2011). Similarly Yang et al. (2015b) found that the presence of workplace restrictions resulted in *more* home smoking restrictions on smoking and hence less smoking in the home. HSRs were more common when there were regional workplace restrictions, but there was no relationship with smoking restrictions in public places. Nazar et al. (2014) for 15 LMICs, including China, showed that participants who were employed in a smokefree workplace, compared to a workplace where smoking was allowed, were more likely to live in a smokefree home. The relationship was significant for 13/15 countries and strongest for China (OR = 2.29). More recently evidence for India (Lee et al. 2014), Nigeria (Kaleta et al. 2015) and Malaysia (Lim et al. 2019) confirm the broad conclusions of the Nazar study.

On the other hand, other studies suggest that public smokefree policies have had a negative impact on the home. Ho et al. (2010), using survey data before and after the 2007 smokefree legislation in Hong Kong, showed that the prevalence of SHS exposure increased both at home (10.2– 14.1%) and outside the home (19.8– 27.2%) following the legislation. Similarly Yao et al. (2009) showed that the adoption of smokefree regulations in Chengdu increased smoking in the home. In a qualitative study of pregnant women in 2006, it was found that 69.1% participants reported that their husbands had shifted their smoking location from public places and workplaces to the home following the adoption of public place and workplace restrictions. Some reports from focus group discussions included (Yao et al. 2009, p.S162):

> *He had received a warning by the security guard when he was smoking in the shopping mall. No he prefers to smoke at home. He said "I feel safer" because there is no need to worry about penalties.*
>
> *Usually, he (husband) smoked on the bus or taxi to home, but since he can't smoke in the worksite or on public transportation anymore, our home became the ideal smoking place for him as there is no restriction on smoking at all.*
>
> *It seems that he was anxious about something and he began to smoke immediately as soon as he returned home because he had refrained from smoking during his work for the whole day. To avoid the constraints of the smokefree policy in public places he would rather stay at home on weekends or holidays.*

Another study reported that although smokefree legislation appeared to have affected the home, the effects were context specific. Zhao et al. (2018) conducted surveys before (2006) and after (2009) the 2008 Summer Olympics in Qingdao. SHS exposure at home increased for urban residents but decreased for rural residents. The authors speculated that this most likely reflected the fact that most facilities affected by the temporary smokefree policy were in urban districts. Following the implementation of the smoking ban, which resulted in reduced consumption especially among lighter smokers, in urban areas heavier smokers smoked more at home than before, whereas in rural districts with alternative smoking spaces, such displacement was less likely.

It is difficult to reconcile these conflicting findings. One problem with many of the early studies relates to the quality of the data. They deal with early smokefree regulations which were often not enforced. In Guangzhou, for instance, there was only a partial ban, which allowed for smoking rooms. This applied to workplaces and places of entertainment, and no one was fined for policy infringements, so the issue of displacement of smoking from public to private spaces was unlikely. Ye et al. (2015) report that no fines were imposed on patrons and managers until 9 months after the legislation and high rates of smoking continued to occur in public places. Yang et al.'s (2015b) study showed that only workplace smoking restrictions had an effect on HSR and not public place restrictions. However, smokers who indicated that they continued to smoke at smokefree venues reported that they were more than twice as likely to face home restrictions. This perhaps suggests reverse causation of home HSRs displacing smokers into the community, a likely outcome when smokefree legislation is poorly enforced. Similarly the problem with the Nazar et al. (2014) study is that the authors provide no rationale as to why China should exhibit the strongest relationship between workplace smoking bans and HSRs and why there should be a positive relationship in LMICs. Although the relative odds ratio was strongest for China, the proportion of workplaces which were smokefree was the smallest of the 15 countries studied. Clearly more research is needed to uncover the implications of public policies for private spaces, the extent to which these are dependent on policy enforcement and how public policies change social norms to the extent that home restrictions are a likely outcome.

9.7 Conclusion

This chapter has examined the development of smokefree initiatives in China since it became a signatory of the FCTC in 2005. In reviewing developments since that time, a number of key conclusions can be drawn not only about the nature of policy development but also about current research on smokefree initiatives and policy implementation.

First it is clear that, although there have been marked improvements in recent years, SHS exposure in China remains high particularly in certain kinds of urban spaces where policy implementation remains weak. From a policy perspective, China lags behind other large LMIC countries, such as Brazil and India, and has yet to develop a comprehensive national smokefree policy. This is in spite of various surveys which have found high support for smokefree policies, even among smokers (Li et al. 2010). In 2015 the WHO report, *Smokefree Policies in China*, noted in comparison with the 15 other countries included in the ITC Project (2007–2012), China had the highest level of smoking in workplaces, the second highest level of smoking in restaurants and bars and the lowest percentage of smokers with smokefree homes (WHO 2015). Although there have been improvements since that date, particularly in public transport, government buildings and healthcare facilities, exposure remains high (China CDC 2019; Jiang 2020).

Second, the development of smokefree policies in China has been largely a local urban phenomenon rather than having a regional or national focus. The 'bottom-up' development of policy has numerous origins but is partly due to the impact of globalisation and the development of anti-smoking norms in particular Chinese cities, usually provincial capitals. The impact of global events such as the Olympics, or the effect of philanthropic activities, such as the city initiatives stimulated by the Bloomberg and Gates Foundations, cannot be overstated. But there are other global connections too such as the impact of Hong Kong, long a smokefree island in a 'sea of smoke', on its immediate hinterland or global tourist flows to places like Beijing or Xi'an which highlight the importance of the global-local nexus. In part they reflect local public health initiatives, but they also represent China putting its best face to the world. Initially the patchwork of local smokefree policies was relatively ineffective and poorly enforced with partial smoking bans having little impact in limiting SHS exposure. Recent developments of more comprehensive smoking bans in some major cities, most notably Beijing and Shanghai, suggest that the tide may be changing. It will be interesting to see the extent to which other provincial capitals begin to adopt smokefree policies and whether these begin to diffuse down the urban hierarchy to smaller cities and towns. At the provincial level, despite some advocacy activities in Shandong Province (WHO 2018/2019), it seems unlikely that smokefree provinces will become a reality.

Third, recent evidence suggests that, given China's pro-smoking culture, issues of compliance will still arise, even with comprehensive bans, particularly in places of entertainment that traditionally cater for male smokers. This will require management training and greater enforcement from local agencies and co-ordination between them. The development of greater citizen input will be necessary to promote smokefree norms. In western countries the development of smoking bans accelerated the process of denormalisation of smoking by restricting its geography in terms of where it could be practised. This is most likely to occur when smokefree legislation is enforced and anti-smoking norms begin to replace pro-smoking ones. This has already begun to occur in China with bans on outdoor advertising, but a range of strategies, including effective smokefree legislation, is necessary.

Fourth, smokefree policies in the home are largely an urban phenomenon but, if the experience of Europe is any guide (Ferketich et al. 2016), are likely to become more widespread as China adopts stronger national tobacco control measures. Despite increased interest in smoking in private spaces, little is known about the behaviour of Chinese smokers when they are at home, particularly whether they smoke both inside and outside their places of residence. Only one study has specifically examined socio-economic variations in home smoking and how smoking practices vary as a result. Also, with few exceptions, no studies have considered the impact of how Chinese urban development is affecting home smoking practices. Given the recent relaxation of the one-child policy and the change in size of some households, a greater understanding of home smoking practices and how they vary with both socio-economic status and home location will be a high priority.

Finally, in terms of research developments, the problem of SHS exposure in China has been well documented, but this is much less true of the policy process. More attention needs be devoted to the geography of smokefree implementation in China and to the particular sets of factors that have been important in leading in policy innovation in different cities. Apart from the ITC surveys, there has been little effort to understand why policy outcomes have varied between different places and how the political and socio-economic environment has been important.

At the international level, recent research has begun to examine the determinants governing countries' enactment of smoking bans in public places and to examine the factors determining their ability to successfully implement these. Perkins and

Neumayer (2014), for instance, found that national income per capita was an important determinant of adoption and compliance, but that the latter was also affected by tobacco industry reliance and levels of health expenditure. Peruga et al. (2018) also examined compliance and found two factors to be important: the involvement of local jurisdictions in providing training and/or guidance regarding inspections of violations and the extent to which countries acted strongly in containing political corruption.

The factors listed above are relevant to exploring the issue of compliance in China and, given the geographically uneven existence of smoking bans, provide a fruitful source of new ideas to explore in more nuanced analyses at both regional and local levels. As Mlinarić et al. (2019) have noted with respect to the USA, where the adoption and implementation of smokefree policies has been very much a local process, it is important to develop an evidence-based theoretical framework to explain why smokefree interventions have worked well in some contexts but poorly or not at all in others. They contend that establishing trust between tobacco control advocates and local authorities, increasing priorities for tobacco control and limiting the power of opposing interests are three key factors in the policy process. A logical extension, therefore, is to apply, and refine, such ideas in non-western contexts. Too many studies in China have uncritically adopted large multi-city samples, yet have not explored the key factors underpinning local variations in policy outcomes. China is a diverse country and place matters, so more subtle and insightful studies are needed to understand why, in a country not noted for pluralistic political processes, this is the case. Furthermore most studies of smokefree policy have largely had a quantitative focus, and few have provided more subtle qualitative insights into the significance of different processes.

References

Abdullah AS, Hua F, Xia X, Hurlburt S, Ng P, MacLeod W, Siegel M, Griffiths S, Zhang Z (2012) Second-hand smoke exposure and household smoking bans in Chinese families: a qualitative study. Health Soc Care Community 20(4):356–364

Astell-Burt T, Zhang M, Feng X, Wang L, Li Y, Page A, Zhou M, Wang L (2018) Geographical inequality in tobacco control in China: multilevel evidence from 98,058 participants. Nicotine Tob Res 20(6):755–765

Au WW, Ma WQW, Zhu Q, Tang L (2016) Problems with cigarette smoking and attitudes towards the ban of smoking in Shantou, China. Public Health 134(May):46–53

Barnett R, Moon G, Pearce J, Thompson L, Twigg L (2017) Smoking geographies space, place and tobacco. Wiley Blackwell, Oxford, 301p

Berg CJ, Zheng P, Kegler MC (2015) Perceived benefits of smoke-free homes, the process of establishing them, and enforcement challenges in Shanghai, China: a qualitative study. BMC Public Health 15(1):89

Brennan E, Cameron M, Warne C, Durkin S, Borland R, Travers MJ, Hyland A, Wakefield MA (2010) Secondhand smoke drift: examining the influence of indoor smoking bans on indoor and outdoor air quality at pubs and bars. Nicotine Tob Res 12(3):271–277

Cai L, Cui W, He J, Wu X (2014) The economic burden of smoking and secondhand smoke exposure in rural South-West China. J Asthma 51(5):515–521

Cai L, Wu X, Goyal A, Han Y, Cui W, He J, Xia X, Zhao K, Jiao F, Song Y (2013) Multilevel analysis of the determinants of smoking and second-hand smoke exposure in a tobacco-cultivating rural area of southwest China. Tob Control 22:ii16–ii20

Chen Y, Chen H, Li Z (2018) Effect of characteristics and life in cities on residents' smoking behaviour. J Int Med Res 46(10):4226–4234

CATC (China Association on Tobacco Control) (2018a) Designated smoke areas harm efforts at reduction. CATC domestic news 9 December. CATC, Beijing. http://www.catcprc.org.cn/index.aspx?menuid=25&type=articleinfo&lanmuid=186&infoid=10898&language=en

CATC (China Association on Tobacco Control) (2018b) City smokers still flouting ban in public indoor spaces. CATC domestic news 3 august. CATC, Beijing. http://www.catcprc.org.cn/index.aspx?menuid=25&type=articleinfo&lanmuid=186&infoid=9992&language=en

CATC (China Association on Tobacco Control) (2019) Fancy giant ashtrays 'stubbed out' in Beijing's 'Fifth Avenue'. CATC Domestic News 12 January. CATC, Beijing. http://www.catcprc.org.cn/index.aspx?menuid=25&type=articleinfo&lanmuid=186&infoid=11012&language=en

China CDC (Center for Disease Control and Prevention) (2015) Report of the China City adult tobacco survey 2013–14. China CDC, Beijing

China CDC (Center for Disease Control and Prevention) (2019) GATS global adult tobacco survey, fact sheet, China 2018. China CDC, Beijing

China Daily (2010) Controversy clouds debate on smoking law. China Daily 23 December, p 5. http://www.chinadaily.com.cn/cndy/2010-12/23/content_11742353.htm

Cho YN (2006) The politics of lawmaking in Chinese local People's congresses. China Q 187(September):596–609

Christakis NA, Fowler NH (2008) The collective dynamics of smoking in a large social network. N Engl J Med 358(21):2249–2258

Cui X, Cui X, Li C (2009) Analysis of the changes of smoking status of residents before and after the implementation of several provisions on the scope of smoking in public places in China. J Cardiopulm Vasc Dis 28(3):198–202

Emory K, Saquib EA, Gilpin EA, Pierce JP (2010) The association between home smoking restrictions and youth smoking behaviour: a review. Tob Control 19(6):495–506

Emory University (2013) Building partnerships for a healthier tomorrow. Emory University, Atlanta, Georgia, Emory Global Health Institute

Emory University (2018) Expanding partnerships for a healthier tomorrow. Emory University, Atlanta, Georgia, Emory Global Health Institute

Fallin A, Goodin A, Rayens MK, Morris S, Hahn EJ (2014) Smoke-free public policy implementation: theoretical and policy implications. Policy Polit Nurs Pract 15(3–4):81–92

Ferketich AK, Lugo A, La Vecchia C, Fernandez E, Boffetta P, Clancy L, Gallus S (2016) Relation between national-level tobacco control policies and individual-level voluntary home smoking bans in Europe. Tob Control 25(1):60–65

Fong GT, Sansone G, Yan M, Craig L, Quah ACK, Jiang Y (2015) Evaluation of smoke-free policies in seven cities in China: longitudinal findings from the ITC China Project (2007-2012). Tob Control 24(Supp 4):iv14–iv20

Frohlich KL, Mykhalovskiy E, Poland BD, Haines-Saar R, Johnson J (2012) Creating the socially marginalised youth smoker: the role of tobacco control. Sociol Health Illn 34(7):978–993

Fu H, Feng D, Tang S, He Z, Xiang Y, Wu T, Wang R, Shao T, Liu C, Shao P, Feng Z (2017) Prevalence of tobacco smoking and determinants of success in quitting smoking among patients with chronic diseases: a cross-sectional study in rural Western China. Int J Environ Res Public Health 14(2):167

Galea G (2018) Law must ban smoking in public places without exemptions. World Health Organisation Representative Office for China, November, 2018. https://www.who.int/china/news/commentaries/detail/law-must-ban-smoking-in-public-places-without-exemptions

Gan Q (2018) Protecting China from secondhand smoke. In: Yang G (ed) Tobacco control in China. Springer, Singapore, pp 167–182

Gao C, Yao Y, Niu H, Li L, Li M, Qu Y, Wang R, Zhang P, Li M, Yan S, Li B (2019) Urban-rural differences in related factors of second-hand smoke exposure: a cross-sectional study of adult non-smokers in Northeast China. J Public Health 41(2):321–328

Guo J, Zhu Y, Fang L, Li M, Fu M, Liu C, Wu Y, Liu C (2020) Husband's smoking behaviours associated with women's depression in China: the importance of a smokefree household. J Subst Abus 25(2):194–200

Hajizadeh M, Nandi A (2016) The socioeconomic gradient of secondhand smoke exposure in children: evidence from 26 low-income and middle-income countries. Tob Control 25(e2):e146

He F, Li T, Lin J, Li F, Zhai Y, Zhang T, Gu X, Zhao G (2020) Passive smoking exposure in living environments reduces cognitive function: a prospective cohort study in older adults. Int J Environ Res Public Health 17(4):1402

He Q, Wang J, Zhu Y et al (2011) Prevalence of passive smoking among inland residents in China: a systematic analysis of 2000-2009 studies. Zhonghua Liu Xing Bing Xue Za Zhi 32(2):159–163

Hernandez EM, Vuolo M, Frizzel LC, Kelly BC (2019) Moving upstream: the effect of tobacco clean air restrictions on educational inequalities in smoking among young adults. Demography 56(5):1693–1721

Ho SY, Wang MP, Lo WS, Mak KK, Lai HK, Thomas GN, Lam TH (2010) Comprehensive smoke-free legislation and displacement of smoking into the homes of young children in Hong Kong. Tob Control 19(2):129–133

Hoe C, Cohen JE, Yang T, Peng S, Zhang W (2021) The association of cigarette production and tobacco retailer density on secondhand smoke exposure in urban China. Tobacco Control (forthcoming: tobaccocontrol-2021-056655.R1)

Hu T-W (2008) Tobacco control policy analysis in China: economics and health. World Scientific Publishing, Singapore

Huang K, Chen H, Liao J, Nong G, Yang L, Winickoff JP, Zhang Z, Abdullah AS (2016) Factors associated with complete home smoking ban among Chinese parents of young children. Int J Environ Res Public Health 13(2):161

Hyland A, Barnoya J, Corral JE (2012) Smoke-free air policies: past, present and future. Tob Control 21(2):154–161

Ji M, Ding D, Hovell MF, Xia X, Zheng P, Fu H (2009) Home smoking bans in an urbanising community in China. Am J Prev Med 37(2):132–136

Jiang L, Sriplung H, Chongsuvivatwong V, Li T, Xiao Y (2010) Attitudes towards a smoking ban in restaurants of managers, employees and customers in Kunming City, China. Southeast Asian J Trop Med Public Health 41(5):1258–1266

Jiang S, Yang X, Yang T (2019) China: City smokefree policy challenge highlights China's FCTC quandary. Tob Control 28(3):241

Jiang Y (2020) Perspectives: urgent need to ratify national legislation banning smoking in public places. China CDC Weekly 2(23):426–427. http://weekly.chinacdc.cn/en/article/doi/10.46234/ccdcw2020.114

Jin Y, Wang L, Lu B, Ferketich AK (2014) Secondhand smoke exposure, indoor smoking bans and smoking-related knowledge in China. Int J Environ Res Public Health 11(12):12835–12847

Kaleta D, Polanska K, Usidame B (2015) Smoke-free workplaces are associated with protection from second-hand smoke at homes in Nigeria: evidence for population-level decisions. Biomed Res Int 2015(1):1–8

Kaufman MR, Merritt AP, Rimbatmaja R, Cohen JE (2015) 'Excuse me, sir. Please don't smoke here' a qualitative study of social enforcement of smoke-free policies in Indonesia. Health Policy Plan 30(8):995–1002

Kegler MC, Hua X, Solomon M, Wu Y, Zheng PP, Eriksen M (2014) Factors associated with support for smoke-free policies among government workers in six Chinese cities: a cross-sectional study. Public Health 14(1):1130

Lee JT, Agrawal S, Basu S, Glantz SA, Millet C (2014) Association between smoke-free workplace and second-hand smoke exposure at home in India. Tob Control 23(4):308–312

Lee YH, Chng YC, Lee YTH, Liu CT, Shelly M (2018) Are older adults' home exposures to secondhand smoking during earlier life stages associated with current health status and life satisfaction in China? A multilevel approach. J Ethn Subst Abus 17:1–15
Leis M (2015) Tobacco smoking among Peking university students – smoking as a resistance, lack of information on health problems, or a societal norm? Case study in Peking University with students aged 18–28. Master's Thesis, Lund University, Lund, Sweden
Li Q, Hyland A, O'Connor R, Zhao G, Du L, Li X, Fong GT (2010) Support for smokefree policies among smokers in six cities in China: ITC China survey. Tob Control 19(Supp 2):i40–i46
Li X, Zheng P, Fu H, Berg C, Kegler M (2013) Results from an evaluation of tobacco control policies at the 2010 Shanghai World Expo. Tob Control 22(Suppl 2):ii21–ii26
Li Z, Yao Y, Han W, Liu Y, Tao Y, Kou C, Jiang L, Sun Q, Yin Y, Zhang H, Li B (2015a) Smoking prevalence and associated factors as well as attitudes and perceptions towards tobacco control in Northeast China. Int J Environ Res Public Health 12(7):8606–8618
Li Z, Yao Y, Yu Y, Shi J, Liu Y, Tao Y, Kou C, Zhng H, Han W, Yin Y, Jiang L, Li B (2015b) Prevalence and associated factors of passive smoking among women in Jilin Province, China: a cross-sectional study. Int J Environ Res Public Health 12(11):13970–13980
Liang X (ed) (2015) Report of China City Adult Tobacco Survey 2013–14: A 14-City Experience. Center for Disease Control (CDC) Foundation, Atlanta
Lim KH, Lim HL, The CH, Ghazali SM, Kee CC, Heng PP, Shaharuddin R, Lim JH (2019) Is the implementation of smoke-free polices at workplaces associated with living in a smoke-free home?: findings from a national population-based study in Malaysia. Tob Induc Dis 17(51):51
Lin H, Chang C, Liu Z, Zheng Y (2019) Subnational smoke-free laws in China. Tob Induc Dis 17(November):78
Liu RL, Jiang Y, Li Q, Hammond SK (2014a) An assessment of health risks and mortality from exposure to secondhand smoke in Chinese restaurants and bars. PLoS One 9(1):e84811
Liu RL, Jiang Y, Li Q, Hammond SK (2014b) Assessing exposure to secondhand smoke in restaurants and bars 2 years after the smoking regulations in Beijing, China. Indoor Air 24(4):339–349
Liu RL, Jiang Y, Travers MJ, Li Q, Hammond SK (2014c) Evaluating the efficacy of different smoking policies in restaurants and bars in Beijing, China: a four year follow-up study. Int J Hyg Environ Health 217(1):1–10
Liu RL, Yang Y, Travers MJ, Fong GT, O'Connor RJ, Hyland A, Li L, Nan Y, Feng GZ, Li Q, Jiang Y (2011) A cross-sectional study of levels of secondhand smoke in restaurants and bars in five cities in China. Tob Control 20(6):397–402
Lu W, Wu H (2019) Does passive smoking at home lead to depression among older people in China? Int J Health Plan Manage 35(2):494–505
Luo B, Wan L, Liang L, Li T (2015) The effects of educational campaigns and smoking bans in public places on smokers' intention to quit smoking: findings from 17 cities in China. BioMed Res Int 2015:Article ID 853418 9pp
Ma J, Apelberg BJ, Avila-Tang E, Yang G, Ma S, Samet JM, Stillman FA (2010) Workplace smoking restrictions in China: results from a six county survey. Tob Control 19(5):403–409
Mao A (2013) Space and power: Young mother's management of smoking in extended families in China. Health Place 21(C):102–109
Mao A, Robinson J (2015) Home smoking restrictions before, during and after pregnancy – a qualitative study in rural China. Health Promot Int 31(3):606–613
Mbulo L, Palipudi KM, Andes L, Morton J, Bashir R, Fouad H, Ramanandraibe N, Caixeta R, Dias RC, Wijnhoven TMA, Kashiwabara M, Sinha DN, d'Espaignet ET (2016) Secondhand smoke exposure at home among one billion children in 21 countries: findings from the. Global Adult Tobacco Survey (GATS) Tobacco Control 25(E2):e95–e100
Mehta N, Zhang C, Hua X, Redmon P, Eriksen M, Koplan J, Ali M (2014) Tobacco smoking among government employees in six cities in China. Heart Asia 6(1):179–183
Mills AI, Messer K, GilpinEA PJP (2009) The effect of smoke-free homes on adult smoking behaviour: a review. Nicotine Tob Res 11(10):1311–1141

Mlinarić M, Hoffman L, Kunst AE, Schreuders M, Willemsen MC, Moor I, Richter M (2019) Exploring mechanisms that influence smokefree implementation at the local level: a realist view of smoking bans. Nicotine Tob Res 21(12):1609–1620

Mons U, Nagelhout GE, Allwright S, Guignard R, van den Putte B, Willemsen MC, Fong GT, Brenner H, Pötschke-Langer M, Breitling LP (2013) Impact of national smoke-free legislation on home smoking bans: findings from the international tobacco control policy evaluation project Europe surveys. Tob Control 22(e1):e2–e9

Monson E, Arsenaut N (2017) Effects of enactment of legislative (public) smoking bans on voluntary home smoking restrictions: a review. Nicotine Tob Res 19(2):141–148

Nagelhout G, de Korte-de Boer D, Kunst AE, van der Meer RM, de Vries H, van Gelder BM, Willemsen MC (2012) Trends in socioeconomic inequalities in smoking prevalence, consumption, initiation, and cessation between 2001 and 2008 in the Netherlands. Findings from a national population survey. BMC Public Health 12(1):303

Nazar GP, Lee JT, Glantz SA, Arora M, Pearce M, Millet C (2014) Association between being employed in a smoke-free workplace and living in a smoke-free home: evidence from 15 low and middle income countries. Prev Med 59(100):47–53

Perkins R, Neumayer E (2014) Adoption and compliance in second-hand smoking bans: a global econometric analysis. Int J Public Health 59(5):859–866

Peruga A, Hayes LS, Aguilera X, Prasad V, Bettcher DW (2018) Correlates of compliance with national comprehensive smoke-free laws. Tob Control 27(6):608–613

Qian J (2017) Tobacco control in China. Institutions, bureaucratic noncompliance and policy ineffectiveness. In: Brødsgaard KE (ed) Fragmented authoritarianism. Earthquakes, energy and environment. Routledge, Abington, Oxford, pp 56–76

Qian X, Gu H, Wang L, Wang X, Xuan Z, Zheng P, Fu C (2018) Changes in smoking prevalence after the enforcement of smoking control regulations in urban Shanghai, China: findings from two cross-sectional surveys. Tob Induc Dis 16(June):27

Redmon P, Chen LC, Wood JL, Li S, Koplan JP (2013) Challenges for philanthropy and tobacco control in China (1986–2012). Tob Control 22(Suppl 2):ii4–ii8

Redmon P, Koplan J, Eriksen M, Li S, Kean W (2014) The role of cities in reducing smoking in China. Int J Environ Res Public Health 11(10):10062–10075

Rogers EM (1962) Diffusion of innovations. Free Press, New York

Song S, Li W, Leng J, Wang L, Li W, Shi F, Liu G, Zhou J, Yang X (2019) Passive smoking and postpartum depression among Chinese women: a prospective cohort study in Tianjin, China. J Women Health 59(3):281–293

Stillman F, Navas-Acien A, Ma J, Ma S, Avila-Tang E, Breysse P, Yang G, Samet J (2007) Secondhand tobacco smoke in public places in urban and rural China. Tob Control 16(4):229–234

Sureda X, Martinez-Sanchez JM, López MJ, Fu M, Aguero F, Salto E, Nebot M, Fernandez E (2012) Secondhand smoke levels in public building main entrances: outdoor and indoor $PM_{2.5}$ assessment. Tob Control 21(6):543–548

Sureda X, Fernández E, López MJ, Nebot M (2013) Secondhand tobacco smoke exposure in open and semi-open settings: a systematic review. Environ Health Perspect 121(7):166–173

Wan X, Stillman F, Liu H, Spires M, Dai Z, Tamplin S, Hu D, Samet JM, Yang G (2013) Development of policy performance indicators to assess the implementation of protection from exposure to secondhand smoke in China. Tob Control 22 (Suppl 2):ii9–ii15

Wang C-P, Ma SJ, Xu XF, Wang J-F, Mei C-Z, Yang G-H (2009) The prevalence of household second-hand smoke exposure and its correlated factors in six counties of China. Tob Control 18(2):121–126

Wang H, Li F, Zhang Y, Jiang F, Zhang J (2019) The association between exposure to secondhand smoke and psychological symptoms among Chinese children. BMC Public Health 19(1):923

Wang X, Ning Y (2019) Designated smoking area draws ire. China Daily 11 January http://global.chinadaily.com.cn/a/201901/11/WS5c37e8e3a3106c65c34e3d1b.html

Wang X, Wang MP, Viswanath K, Wan A, Lam TH, Chan SS (2016) Smoking and secondhand smoke exposure at home were associated with poor perceived family well-being. Findings of FAMILY project. PLOS One 11(8):e0161761

Wei X, Zhang Z, Song X, Xu Y, Wu W, Lao X, Ma W (2014) Household smoking restrictions related to secondhand smoke exposure in Guangdong, China: a population representative survey. Nicotine Tob Res 16(4):390–396

Wei Y, Borland R, Zheng P, Fu H, Wang F, He J, Feng Y (2019) Evaluation of the effectiveness of comprehensive smoke-free legislation in indoor public places in Shanghai, China. Int J Environ Res Public Health 16:4019

WHO (World Health Organisation) (2018–19) China- WHO - Programme Budget Web Portal. World Health Organisation, Western Pacific Region, Manila. https://open.who.int/2018-19/country/CHN

WHO (World Health Organisation) Western Pacific Region and University of Waterloo, ITC Project (2015) Smoke-free polices in China. Evidence of effectiveness and implications for action. World Health Organisation, Western Pacific Region, Manila

Xia C, Zheng R, Zeng H, Zhou M, Wang L, Zhang S, Zou X, Sun K, Yang Z, Li H, Parascandola M, Islami F, Chen W (2020) Provincial-level cancer burden attributable to active and second-hand smoking in China. Tob Control 28(6):669–675

Xiao H, Zhang H, Wang D, Shen C, Xu Z, Zhang Y, Jiang G, Yang G, Wan X, Naghavi M (2020) Impact of smoke-free legislation on acute myocardial infarction and stroke mortality: Tianjin, China, 2007-2015. Tob Control 29(1):61–67

Xiao L, Jiang Y, Liu X, Gan Q, Liu F (2017) Smoking reduced in urban restaurants: the effect of Beijing smoking control regulation. Tob Control 26(e1):e75–e78

Xu Y, Xu S, Wu Q, Guo Y (2016) Association between secondhand smoke exposure at home and cigarette gifting and sharing in Zhejiang, China: a repeat cross-sectional study. BMJ Open 6(3):e010058

Yang JA, Tsou MH, Janowicz K, Clarke KC, Jankoski P (2019) Reshaping the urban hierarchy: patterns of information diffusion on social media. Geo-Spat Inf Sci 22(3):149–165

Yang L, Tong EK, Mao Z, Hu R-W (2010) Exposure to secondhand smoke and associated factors among non-smoking pregnant women with smoking husbands in Sichuan Province, China. Acta Obstet Gynecol Scand 89(4):549–557

Yang T, Abdullah AS, Li L, Rockett IRH, Lin Y, Ying J, Guo W, Wu D, Li M (2013) Public place smoke-free regulations, secondhand smoke exposure and related beliefs, awareness, attitudes, and practices among Chinese urban residents. Int J Environ Res Public Health 10(6):2370–2383

Yang T, Barnett R, Jiang S, Yu L, Xian H, Ying J, Zheng W (2016a) Gender balance and its impact on male and female smoking rates in Chinese cities. Soc Sci Med 154(April):9–17

Yang T, Cao C, Cottrell RR, Wu D, Yu L, Lin H, Jiang S, Young KJ (2015a) Second hand smoke exposure in public venues and mental disorder: a representative nationwide study of China. Tob Induc Dis 13(1):18

Yang T, Jiang S, Barnett R, Peng S, Yu L (2015b) Individual and city-level determinants of second-hand smoke exposure in China. Int J Health Geogr 14:36

Yang T, Jiang S, Barnett R, Oliffe JL, Wu D, Yang X, Yu L, Cottrell RR (2016b) Who smokes in smoke-free public places in China? Findings from a 21 city survey. Health Educ Res 31(1):36–47

Yang T, Rockett IRH, Li M, Xu X, Gu Y (2012) Tobacco advertising, environmental smoking bans, and smoking in Chinese urban areas. Drug Alcohol Depend 124(1–2):121–127

Yang T, Xu X, Rockett IRH, Guo W, Zhou H (2011) Effects of household, workplace, and public place smoking restrictions on smoking cessation. Health Place 17(4):954–960

Yang T, Yu L, Jiang S, Feng X, Xian H, Cottrell RR, Rockett IRH (2015c) Household smoking restrictions among urban residents in China: individual and regional influences. Int J Public Health 60(4):479–486

Yang Y, Liu F, Wang L, Li Q, Wang X, Chen JC, Wang Q, Shen H, Zhang Y, Yan D, Zhang M et al (2017) Association of husband smoking with wife's hypertension status in over 5 million Chinese females aged 20 to 49 years. J Am Heart Assoc 6(3):e004924

Yang Y, Yang XY, Yang T, He W, Peng S, Rockett IR (2021) Social deprivation and secondhand smoke exposure among urban male residents: A nationwide study of China. Tob Induc Dis 19(March):21

Yao T, Lee AH, Mao Z (2009) Potential unintended consequences of smoke-free policies in public places on pregnant women in China. Am J Prev Med 37(2S):S159–S164

Yao T, Sung H-Y, Mao Z, Hu T-W, Max W (2012) Secondhand smoke exposure at home in rural China. Cancer Causes Control 23(Suppl 1):109–115

Ye X, Chen S, Yao Z, Gao Y, Xu Y, Zhou S, Zhu Z, Wang L, Yang Y (2015) Smoking behaviors before and after implementation of a smoke-free legislation in Guangzhou, China. BMC Public Health 15(1):982

Ye X, Yao Z, Gao Y, Xu Y, Xu Y, Zhu Z, Chen S, Yang Y (2014) Second-hand smoke exposure in different types of venues: before and after the implementation of smoke-free legislation in Guangzhou, China. BMJ Open 4(2):e004273

Yuan J (2019). Time is ripe to boost national smoke-free city campaign. China Daily 28 January. http://global.chinadaily.com.cn/a/201901/28/WS5c4e97dda3106c65c34e6d92.html

Zeng J, Yang S, Wu L, Wang J, Wang Y, Liu M, Zhang D, Jiang B, He Y (2016) Prevalence of passive smoking in the community population aged 15 years and older in China: a systematic review and meta-analysis. BMJ Open 6(4):e009847

Zhang J, Cui X, Liu H, Zhang J, Cui X, Liu H, Han H, Cao R, Sebrie EM, Yin X (2019) Public mobilization in implementation of smoke-free Beijing: a social media complaint platform. Tob Control 28(6):705–711

Zhang J, Jiang Y, Wang Y, Xiao L (2018) The study of the relationship between the construction of a smoke-free environment and smoking cessation behaviour. Chin J Prev Control Chronic Dis:2018–2014. http://en.cnki.com.cn/Journal_en/E-E000-ZMXB-2018-04.htm

Zhao J, Li X, Stewart SL, Gao W, Qi F, Zhang L, Pang Z, Qiao Q, Ning F, Tong E (2018) Cigarette smoking and secondhand smoke exposure before and after a tobacco-free Olympic policy period: Qingdao, China. Nicotine Tob Res 25(11):1531–1538

Zheng P, Berg CJ, Kegler MC, Fu W, Wang J, Zhou X, Liu D, Fu H (2014a) Smoke-free homes and home exposure to secondhand smoke in Shanghai, China. Int J Environ Res Public Health 11(11):12015–12028

Zheng P, Fu H, Li G (2009) Smoke-free restaurants in Shanghai: should it be mandatory and is it acceptable? Health Policy 89(2):216–224

Zheng P, Kegler MC, Berg CJ, Fu W, Wang J, Zhou X, Liu D, Fu H (2014b) Correlates of smoke-free home policies in Shanghai, China. Biomed Res Int 2014: Article ID 249534 8pp

Zheng Y, Ji Y, Dong H, Chang C (2018) The prevalence of smoking, second-hand smoke exposure, and knowledge of the health hazards of smoking among internal migrants in 12 provinces in China: a cross-sectional analysis. BMC Public Health 18(1):655

Zhou L, Niu L, Jiang H, Jiang C, Xiao S (2016) Facilitators and barriers of smokers' compliance with smoking bans in public places: a systematic review of quantitative and qualitative literature. Int J Environ Res Public Health 13(12):1228

Chapter 10
Conclusions: Environments and Smoking in China

Ross Barnett, Tingzhong Yang, and Xiaozhao Y. Yang

Abstract This concluding chapter draws upon the authors' own empirical work, and a synthesis of existing literature to show how a knowledge of economic, social and cultural environments is needed to fully understand changing patterns of smoking. As China is the world's largest producer and consumer of cigarettes, it is important to understand the significance of tobacco and the wider set of factors influencing its consumption. The chapter concludes by examining some of the substantive conceptual and methodological challenges facing the Chinese academic public health community in addressing smoking and ways in which this community might contribute to more effective approaches to tobacco control. While western ideas of a tobacco endgame have not been seen as particularly relevant in China, there are, nevertheless, substantial research opportunities that may help contribute to that goal.

Keywords Conceptual and methodological challenges · Environmental (contextual) effects · Framework Convention on Tobacco Control (FCTC) · Public health · Research priorities · Stages of tobacco regulation · Tobacco endgame

R. Barnett (✉)
School of Earth and Environment, University of Canterbury, Christchurch, New Zealand
e-mail: ross.barnett@canterbury.ac.nz

T. Yang
School of Medicine, Zhejiang University, Hangzhou, China
e-mail: tingzhongyang@zju.edu.cn

X. Y. Yang
School of Sociology and Anthropology, Sun Yat-sen University, Guangzhou, China
e-mail: yangxzh27@mail.sysu.edu.cn

R. Barnett et al. (eds.), *Smoking Environments in China*,
Global Perspectives on Health Geography,
https://doi.org/10.1007/978-3-030-76143-1_10

10.1 Introduction

This book has examined environmental factors which have contributed to the smoking epidemic in China. Drawing upon the authors' own empirical work, and a synthesis of existing literature, we have shown how an understanding of economic, social and cultural environments is needed to fully understand changing patterns of smoking. We cannot understand smoking unless we appreciate its complexity: as an addictive behaviour, as a relief from stress, as a social and cultural phenomenon, as an economic commodity or as a symbol which may be politically valued as an important part of national and local identity. Understanding this complexity and how different individual and structural factors interact in China to influence smoking has been the focus of this book. China is not the West, and, as the world's largest producer and consumer of cigarettes, it is important to understand the significance of tobacco and the wider set of factors influencing its consumption. It would be a mistake, however, to assume that the societal forces affecting smoking in China are unique, compared to other LMICs or even western countries, for there are certain similarities, such as the effect of the industry in maintaining consumption. Nevertheless, there are also important societal differences, both within China and between China and the rest of the world, and understanding the significance of these has been an important task.

Our work has drawn on a wide range of disciplines. This has been necessary because of the disparity of research interests and the limited attention paid by some social sciences, such as health geography, to certain 'drivers' of the global tobacco epidemic and how these have played themselves out in LMICs. With this in mind, we have structured this final chapter into six parts, which revisit the main issues covered in this book. The first three sections (10.2, 10.3 and 10.4) provide a brief summary of the main environmental factors, economic, cultural and social, which are fundamental to understanding the smoking epidemic in China and the way it has evolved. In the fourth section (10.5), we address issues relating to the political and policy process. We re-emphasise that it is important to understand the barriers to policy development and why important changes have occurred in some locations but not in others. In Sect. 10.6 we discuss some of the substantive conceptual and methodological challenges facing the Chinese academic public health community in addressing smoking and ways in which this community might contribute to more effective approaches to tobacco control. While western ideas of a tobacco endgame have not been seen as particularly relevant in China, in the final section (10.7), we highlight some of the substantial research opportunities which may help contribute to that goal.

10.2 The Economic Environment and Its Impacts

Greater understanding of the economic environment is crucial to giving meaning to patterns of consumption and tobacco control. As the largest producer and consumer of tobacco products in the world, the tobacco industry has become an integral part of the national economy of China. China is one of the few LMICs not dominated by transnational tobacco companies, with the state tobacco company, the CNTC, viewed as an important symbol of resistance to foreign interference and a source of national pride. Thus it is hardly surprising that the Chinese political leadership has been ambivalent to tobacco control. Tobacco growing and cigarette production were important in the early days of the rise of the Chinese Communist Party, and the expansion of the industry since then has been nothing short of phenomenal.

The economic geography of tobacco has been an under-researched theme in geography, and undertaken largely by other social sciences, most notably anthropology, political science and sociology. This is unfortunate since the evolution of the Chinese tobacco industry reflects many contemporary geographical processes such as regional specialisation and multiplier effects or broader processes such as the globalisation of commodity chains and networks of production. Both have been evident in China. Within China tobacco growing, although relatively widespread, has become concentrated into a smaller number of provinces. Never popular with peasants, and often only grown through coercion, tobacco is increasingly being substituted for other crops partly reflecting national imperatives for increased food security and the replacement of domestic sources of production with foreign imports. China has increasingly become a 'resource seeker' with the CNTC securing high-quality sources of tobacco from abroad as it seeks to improve product quality for both internal and overseas markets.

China's cigarette manufacturing industry likewise, has seen many changes, evolving from a locally protected industry to one that is increasingly globalised. Despite the formation of a central state monopoly in the Maoist years, the most significant drivers of change in the industry occurred largely at the provincial level. The particular nature of central-provincial government tax arrangements led to the growth of provincial protectionism of manufacturing enterprises and controls over competition from non-local plants. To some extent central government lost control of the industry which became increasingly inefficient with many marginal enterprises producing poorer-quality products. It was not until after the 1978 economic reforms, and particularly the entry of China into the WTO in 2001, that the industrial structure began to change as a result of new central-local tax arrangements and industry restructuring aimed at increasing market efficiency. These changes reflected wider trends in the 1980s, both in capitalist and former Communist countries, where neoliberal economic reforms ushered in new economic realities. In this respect China was no different. What was different, however, was that the economic reforms which affected tobacco manufacturing did not dismantle the state monopoly, which remained.

Chapter 4 considered the effects of the industry on consumption within China. Market saturation has resulted in the creation of many new products especially electronic cigarettes, most of which are sold outside China, but also premium low-tar and slim cigarettes designed to encourage consumption among youth and women. Consumption patterns also reflected the regional importance of the tobacco industry with smoking prevalence being greatest in regions with higher levels of cigarette production. Important here has been company engagement in a variety of philanthropic projects which have helped promote a positive public image of corporate social responsibility. Such practices, by promoting tobacco as a normal part of everyday life, have subverted attempts at denormalisation by tobacco control advocates and made their task much more difficult.

An important part of this chapter was to consider other important parts of the commodity chain, most notably the tobacco retail landscape. As other forms of tobacco advertising have become more constrained through the enactment of stricter tobacco control laws, point of sale, together with social media, advertising has become more important. However, unlike western research, where such themes have received much attention, there has been remarkably little research on the tobacco retail landscape and how this is evolving in response to industry trends at the macro-level.

10.3 Cultural Environments and Smoking

In Chap. 2 we highlighted the important changes that have taken place in the social and gender distribution of smoking in China and the uncertain relevance of conventional models of the smoking transition. While such models have provided an important conceptual framework for understanding smoking transitions in western countries, there has been little attention to geographic differences in the pace of change within particular national contexts and how such changes have been affected by the persistence of traditional cultural norms.. In China, as in many other Asian countries, traditional cultural norms vary not only by region but also by urban-rural residence. As we saw in Chap. 5, these relate not only to the role of patriarchy but also to a variety of other cultural factors that influence smoking.

Although culture is a diverse subject and hard to define, certain cultural traits, such as patriotism, patriarchy, self-exempting beliefs and cigarette gifting, reflecting economic, political and social processes, can nevertheless be identified. Many of these are not unique to China, but others, such as the authoritarian *hukou* system, form a relatively distinct part of Chinese culture. In China urbanisation has been a managed process in the sense that the *hukou* system has provided institutional hurdles preventing rural migrants from freely accessing urban services, thus limiting their status and affecting health outcomes (Song and Smith 2019). While such arrangements have benefitted cities by providing a ready source of rural labour, they have helped alter the social fabric by creating disadvantaged class of urban migrants less likely to adopt healthier lifestyles. However, the implications of *hukou* policy remain unclear. Since urban policies regarding *hukou* were decentralised to local governments in the 1990s, there is likely to be greater geographic variation in *hukou* policy making. The same could be said for other practices, such as cigarette gifting, where the evidence is still relatively unclear about how urbanisation and changing social values are altering its cultural significance.

Chinese culture has also emphasised strong disapproval of female smoking. While this is not unique to China, it will be interesting to see how gender norms are likely to change in the future as women gain new freedoms as a result of urbanisation and economic growth. To what extent will China be like other LMICs where women are smoking more like men? This has been especially true in countries which have experienced a growth in political freedoms but less true in places where a greater proportion of men work relative to women, where levels of urbanisation are lower and where gender differences in educational attainment remain high (Lillard and Christopoulu 2015). In China many studies reaffirm the importance of gender differences in smoking, particularly in rural areas, but there have been few attempts to identify the contexts in which patriarchal values are changing. Thus we are still left with the puzzle of understanding the types of community conditions in which female smoking will increase. Certainly the CNTC has promoted new brands to entice more women to smoke, but it is unclear whether such trends depend upon individual social status or on the characteristics of the communities where women reside.

10.4 The Social Environment: Urbanisation and Smoking

One of the important contributions of this book has been to locate smoking in the context of urbanisation and socio-economic change. In the light of rapid urbanisation and the marked socio-economic transition that has characterised China since 1978, it is important to highlight the impact of such transitions and their effects on health behaviour. Although Chinese scholars have increasingly engaged with the health consequences of urbanisation in a general sense (Gong et al. 2012; Li and Wu 2016; Yuan et al. 2018), with the exception of work on obesity (Hou et al. 2019; Miao and Wu 2016; Zhou et al. 2017), there has been little attention directed to the how smoking and other health behaviours have been affected by community economic and social developments.

In Chap. 6 we identified two key processes that were likely to alter the distribution of smoking: the effects of income inequality and changes in the urban environment. Of the first, income inequality has become increasingly important in China, attracting considerable attention not so much for its health consequences but for its fiscal ramifications. As in western countries, increased inequality has generated a range of negative health impacts, including poor mental health, decreased happiness and well-being and adverse health behaviours, including smoking. Also important was that many of the health impacts of increased inequality, as in the western case, were mainly felt among the poor. This is not just the migrant poor as the effects of income inequality are felt more widely.

Changes in the built and social urban environment can also be implicated in the urban distribution of smoking. The growth in income inequality is reflected in changes in the housing market, not just in the creation of migrant enclaves but also in changes in housing tenure which have produced high levels of urban polarisation. Urban neighbourhoods, initially planned and constructed by central government agencies, have slowly given way to more private housing construction, albeit still constrained by planning and zoning. While urban neighbourhood social distinctions are less apparent than in western cities, they both exist and affect health (Liu et al. 2017). Although data is scarce, we speculated that China's current pattern of urban socio-economic development is likely to lead to increased social segregation and the emergence of area deprivation (and associated stigmatisation) now typical of some other East Asian cities (Yun et al. 2015). As in the case of income inequality, such trends, along with the socially selective impact of urban tobacco control measures, are likely to help alter the social distribution of smoking in major Chinese cities and perhaps in nearby rural areas.

Consistent with expectations, we found that the income pattern of smoking, while positive at the individual level, was changed by community-level social development. In other words, as a community developed socially, more affluent people smoked (and drank) less; but richer people living in less developed communities still consumed more alcohol and cigarettes relative to the poor. It was also found that in more urban socio-economically developed communities, traditional cultural constraints were lessened with women smoking and drinking more. Such findings are partly consistent with those reported in Korea (Yun et al. 2015). These trends and national differences are intriguing and in China deserve much more attention than they have received to date.

The findings reported above are significant in a number of respects. First, it is clear that the socio-economic and gender patterning of smoking depends very much on the level of community socio-economic development. Such patterns indicate that health transition models need to examine within country variations and to more clearly specify the community variables which have been important. Second, the social patterning of smoking in more economically and socially developed areas of China is beginning to resemble patterns found in the West, where historically there has been an inversion of the social gradient in smoking as societies changed over time. Thus, while China is not unique in this respect, the speed with which the process has taken place in some areas is remarkable. This perhaps can be linked with changing social norms and the beginnings of denormalisation of smokers, greater awareness of health risks and perhaps a lower availability of smoking outlets in more affluent communities. Third, the findings have implications for tobacco control policies. They suggest that tobacco control advocates need to pay more attention to differences in community contexts in the implementation of such policies as clearly 'one size does not fit all'. Smoking remains higher in the countryside and is likely to remain so given that economic and social developments in cities have outpaced those in rural areas. Clearly more tailored initiatives are needed especially in poorer rural areas and in areas dominated by the tobacco industry where strong pro-tobacco norms dominate social interactions within communities.

10.5 Environments of Smoking Cessation and Tobacco Control

An important goal of this book has been to add a geographical lens to the development of tobacco control policies in China by focusing on the range of policies that have developed at the national, provincial and local (urban) scale. Our incentive for doing so is that much Chinese smoking research to date has focused on individual decision-making and unassisted smoking cessation. As we showed in Chap. 8, Chinese smokers have been reluctant to engage with official sources of help when attempting to quit, and consequently much research has focused on why this is the case and the avenues used by individual smokers. These results stand in contrast to most western countries, where there has generally been greater proportion of smokers receiving cessation assistance. The reasons clearly are not the absence of such services in China, for these exist in most large cities, but the fact that such services are not used. Many Chinese men think that smoking cessation services are unnecessary, an intrusion into one's privacy, a source of potential embarrassment and loss of face, an affront to their gender identity and in conflict with a culture of self-reliance so typical of much of Chinese society.

While such research is of value in understanding individual behavioural decisions, with few exceptions, it has not considered the environmental factors involved in quitting other than the effects of anti-tobacco health campaigns or environmental smokefree policies. Tobacco control has lacked research that has increased our understanding of the role of environmental factors affecting tobacco control policy and its implementation. Chapters 7, 8 and 9 dealt with these considerations: Chap. 7 set the scene by considering the national institutional context of tobacco control, Chap. 8 reviewed tobacco control instruments and policy outcomes, while Chap. 9 provided a more detailed case study of smoking bans and how geography played a role in their implementation.

What is clear is that at the international scale, China has performed poorly compared to other LMICs in implementing the provisions of the FCTC. In addition to an initial reluctance to engage with the WHO, subsequent policy development has been characterised by tardiness, in some cases deception, but above all a clear message that China is only doing the minimum required in order to be part of the international tobacco control community. Centrally directed changes, such as the 2009 tax increase, or the introduction of stronger smokefree policies associated with the holding of international events, have largely been cosmetic and designed to present a positive image to the world and provide evidence of China's commitment as a signatory of the FCTC. Similarly either tax increases have not been passed on to consumers, as in 2009, or the scale of the tax increase in 2015 was insufficient to significantly affect consumption and was criticised for having an economic rather than a social goal. The Chinese government has long been ambivalent about the FCTC and the implementation of more effective tobacco control policies. This reflects the nature of the institutional framework and the continued influence of the CNTC relative to public health advocates in the policy-making process. While in other low- and middle-income countries TTCs have interfered with tobacco control policy, it is ironic that in China, it is the state-owned CNTC that has undermined government initiatives in tobacco control.

The reluctance of central government to implement stronger tobacco control policies was evident in Chap. 9. The important conclusion reached is that policy development has largely been a subnational process which has occurred in particular types of cities, usually provincial capitals, but also in larger cities and especially in global tourist cities such as Xi'an or Shanghai. The localised pattern of smokefree policies in particular types of cities is significant both because it signals the impact of urban competition for international tourists and because it is reminiscent of the past importance of international events in changing attitudes towards tobacco control. While there is evidence that regional clusters of cities have begun to develop similar policies, it is unclear whether this reflects the rising power of local public health advocates or evidence of policy diffusion between political elites in adjacent cities.

It is difficult to speculate on the future path of tobacco control in China and of the implications of the *Healthy China 2030* goals in reducing smoking prevalence. At the regional and local levels, we considered the significance of local political elites in preserving the status quo, especially in regions more reliant on the tobacco industry, and provided an analysis of antitobacco control strategies used by the industry. The evidence suggests that the tide may have turned with central government now being more aware that the health costs of tobacco consumption outweigh any economic gains for the country. But the tobacco industry is also a source of pride and has been seen as a source of political stability. In an authoritarian society such as China, both are important and likely to remain so for some time. On the other hand, however, while recent trends may suggest otherwise, the power of civil society groups has grown and tobacco control advocates are now more significant than before.

10.6 Challenges for Future Research in China

By adopting a largely environmental lens, this book has aimed to shift the emphasis of smoking research in China and to consider the interaction of structural and individual factors and the importance of different types of environments on smoking prevalence. In this last section, we examine some of the key research issues we see as important in improving both our understanding of individual smoking behaviour in China and how it is shaped by context, but also how place effects have been important in policy making and its outcomes. We focus on three issues: resource challenges of undertaking research on tobacco control in China (10.6.1), conceptual questions relating to the practice of tobacco control (10.6.2) and methodological issues associated with work to date (10.6.3).

10.6.1 The Big Picture: Resource Challenges and the Academic Role of Public Health

Since 1978 China has been transitioning from a centralised to a market-based economy. The transition has improved living standards and markedly increased choices in consumer goods, education, health and employment. Until recently, governments, particularly local governments, have prioritised cigarette production as an economic development strategy and consequently attached little importance to the prevention of smoking and tobacco control or investment in tobacco control research. It is still unclear whether the *Healthy China 2030* initiative and its follow-up Action Plan will have a positive and significant impact on tobacco control and public health outcomes. Experience in western countries shows that the most effective tobacco control comes about when there are effective community, professional and state coalitions. The question remains whether community advocacy and NGOs will be allowed to play an effective role in conjunction with the state apparatus in China. As many recent initiatives have originated in local urban settings, this highlights the need for provincial governments to play a much more active role in tobacco control policy.

Tobacco control falls within the professional discipline of public health, but in China the scope and practice of public health is not well-understood, particularly among government officials and managers. Public health is an applied health discipline with strong population and social dimensions, but in China it is often thought of as general medicine, including basic, preventative and clinical medicine, and evaluated according to individual and biological criteria. State funding for public health research is often directed towards individual health and biological topics, with less recognition of population health issues and strategies. This means that research funding for tobacco control is not easy to acquire and public health professional education in universities does not receive attention adequate for work in smoking prevention and tobacco control. In the past, resources for tobacco control research and policy development have come largely from intergovernmental agencies such as the WHO, or philanthropic organisations such as the Gates and Bloomberg Foundations. Given the magnitude of the smoking problem in China, however, such an important public health and strategic issue should not be addressed, and cannot be solved, by reliance on external expertise and funds. At present, public health professional education in universities is far removed from social needs, and this has seriously affected the quality of disease control and prevention work. Consequently, public health preparedness has been very poor as evident in the lack of a public health input into managing the current COVID-19 epidemic.

10.6.2 Conceptual Challenges

The development of tobacco control in China is in its infancy, and, to date, there is little evidence of substantial progress or of attempts to monitor progress in reducing smoking prevalence other than the broad 2030 smoking goals. The common problems are the generally low quality of the activities and the lack of a theoretical framework to assess policy outcomes. Most research on tobacco control and preventive programs in China are not based on any theoretical model of the factors that influence peoples' behaviour, nor of the role and influence of state actors in attempting to change behaviour. There is a need for more concise model development that links economic, social, educational, cultural and political concerns and their interlinkages.

So far, the vast majority of health behaviour theories used in China come from western countries, and few studies have considered the significance of cultural factors in perpetuating smoking. We have argued that, while many of these are not necessarily unique to China, they need to be acknowledged in tobacco control policy. WHO documents state that there are multiple interventions that can reduce smoking prevalence on the population level. For example, if society increases the cigarette price, implements smokefree policy in public places and provides assistance in quitting, smoking prevalence will generally decline. However, how to do this in a particular culture is not always clear. For example, the fact that the Chinese government owns the tobacco industry makes it difficult to apply many of the known effective tobacco control strategies. Moreover, Chinese culture, to a large extent, is still adhering to agrarian social mores. This suggests that the social values may frame people's behavioural response models differently from those of western countries (Yang 2007, 2017). However, many of the approaches used in tobacco control have been directly copied the western experience with results that are far from satisfactory. This suggests that tobacco control policies need more nuanced appreciation of Chinese cultural traditions.

It is important, therefore, to explore the unique social-behavioural rules and adapt appropriate measures in tobacco control campaigns. Previous studies have already indicated that Chinese smokers have unique social psychological motivations, and thus it is necessary to adopt culturally appropriate measures for tobacco control campaigns to be effective. However, the majority of current tobacco control professionals in China are medically trained with little background in the social sciences and unfamiliar with ways of understanding culture and how to use this in tobacco control. An important step is for public health professionals to have more social science literacy and to understand that multidisciplinary collaboration is essential for tobacco control research.

10.6.3 Methodological Issues

There are a number of methodological challenges in research on smoking and health behaviour in China. To date research has been largely quantitative in focus with an emphasis on using 'big data', usually large-scale surveys of individuals stratified at the provincial or city level. The advantage of such approaches is that they have the potential to capture the diversity of sociocultural and economic environments in China at the macroscale and then to link these to environmental differences to patterns of individual behaviour. Clearly a country as large, diverse and dynamic as China requires a range of research approaches to understand overall patterns of change and their effects on health behaviour and health, but also a variety of more nuanced approaches which take into account the significance of differences in local conditions. While there is a significant social science and public health literature on smoking in China which has produced many new insights into smoking prevalence and policy issues, we highlight five main methodological problems of studies to date.

First, many of the studies that have included environmental factors have lacked a theoretical focus with such variables often included in multi-level models as confounders, rather than as key factors in a conceptual model. Thus research often has had a narrow inductive focus searching for important relationships rather than theorising these at the outset. More thought needs to be given to the nature of different contextual factors, why they are important and the most appropriate scale of measurement. This is especially important for smoking where processes operate at a variety of scales, such as the power of the tobacco industry at the provincial level, how this has shaped attitudes towards tobacco control at the provincial and urban levels or the importance of POS advertising at the local neighbourhood level. Both the type of contextual variable and the interconnections between different scales are important factors to consider.

Second, in examining environmental effects on behaviour, more thought needs to be given to levels of exposure to different environmental norms or stressors that will be important for certain groups. This is particularly important at the urban scale where migrants may be exposed to constraints at the workplace but not in the home. On the other hand, neighbourhood patterns of exposure to sets of norms, patterns of social capital and neighbourhood resources may be very stable for older residents compared to younger people who live in newer privately developed housing estates. Patterns of movement and exposure to 'environmental bads' are poorly understood in Chinese cities, so new research approaches are needed to capture the significance of these new patterns of urban living on smoking and other unhealthy behaviours.

Third, with few exceptions (e.g. Song and Smith 2019), much of the research which has been undertaken has been cross-sectional in nature. This allows conclusions to be drawn for particular places and times, but provides few insights to the impact of important processes that unfold over time, such as urbanisation, socioeconomic transition or the impacts of rural-urban migration. More longitudinal studies, therefore, are needed in order to gain a greater appreciation of the effect of different factors at different points in the urbanisation curve or how these unfold through the life course. For example, the migration literature indicates that the prevalence of smoking among females in China is low in the early stages of migration, but rises with an increased length of urban residence (Chen et al. 2004). Similarly the prevalence of unhealthy lifestyle behaviours among different social groups depends upon the level of urbanisation and social development. Such findings point to the significance of acculturation but also to the impact of different social processes operating in urban and rural environments. The study of urbanisation and acculturation effects on health behaviour and health outcomes has a long history in western countries, but in China more theoretically informed and methodologically sophisticated analyses are needed.

Fourth, an overly quantitative focus of Chinese health behaviour research has meant that few studies of smoking have adopted qualitative or mixed methods approaches. These have the potential to provide added insights into attitudes towards smoking and the extent to which these vary among different groups, including women, recent migrants and youth. For example, while the literature suggests that many Chinese smokers are strongly in support of tobacco control, there is also a certain fatalism on the part of individuals about how much they are able to change their behaviour in the light of the current societal pressures. In Foucauldian terms, the general power of a national smoking culture makes it difficult for individual smokers to resist such forces and change their behaviour. Sustained qualitative exploration of why smoking persists among particular social groups, or in particular localities, would help provide more insightful interpretations of place or smokers' localised responses to different tobacco control measures such as smokefree policies or tax increases.

Finally, most work on Chinese smoking usually considers consumption and behaviour according to population elements, such as age and gender. As in western research, however, it has been rare that studies have considered the differential effects of place on such groups, or in multi-level modelling terms, considered patterns of interaction. Analyses which test for subgroup effects offer a greater opportunity for more insightful interpretations of the different environmental effects on smoking and the groups most affected. For example, in the Chinese cultural context, the differential gender, socio-economic or ethnic effects of the stresses of urbanisation and migrant status would seem to be particularly important.

10.7 Research Priorities: A Guiding Framework, Erickson's Model

Despite many positive developments in smoking research in China, we can identify a number of substantive challenges for future research. These challenges need to be considered in the light of the nature of the academic community in China, the extent to which it has influence on government and the degree to which the tobacco industry has infiltrated academic research programmes to cast doubt on the dangers of smoking and the need for tobacco control. All factors have undoubtedly influenced the nature of tobacco control research in China and the type of policy research which has been undertaken. While Chinese universities are relatively open institutions in terms of teaching and research, Chinese academics nevertheless face a number of constraints that need to be acknowledged as influencing the types of research and the nature of the conclusions drawn. Hence, unlike their western counterparts, there has been a strong research focus on individual behaviour and limited criticism of structural factors associated with state policy. The latter has usually only occurred among senior academics with established reputations or by China scholars outside China.

With these caveats in mind, we identify substantive challenges for future geographical research on smoking and tobacco control in China. We use a modified version of a model of stages of tobacco regulation proposed by Eriksen et al. (2015) as an organising framework. Use of this framework was helpful in undertaking a similar analysis of key spatial issues associated with the more general international issue of the tobacco endgame (Moon et al. 2018). Erickson's model focuses on key stages of the lifecycle of the cigarette starting with tobacco growing to eventual disposal. We supplement Erickson's model by separating processes involved in the supply chain from policy processes concerned with tobacco control at various scales. This provides a method to identify key research questions associated with tobacco supply and demand, along with key policy issues seen as important (Table 10.1). Many of these, such as the globalisation of CNTC, are already receiving increased attention in China. Earlier Koplan et al. (2013) stressed the value of research as a component of successful tobacco control in China, but their brief analysis was almost devoid of environmental considerations. Guided by the potential research questions set out in Table 10.1, we identify five selected issues deemed of high research priority: greater understanding of the policy process at regional and local levels; the role of neighbourhood-level factors in promoting smoking; cultural limitations on smoking cessation; greater consideration of geographic variations in the smoking transition; and developing GIS approaches and better information systems.

10.7.1 Geographic Variations in the Policy Process

As in western countries, much of the literature on tobacco control in China has focused on the nature of different types of policy interventions and their effectiveness. Less attention has been paid to national and local variations in policy interventions and enforcement. Most studies of tobacco control initiatives, such as tobacco taxation and pricing, are aspatial in their emphasis despite evidence suggesting that considerable variations in local pricing exist. In Europe now, for example, mainly as a result of the pioneering work of Joossens and Raw (2006), there is a good understanding of the extent of national variation in policy implementation. More recently

Table 10.1 Potential research questions important in further understanding of smoking and tobacco control in China

Stages of the tobacco supply chain	
Processes	Policy
Growing	
To what extent has the dependence on local sources of tobacco production been replaced by tobacco imports from other LMICs? What are the regional implications of increased foreign imports of tobacco for China's tobacco-growing regions? To what degree do local farmers have a choice in growing tobacco under the state monopoly system? To what extent is regional specialisation occurring in crop types and how is this related to changes in tobacco pricing?	How does tobacco growing fit with UN Sustainable Development goals? What key strategies have and can be used in tobacco-growing regions to reduce tobacco growing? What have been some of the environmental implications of tobacco growing? What are some implications of China's encouragement of tobacco growing in other LMICs? How does external tobacco investment fit in with China's 'Belt and Road' policy
Manufacturing	
What are the locational implications of the amalgamation of tobacco companies? How has the CNTC responded to the changing social dynamic of China in terms of its marketing strategies? What companies have been most involved in globalisation and why? How does the pattern of CNTC overseas expansion compare with the past globalisation of TTCs? To what extent is the counterfeit tobacco industry still important?	To what extent does provincial competition for tobacco manufacturing still exist? To what degree has industry presence affected the policy process and development of particular provincial and local tobacco control policies? In what ways has the industry promoted 'corporate social responsibility' in different provinces?
Marketing	
How much variation exists in cigarette pricing under the state monopoly system? Despite increased regulation, to what extent does tobacco advertising still have a marked effect on consumption? To what extent does tobacco advertising vary between different types of retail outlets? To what degree are tobacco outlets clustered in poorer neighbourhoods and does the degree of clustering depend upon the economic reliance of a city on cigarette manufacturing? To what extent is POS outlet density and advertising related to smoking initiation and pro-smoking norms in different types of neighbourhoods? To what extent are Chinese tobacco companies increasing their domestic marketing of e-cigarettes as an alternative to conventional forms of smoking?	To what extent is the impact of tobacco marketing counteracted by public health anti-smoking advertising? Why has China, along with many other nations, resisted implementing POS advertising restrictions? To what extent are POS marketing restrictions being enforced? What POS monitoring data needs to be routinely collected and at what spatial scale? How can the development of GIS information systems aid in any such developments?

(continued)

Table 10.1 (continued)

Stages of the tobacco supply chain	
Processes	Policy
Consumption: Product use	
To what degree do gender and socio-economic differences in smoking prevalence reflect differences in urban development and neighbourhood social status? To what extent is *hukou* status related to urban inequality, psychological stress and smoking rates among different migrant groups? What particular aspects of Chinese urban development have contributed to increased stress and adverse health behaviours? To what extent are patterns of female smoking affected by city type and neighbourhood social status? The urban-rural divide: What have been the main socio-economic processes leading to reductions of smoking in the countryside?	Why do most smokers use unassisted forms of cessation? To what extent has smokefree legislation in public places displaced smoking into private spaces? In what types of places has this been most prevalent? What types of places, cities and neighbourhoods have been most resistant to cessation initiatives? How is the CNTC responding to the 2030 national goal of reducing smoking prevalence?
Regulation and enforcement	
Who ignores smokefree policies? What types of cities have been most proactive in implementing tobacco control initiatives? What types of anti-smoking coalitions have developed in these cities? How do provinces and cities compare in the implementation of tobacco control measures? What examples are there of 'best practice' that can be emulated by other communities?	What bureaucratic constraints have limited the effectiveness of enforcement? Why have no smokefree policies been developed at the provincial level? Could a European-style Joossens index be developed for China? Given China's cultural and economic diversity, how can tobacco control policies be adapted to different local circumstances?

the Southeast Asia Tobacco Control Alliance (SEATCA) has developed a tobacco industry interference index (Assunta and Dorotheo 2015) which has enabled ASEAN country comparisons to be undertaken on the degree of interference and its effects on policy implementation and smoking prevalence. Not surprisingly, countries with high scores on the index were more likely to have postponed tobacco control legislation or limited its impact.

In China, where subnational governments play a significant role in policy development, these issues will be doubly important. There has been research on how the tobacco industry publications (Yiqun et al. 2018) and other strategies were designed to subvert the implementation of the FCTC (Guo 2018), but there has been a lack of structured policy analysis particularly at the provincial level. Li's (2018) insightful analysis of political corruption and policy making in Yunnan is perhaps a beginning to a more systematic comparison of policy development in provinces heavily reliant on the tobacco industry compared to those which are not. Similarly the differential development of smokefree policies deserves much more attention, as do variations in policy enforcement. Such analyses would provide greater insights into the extent to which local officials have adequate knowledge of FCTC provisions and access to sufficient resources to implement local policies, or even whether they see these as a high priority (Mohan et al. 2013). This would lead to a greater understanding of the relative power of key interest groups in tobacco control and how this power is affected by the presence of the industry. Pro-tobacco interests are not always a monolithic bloc, and such alliances may be fractured if one actor pursues an independent path which affects the well-being of other members of the coalition. For example, in tobacco-growing states in the USA, rapid progress in the implementation of smokefree laws occurred after tobacco companies increasingly switched to imported tobacco. This affected not only the well-being of local growers but also hospitality associations who began to stop opposing smokefree legislation (Fallin and Glantz 2015). Thus it is important to monitor the effects of the growth of tobacco imports into China on tobacco-growing regions and groups within them.

Policy implementation at the provincial level poses challenges in the future. Given China's vast territory and different cultures and levels of social and economic development, one of the requirements for tobacco control is to produce tailored programmes best fitted to the needs of particular places. For example, in the North smoking is often more of a group activity than in the South where it is more common for people to smoke individually (Yang 2007). Geographic differences in policy application also need to take account of differences in levels of urbanisation and economic development, in the awareness of health harms of smoking and its social distribution. In a similar manner, geographic and urban-rural differences in tobacco sharing and gifting need to be acknowledged in tobacco control policies.

10.7.2 Neighbourhoods and Pro-smoking Cultures

A major gap of smoking research in China has been an absence of research on the effects of neighbourhood-level processes on smoking and other health behaviours. In western countries neighbourhood-level factors are known to be important contextual determinants of smoking with neighbourhood 'liveability' and stressors, along with the impacts of social capital and tobacco retailing and availability, receiving much attention. Local place-based practices have formed a critical component in the wider conceptualisation of contextual effects on smoking and have contributed to a greater understanding of both the initiation of smoking and its continuance. Important here has been work, drawing on Bourdieu, which has attempted to develop an understanding of the social practices underpinning persistently high levels of smoking among groups with lower socio-economic status (Dixon and Banwell 2009) and the importance of smoking as a marker of neighbourhood identity (Stead et al. 2001).

Many of these ideas have applicability in China, but remain unexplored. For instance, social ties and bonds within migrant communities may be a critical element of social capital, with smoking forming an important part of collective social identity. This may also occur in affluent urban communities, where cigarette gifting forms an integral part of business interactions. Other affluent communities, however, may have collective characteristics, such as higher levels of education, or a greater voluntary enforcement of home smoking bans that bring about lower levels of smoking. Similarly, there has been little work on tobacco retailing in China and particularly on the neighbourhood social distribution of tobacco retail outlets and their location near schools. It is likely that local neighbourhood patterns vary depending on the importance of the tobacco industry at the urban and provincial level and how this has affected the degree of enforcement of local regulations. More investigation is needed on how retail availability affects local cultures of smoking, and specially the extent to which these are related to smoking initiation among youth and other at-risk groups.

Given differences in western and Chinese patterns of urban development, it is difficult to transfer western ideas of neighbourhood research to China. In western countries neighbourhood-level health research has frequently been criticised because it did not take account of the structural conditions that created disadvantaged neighbourhoods in post-industrial cities. By contrast, few Chinese cities are post-industrial and neighbourhood developments in the past reflected the pre-eminence of state planning rather than market processes, although this has clearly changed in recent decades. In the light of such trends, we argue that it would be fruitful to consider more closely the links between types of urban development upon urban stress and health behaviours.

10.7.3 Better Understanding of Cultural Practices Limiting Smoking Cessation

Ecological theory believes that human behaviour is influenced by both personal and environmental attributes. Norms in the social environment affect the behaviour of smokers and those around them. On one level pro-smoking norms are strong in countries, such as China, Japan and South Korea, which emphasise collectivism. Despite the importance of collective norms, these have received little attention, especially how these have been altered by key role models in different social and cultural environments. It is somewhat ironic that, despite the widespread importance of collective Confucian values, and the presence of different regional cultures such as the 'collective' North, compared to the more 'individualistic' South, most Chinese smoking research remains strongly individualistic.

Tobacco control research and policy interventions in China, therefore, should pay greater attention to local social practices and how 'viewpoint leaders' generate group effects through 'innovative promotion' of healthy behaviours. In this way individual behaviour change can be more effectively located in models of the smoking transition which emphasise the group diffusion of anti-smoking norms (e.g. Christakis and Fowler 2008) and the extent to which these are becoming more important in different types of communities. The absence, however, of a strong locality research tradition in China has so far impeded the development of this type of research.

10.7.4 Geographic Variations in Smoking Transitions

In China, much has been written about the urban-rural, and to a lesser extent regional, variations in smoking, but there has been less attention on how the smoking transition model is playing itself out in different types of communities. Rural areas continue to suffer deprivation as migrants to urban areas are themselves disadvantaged by the *hukou* system and are often unable to fulfil their obligations to support those left behind. In addition, although urban culture has permeated the countryside, traditional social practices persist, particularly regarding the importance of family and the role of women. Social distinctions remain strong and have different relationships to health behaviour than in larger cities.

Given differences in regional and urban-rural cultures, levels of economic development and reliance on the tobacco industry, we suggest that more attention needs to be paid to such variations. Particularly important here is the inversion of the social gradient in smoking between rural and urban areas and how social practices have altered as a result of urbanisation and socio-economic change. This trend reflects broader differences between HICs and LMICs in the social distribution of health behaviours, particularly the changing social distribution of obesity which has captured most attention. In China it is unclear how far the smoking transition has occurred at different points along the urbanisation continuum. This has implications for tobacco control in different places and the extent to which different groups could be targeted by educational programmes and antitobacco messages.

There are research opportunities to investigate social variations in smoking in different types of rural locales, how social practices differ and the barriers faced by different groups of rural smokers who wish to quit (Wang et al. 2014). A more nuanced approach, which examines smoking prevalence and cessation in different types of rural areas, whether in terms of their location relative to larger cities, the importance of tobacco growing in the rural economy or the availability of health services, would be fruitful.

10.7.5 Monitoring Tobacco Control: Developing Better Information Systems

The development of surveillance and evaluation systems is one of the obligations of the parties of the WHO FCTC. Wan et al. (2018) have concluded that current surveillance system in China is in need of improvement as it is unable to effectively monitor the tobacco industry, has trouble with data sharing and data use and is hampered by lack of capacity and funding. Given the development of public health advocacy in the health system, there is an opportunity for public health practitioners to play a greater role in monitoring industry trends and local policy compliance and evaluating health outcomes.

The development of better information systems thus provides an opportunity for the increased application of geographical information system (GIS) methodologies. There are numerous areas where greater monitoring could occur at provincial, urban/rural and local neighbourhood scales. At the provincial scale, the development of a Joossens and Raw (2006) European-type tobacco control scale (based on different policies) would be helpful to quantify the extent to which different provinces are meeting their FCTC obligations. In Europe this has been a useful index which has proved to be an important predictor of quit rates (Schaap et al. 2008). This could be supplemented by a tobacco industry interference index based on stakeholder perspectives similar to that applied by Assunta and Dorotheo (2016) in seven Southeast Asian countries and by the Think-Tank Research Center for Health and Development in Beijing (Jinkui et al. 2018).

At the urban/rural scale, variations in the application of policy performance indicators monitoring the development and enforcement of smokefree policies can be encouraged. These should pay particular attention to variations in group exposure to SHS and how this is changing in different locales. Until a national smokefree law is developed, they should also monitor variations in the enforcement of such policies as is currently occurring in Beijing (see Chap. 9). Given the different cultural environments of urban and rural areas and higher smoking prevalence in the latter, attention should also be directed at support for smokefree legislation in different locales and the extent to which this varies between different population groups.

Finally at the local level, GIS methodologies could be effectively applied to monitor neighbourhood variations in smoking, the location of retail outlets relative to schools and the influence of tobacco outlet density on pro-tobacco attitudes and smoking initiation/cessation. Given the increased importance of POS tobacco advertising in the current marketing strategies of the CNTC, this would be a valuable performance indicator. Given the increased role of public health practitioners in tobacco advocacy, the acquisition of increased GIS skills could enhance the effectiveness of this workforce in monitoring compliance with national laws regarding compliance with smokefree legislation. The new phone-based app in Beijing which records the locations of complaints of smoking ban violations provides an excellent example of one such application (see Chap. 9). It would also provide a much-needed inventory of the location and characteristics of tobacco retailers and particularly sales to minors.

10.8 Conclusion

China presents many challenges for the academic study of smoking and tobacco control practice. Tobacco has become so ingrained into the cultural, political and economic fabric of the country that it is difficult to imagine how things may change in the future. While notions of a tobacco endgame are becoming part of the political rhetoric in richer countries, such ideas almost seem irrelevant to China which, despite signing the FCTC, still remains ambivalent to stronger control measures which might be more effective in combatting the tobacco epidemic.

But maybe the continuing high prevalence of smoking is just a temporary phase in China's economic and social development. Historically China was a world leader in many aspects of economic innovation, but social turmoil, accentuated by decades of foreign invasion and control, weakened societal and civic engagement and the evolution of society at large. While the economic and political changes of the late twentieth century eventually bought a certain political stability, the impact of these wider processes needs to be acknowledged in any discussion of smoking in China and the role of the tobacco industry as an instrument of economic independence and source of national pride. It is likely that in the future, as China develops, tobacco will lose the privileged position it now holds and the central state will become more receptive to promoting stronger measures of tobacco control.

In the meantime the academic public health community in China can continue to emphasise the economic costs of tobacco for the society at large, as well as the individual harms it has produced, which are substantial given China's large smoking population. Smoking prevalence is reflective of a variety of environmental factors, and a greater knowledge of these will be helpful to recommending and promoting various forms of tobacco control. Given China's economic and social diversity, it is evident that 'one size will not fit all', so a more nuanced geographical approach will be essential to producing different sets of recommendations for different types of places, whether these be provinces, cities or rural areas. Tackling issues such as this, but also attempting to have a more outspoken voice in policy making, is a large task, but one in which the academic community has an important role to play.

References

Assunta M, Dorotheo EU (2016) SEATCA tobacco industry interference index: a tool for measuring implementation of WHO framework convention on tobacco control article 5.3. Tob Control 25(3):313–318

Chen X, Li X, Stanton B, Fang X, Lin D, Cole M, Liu H, Yang H (2004) Cigarette smoking among rural-to-urban migrants in Beijing, China. Prev Med 39:666–673

Christakis NA, Fowler JH (2008) The collective dynamics of smoking in a large social network. N Engl J Med 358:2249–2258

Dixon J, Banwell C (2009) Theory driven research designs for explaining health risk transitions: the case of smoking. Soc Sci Med 68:2206–2214

Eriksen M, Mackay J, Schluger N (2015) The tobacco atlas, 5th edn. The American Cancer Society, Atlanta
Fallin A, Glantz SA (2015) Tobacco-control policies in tobacco growing states: where tobacco was king. Milbank Q 93(2):319–358
Gong P, Liang S, Carlton EJ, Jiang Q, Wu J, Wang L, Remais JV (2012) Urbanisation and health in China. Lancet 379:84–852
Guo H (2018) Tobacco product regulation and tobacco industry interference. In: Yang G (ed) Tobacco control in China. Springer, Singapore, pp 201–220
Hou B, Nazroo J, Banks J, Marshall A (2019) Are cities good for health? A study of the impacts of planned urbanisation in China. Int J Epidemiol 48(4):1083–1090
Jinkui L, Tong L, Xiao L, Jiarui W, Yu X (2018) Exploration into and practice in tracking and monitoring the tobacco industry. Sci J Public Health 6(1):15–20
Joossens L, Raw M (2006) The tobacco control scale: a new scale to measure country activity. Tob Control 15(3):247–253
Koplan JP, Eriksen M, Chen L, Yang G (2013) The value of research as a component of successful tobacco control in China. Tob Control 22:ii1–ii3
Li C (2018) Tobacco governance. Elite politics, subnational stakeholders, and historical context. In: Kohrman M, Quan G, Wennan L, Proctor RN (eds) Poisonous pandas: Chinese cigarette manufacturing in critical historical perspectives. Stanford University Press, Stanford, California, pp 179, 308pp–204
Li M, Wu YF (2016) Urbanisation and public health in China. Imperial College Press, London
Lillard DR, Christopoulou R (2015) Life course smoking behavior: patterns and national context in ten countries. Oxford University Press, New York
Liu Y, Dijst M, Faber J, Geertman S, Cui C (2017) Healthy urban living: residential environment and health of older adults in Shanghai. Health Place 47:80–89
Miao J, Wu X (2016) Urbanisation, socioeconomic status and health disparity in China. Health Place 42:87–95
Mohan S, Mini GK, Thankappan KR (2013) High knowledge of framework convention on tobacco control provisions among local government representatives does not translate into effective implementation: findings from Kerala, India. Public Health 127(2):178–181
Moon G, Barnett R, Pearce J, Thompson L, Twigg L (2018) The tobacco endgame: the neglected role of place and environment. Health Place 53:271–278
Schaap MM, Kunst AE, Leinsalu M, Regidor E, Ekholm O, Dzurova D, Helmert U, Klumbiene J, Santana P, Mackenbach JP (2008) Effect of nationwide tobacco control policies on smoking cessation in high and low educated groups in 18 European countries. Tob Control 17(4):248–255
Song Q, Smith JP (2019) *Hukou* system, mechanisms, and health stratification across the life course in urban and rural China. Health Place 58:102150
Stead M, MacAskill S, MacKintosh AM, Reece J, Eadie D (2001) 'It's as if you're locked in': qualitative explanations for area effects on smoking in disadvantaged communities. Health Place 7(4):333–343
Wan X, Hsia J, Yang G (2018) Monitoring tobacco use and implementation of tobacco control policies in China. In: Yang G (ed) Tobacco control in China. Springer, Singapore, pp 141–166
Wang J, Li C, Jia C, Liu Y, Liu J, Yan X, Fang Y (2014) Smoking, smoking cessation and tobacco control in rural China: a qualitative study in Shandong Province. BMC Public Health 14:916
Yang T (2007) Health behavior theory and research. People's Medical Publishing House, Beijing
Yang T (2017) Health research: social and behavioral theory and methods. People's Medical Publishing House, Beijing
Yiqun W, Jinkui L, Yingfa P (2018) Manuals of obstruction. China tobacco blueprints its resistance to the WHO's framework convention on tobacco control. In: Kohrman M, Quan G, Wennan L, Proctor RN (eds) Poisonous pandas: Chinese cigarette manufacturing in critical historical perspectives. Stanford University Press, Stanford, pp 254, 308pp–275
Yuan J, Lu Y, Ferrier RC, Liu Z, Su H, Meng J, Song S, Jenkins A (2018) Urbanization, rural development and environmental health in China. Environ Dev 28:101–110

Yun WJ, Rhee JA, Kim SA, Kweon SS, Lee YH, Ryu SY, Park SW, Kim DH, Shin MH (2015) Household and area income levels are associated with smoking status in the Korean adult population. BMC Public Health 15:39

Zhou M, Feng X, Yong J, Li Y, Zhang M, Page A, Astell-Burt T, Zhao W (2017) Lifting the lid on geographic complexity in the relationship between body mass index and education in China. Health Place 46:1–5

Index

R. Barnett et al. (eds.), *Smoking Environments in China*,
Global Perspectives on Health Geography,
https://doi.org/10.1007/978-3-030-76143-1

Q

R

S

T

GPSR Compliance

The European Union's (EU) General Product Safety Regulation (GPSR) is a set of rules that requires consumer products to be safe and our obligations to ensure this.

If you have any concerns about our products, you can contact us on ProductSafety@springernature.com

In case Publisher is established outside the EU, the EU authorized representative is:

Springer Nature Customer Service Center GmbH
Europaplatz 3
69115 Heidelberg, Germany

Batch number: 10777424

Printed by Printforce, the Netherlands